Industrial Applications of Laser Remote Sensing

Edited by

Tetsuo Fukuchi

Central Research Institute of Electric Power Industry
Japan

&

Tatsuo Shiina

Chiba University
Japan

eBooks End User License Agreement

CONTENTS

FOREWORD

Lasers have been used as remote probes of the environment for close to 50 years, ever since the invention of the laser in 1960. Because of the extremely small divergence and high intensity of a laser beam, active laser remote sensing could be conducted at very large distances compared to other light sources or microwave/radio waves. In particular, an "optical radar" laser beam was bounced off the moon as early as 1962 using a Ruby laser at MIT Lincoln Laboratory, and also used for the detection of water vapor in the atmosphere as early as 1964 by Prof. Richard Schotland. During the past several decades, laser remote sensing (often called lidar, laser radar, or stand-off remote sensing) has become increasingly important in the detection and monitoring of the Earth's ozone hole, global climate change atmospheric gases, and a wide range of environmental trace species. All of these important lidar or laser remote sensing studies have shown that laser beams can be used as sensitive and unique optical spectroscopic probes of the environment and can detect a wide range of chemical and biological substances and targets at ranges out to several kilometers. As such, the use of laser probes in laser remote sensing often can be thought of as a "remote analytical chemistry laboratory" in that the chemical analysis is conducted at the far end of the laser beam.

The unique properties of optical and laser beams that lend themselves to remote sensing applications often use standard optical spectroscopy techniques, such as absorption, fluorescence, Doppler, Raman, and Mie/Rayleigh backscatter, for detection and monitoring of unique trace species and environmental substances. It is important to note that the same optical spectroscopy and laser remote sensing techniques can also be used at much shorter ranges on the order of several meters or less. The only difference between close-in or point optical spectroscopic detection techniques and longer range laser remote sensing is that different optical collection techniques are used to detect the emission optical signal, often using a telescope instead of a single collection lens, and that one usually uses the time-of-flight (i.e. 2-way lidar return delay of 6.6 microseconds for a range of 1000 m) of the returned optical signal as an added discriminator against background noise. As such, laser remote sensing techniques are starting to be applied to a wide range of industrial applications involved in on-line monitoring of chemical species, process control, trace contaminant detection, and a wide range of optical spectroscopy sensing applications.

The book edited by Dr. Tetsuo Fukuchi and Prof. Tatsuo Shiina presents a comprehensive overview of laser remote sensing techniques and how they may be applied to industrial applications at much closer ranges, on the order of tens of meters or less. What is important is that these chapters explain the optical spectroscopic techniques used, and show that they have both remote sensing and close-range industrial applications. Chapters are written by experts in their field and present the fundamental laser spectroscopy and physics involved, show examples of laser remote sensing applications, and explain how this technique can be used in industry and process control applications. Chapters include basic lidar and laser spectroscopy theory, detection of stack exhaust gases, lidar sensing of methane and hydrogen leaks and marine oil spills, use of lasers for wind field mapping near wind power farms and airfields, and use of lasers to monitor plant and tree vegetation, minor trace species , vehicle traffic control, and defects or cracks in concrete structures. As such, the book should be particularly useful to laser remote sensing scientists in developing new laser spectroscopic instruments, for engineers involved in industrial and process control applications, and system engineers interested in the latest advances in this emerging and exciting field.

Dennis K. Killinger
University of South Florida
Tampa, FL
USA

PREFACE

This book covers industrial applications of laser remote sensing. Traditionally, laser remote sensing (lidar) has dealt with atmospheric measurement, with measurement ranges in the order of km. Therefore, lidar systems have been large in size and designed primarily for permanent installation and continuous measurement. However, for industrial applications, the system needs to be mobile or portable so that sensing could be performed at the necessary location at the necessary time, *e.g.* at regular servicing or maintenance, in case of accidents or malfunctions, at occasional environmental inspections.

There exist several books on laser remote sensing, *e.g.* R. Measures, *Laser Remote Sensing* (John Wiley, 1984), T. Fujii and T. Fukuchi, eds., *Laser Remote Sensing* (Dekker, 2005), C. Weitcamp, ed., *Lidar: Range-Resolved Optical Remote Sensing of the Atmosphere* (Springer, 2005). These books mainly cover atmospheric measurement. Recent lidar development has focused on satellite-borne lidar, whose aim is global monitoring of water vapor, ozone, CO_2, wind, clouds, aerosols. On the other hand, a book on applications of laser remote sensing to closer ranges, in the order of m to tens of m, has not been previously published. These ranges require remote sensing because they are too large for *in situ* measurement using conventional sensors or sampling methods, but are too small for applying conventional lidar for atmospheric measurement.

Laser remote sensing has several potential industrial applications in these closer measurement ranges, such as leak gas detection, pollutant detection, environment monitoring, wind profiling, and structural health monitoring. This book aims to provide some specific applications which may be useful to industry, as well as other applications such as marine environment monitoring, vegetation monitoring, and minor constituent monitoring, which are more oriented toward science, but may have applications to industry in the future.

An overview of laser remote sensing and its applications to industry are presented in Chapter **1**. Various kinds of lidar and measurable quantities are described.

Conventional lidar design intended for atmospheric measurement does not allow sensing at very close range, because of insufficient overlap between the transmitted laser beam and the receiver field of view. In order to overcome the problem of insufficient overlap, new concepts on lidar design are under development for near field applications. The design of an in-line type lidar is covered in Chapter **2**.

Lasers have found increased use in industrial applications such as gas leak detection and pollutant detection. The most common technique used for gas leak detection is laser absorption spectroscopy, which is covered in Chapter **3**. Another technique used for gas leak detection is Raman scattering, whose application to hydrogen gas leaks is covered in Chapter **4**. Hydrogen is a gas species which cannot be detected by absorption, and this recently developed technology could find wide use with the increasing introduction of hydrogen energy. Pollutant detection using differential absorption lidar, *e.g.* measurement of gas species in stack emission, is briefly covered in Chapter **3**.

Laser remote sensing has wide applications, which are not limited to detection of gases. The applications to the marine environment, such as bathymetry, oil spill detection, and water quality inspection are presented in Chapter **5**. Application to vegetation monitoring is covered in Chapter **6**. Laser-induced fluorescence from chrollophyl can provide useful information on vegetation growth.

Laser sensing has also found use in safety and security. Although the death rate due to traffic accidents is declining every year owing to safer vehicles and better infrastructure, traffic accidents still rank at the top of the causes of accidental deaths. An example of the application of laser radar to traffic safety is covered in Chapter **7**.

The increase in use of renewable energy sources has led to a rapid increase in electricity generation using wind power. For optimal siting of windmills, profiling of local winds is necessary. The all-fiber laser

Doppler lidar, which has recently been developed, has dramatically decreased the size and power consumption, so that portable wind profiling has become possible. This is covered in Chapter **8**.

An important social issue is the safety of infrastructure such as bridges and tunnels. The use of conventional ultrasound techniques is labor intensive, as it requires contact between the sensor and the object under testing. The application of laser ultrasound provides non-contact testing with a standoff distance of several meters. Recent developments in this field and application to inspection of concrete structures such as railway tunnels are covered in Chapter **9**.

Lastly, remote Laser-Induced Breakdown Spectroscopy (LIBS) for minute concentration detection is covered in Chapter **10**. This method provides the equivalent of Atomic Emission (AE) spectroscopy in a remote configuration. Since no sampling is necessary, the technology could be useful for minute concentration detection in hazardous environments.

The affiliations of the authors of this book are distributed among academic institutions, private and government research institutes, and private companies. The distribution was chosen so that the content will vary from fundamental research to practical applications.

Although laser remote sensing is especially suited to plasmas and combustion fields because of its ability to perform non-contact measurement, the applications to these areas are not covered in this book, because comprehensive texts already exist. The interested reader is requested to refer to these texts, such as K. Muraoka and M. Maeda, *Laser-Aided Diagnostics of Plasmas and Gases* (Institute of Physics Press, 2001), A. Eckbreth, *Laser Diagnostics for Combustion Temperature and Species* (Taylor & Francis, 1996).

The editors hope that this book be a useful addition to the technical library of researchers and engineers interested in laser sensing and its applications.

Tetsuo Fukuchi
Central Research Institute of Electric Power Industry
Japan

Tatsuo Shiina
Chiba University
Japan

List of Contributors

Ando, Toshiyuki
Mitsubishi Electric Corporation
5-1-1 Ofuna, Kamakura, Kanagawa 247-8501, Japan

Asaka, Kimio
Mitsubishi Electric Corporation
5-1-1 Ofuna, Kamakura, Kanagawa 247-8501, Japan

Fujii, Takashi
Central Research Institute of Electric Power Industry
2-6-1 Nagasaka, Yokosuka, Kanagawa 240-0196, Japan

Fukuchi, Tetsuo
Central Research Institute of Electric Power Industry
2-6-1 Nagasaka, Yokosuka, Kanagawa 240-0196, Japan

Hirano, Yoshihito
Mitsubishi Electric Corporation
5-1-1 Ofuna, Kamakura, Kanagawa 247-8501, Japan

Hisamitsu, Yutaka
IHI Corporation
3-1-1 Toyosu, Koto-ku, Tokyo 135-8710, Japan

Kameyama, Shumpei
Mitsubishi Electric Corporation
5-1-1 Ofuna, Kamakura, Kanagawa, 247-8501, Japan

Kobayashi, Takao
Fukui University
3-9-1 Bunkyo, Fukui, Fukui 910-8507, Japan

Kotyaev, Oleg
Institute for Laser Technology
2-6, Yamada-oka, Suita, Osaka 565-0871, Japan

Nagata, Kouichirou
IHI Corporation
3-1-1 Toyosu, Koto-ku, Tokyo 135-8710, Japan

Ninomiya, Hideki
Shikoku Research Institute
2109-8 Yashima-Nishimachi, Takamatsu, Kagawa 761-0192, Japan

SAITO, Kazunori
Shinshu University

4-17-1 Wakasato, Nagano, Nagano 380-8553, Japan

Sasano, Masahiko
National Maritime Research Institute
6-38-1 Shinkawa, Mitaka, Tokyo 181-0004, Japan

Sekimoto, Kiyohide
IHI Corporation
3-1-1 Toyosu, Koto-ku, Tokyo 135-8710, Japan

Shiina, Tatsuo
Chiba University
1-33 Yayoi-cho, Inage-ku, Chiba, Chiba 263-8522, Japan

Shimada, Yoshinori
Institute for Laser Technology
2-6, Yamada-oka, Suita, Osaka 565-0871, Japan

2

CHAPTER 1

Overview of Laser Remote Sensing Technology for Industrial Applications

Takao Kobayashi[*]

Graduate School of Engineering, University of Fukui, 3-9-1 Fukui-shi, Fukui 910-8507, Japan

Abstract: Laser remote sensing systems have been developed as laser radar or lidar (Light Detection And Ranging). The spatial distribution of dust aerosol particles, water droplets, atomic and molecular components of low concentration could be detected efficiently. Various meteorological applications were developed, and effective sensing techniques have been accumulated in the field of atmospheric, oceanic and terrestrial studies, ranging from *in situ*, local, to global sensing. In this chapter, the basic principles, performance, and historical progress of laser remote sensing techniques are briefly introduced. Optical interaction processes used in lidar and other remote sensors are discussed and their detection sensitivities are compared. The basic concept can be applied to the design of laser sensing techniques and systems for various industrial applications. As examples of laser sensors for industrial applications, Mie lidars in the eye-safe near infrared and ultraviolet spectral regions for dust plume monitoring of urban and industrial areas are introduced.

Keywords: Laser remote sensor, laser radar, absorption lidar, Mie lidar, differential absorption lidar, Mie scattering, Raman scattering, fluorescence, absorption, eye safety.

1. INTRODUCTION

Just after the invention of the laser in 1960, it became obvious that the useful application of the laser would be optical radar or laser radar, which is an extension in frequency of coherent electromagnetic radiation of microwave radar used widely in meteorology, aviation, and military applications. This finding is based on potential advantages of the laser over microwave radiation: narrow beam divergence, extremely wide spectral bandwidth, high power output, and a variety of detectable physical and chemical quantities.

The first successful application of laser remote sensing systems developed in the 1960's was Mie scattering laser radar or Mie lidar (Light Detection And Ranging) [1, 2]. High sensitivity detection of small aerosol particles and water droplets by Mie lidar was reported. Various meteorological applications of Mie lidar were found, such as dust pollutant monitoring, sensing of structural and dynamical properties of the atmosphere [2].

In addition, several spectroscopic techniques have been developed for the measurement of molecules and chemical compounds in the environment. Raman lidar or Raman scattering lidar was developed in the 1970's for specific measurement of molecular atmospheric components such as nitrogen and water vapor, and of chemical species in stack plumes [3, 4]. Laser-induced fluorescence sensing systems are also extremely highly sensitive for the detection of atomic and molecular species in the environment.

High sensitive measurement of low concentration molecules has been realized by long-path absorption lidar and differential absorption lidar (DIAL) [5]. In these systems, wavelength tunable lasers, such as diode-lasers, dye lasers, CO_2 gas lasers, and nonlinear optical sources in infrared, visible and ultraviolet spectral regions were employed. These sensing systems have been used widely in environmental and industrial applications.

For measuring wind speed and direction, Doppler lidar was developed in 1970 for detecting the Doppler shift frequency of Mie scattering by aerosol particles in the atmosphere [6]. Doppler lidar systems can be applied to wind sensing in clear weather conditions, which is difficult for microwave radar.

*Address correspondence to Takao Kobayashi: Graduate School of Engineering, University of Fukui, 3-9-1 Fukui-shi, Fukui 910-8507, Japan; Tel: +81-776-27-8558, Fax: +81-776-27-8557, E-mail: t.koba100@gmail.com

Tetsuo Fukuchi and Tatsuo Shiina (Eds)

The use of lidar systems are expanding from compact, ground-based stationary systems and mobile systems for local sensing to airborne and spaceborne systems for global sensing.

Several comprehensive review books have been published covering lidar technology and its applications [7-12].

In this chapter, the basic principles and performance of laser remote sensing techniques are introduced. The optical interaction processes used in laser sensing systems are classified, and the detection sensitivites and detectable quantities of laser sensing systems are compared. These are intended for understanding the fundamental aspects of the laser sensing techniques described in the following chapters of this book.

Finally, eye-safety conditions, which are basically required for industrial laser sensors are described. Two examples of eye-safe lidars are introduced: a compact near infrared system and an ultraviolet Mie lidar system for use in dust monitoring and environmental sensing.

2. PRINCIPLES OF LASER REMOTE SENSING TECHNIQUES

2.1. Principles of Lidar Performance

The basic principle and performance of lidars and other laser remote sensors are introduced in this section [13].

The basic system arrangement of lidar measurement is shown in Fig. (1). The output beam of the laser transmitter is collimated by focusing optics and transmitted into the atmosphere. Backscattered light by a solid target, as shown in Fig. (1a), or distributed targets, as shown in Fig. (1b), is collected by the receiving optics and transmitted through a spectral filter for blocking sky background and selectively transmitting the signal components. The output light is detected by an optical detector, whose output is analyzed by the signal processor and the data is illustrated on the PC display.

The basic system characteristics is analyzed by the relation between the transmitted laser power and the received signal power, which is called the lidar equation or the laser radar equation for two different types of the system.

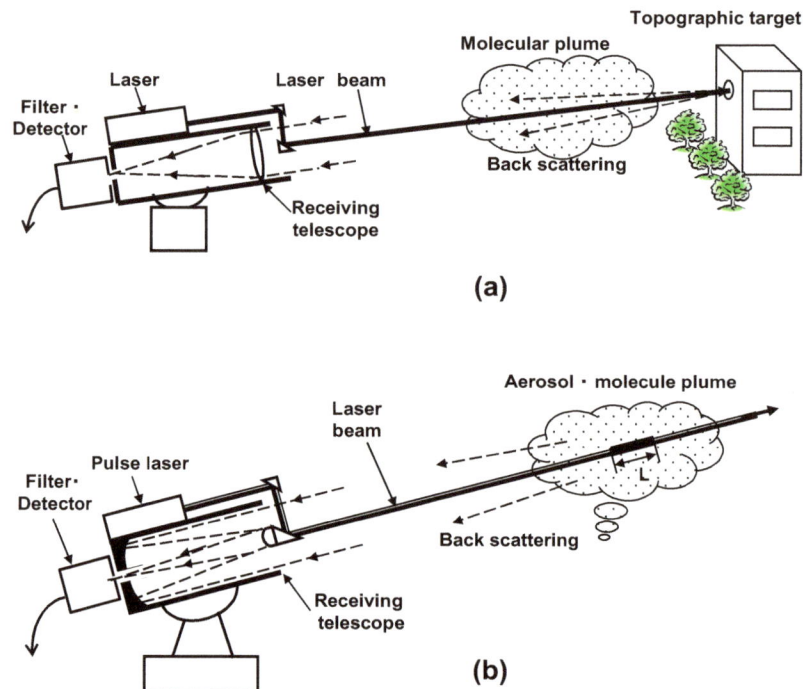

(a)

(b)

Fig. (1). Measurement schemes of the laser remote sensors. **(a)** Laser radar and absorption lidar. **(b)** Lidar and Differential Absorption Lidar (DIAL).

(1) Laser Radar and Absorption Lidar

Laser radar is used to measure the range and the surface quantities of topographic targets such as buildings, ground, and seawater. In absorption lidar, topographic targets are used as reflectors of the laser beam and the total column content of molecules and other substances can be measured by using a tunable laser.

In these systems, reflected power by the target and received by the detector is given by

$$P_r(R) = P_0 KY(R) A \sigma_0 T^2 / R^2 , \qquad (1)$$

where P_0 is the transmitted continuous wave (cw) or pulsed laser power, K is the round trip optical transmittance of the transmitting and receiving optics, $Y(R)$ is the geometrical overlap factor of the laser beam and the receiver field of view at range R, and A is the area of the receiving optics. σ_0 is the scattering coefficient of the target, given by $\sigma_0 = r\cos\theta/\pi$ for rough surface targets called Lambert reflectors with surface reflectance r and angle θ between the surface normal direction and the receiver direction.

(2) Lidar

For the measurement of distributed targets such as aerosols, cloud droplets, atoms and molecules, a pulsed laser is used in the lidar system to resolve the target range and the power received is given by the lidar equation;

$$P_r(R) = P_0 LKY(R) A \beta(R) T^2 / R^2 \qquad (2)$$

where R is the range of the target, L is the scattering depth of the laser beam given by

$$L = c\tau / 2 \qquad (3)$$

which corresponds to half of the laser pulse length, which is given by the product of the laser pulse duration τ and the speed of light c. $\beta(R)$ is the volume backscatter coefficient of the scattering target, and is related to the concentration of the particle or molecule $N(R)$ and the differential backscattering cross section $\sigma(\pi)$ of the particle as

$$\beta(R) = N(R)\sigma(\pi) . \qquad (4)$$

The parameter $T(R)$ is the single-path laser beam transmittance of the atmosphere, given by

$$T(R) = \exp\left[-2\int_0^R \alpha(r)dr \right], \qquad (5)$$

where $\alpha(r) = \alpha_m(r) + \alpha_a(r)$ is the extinction coefficient which is composed of the molecular extinction $\alpha_m(r)$ and the aerosol extinction coefficient $\alpha_a(r)$ at range r.

The range R of the target is measured by the lidar with a pulsed laser by measuring delay time of the signal Δt from the laser pulse

$$R = c\Delta t / 2 . \qquad (6)$$

The volume backscatter coefficient of the distributed target is derived by using the range-corrected received power $P_r(R)R^2$;

$$\beta(R) = kP_r(R)R^2 / T^2(R) \qquad (7)$$

where k is a system parameter. The only unknown parameter in this relation is the atmospheric transmittance $T(R)$. For a short distance, one can assume $T(R)=1$, but for general cases, spectroscopic lidar methods have been used for accurate measurement of $\beta(R)$, as discussed in the following sections.

2.2. Optical Interaction Process and Measurable Quantities

Various optical interaction phenomena can be observed from the target when the laser beam is irradiated. In Table **1**, several optical interaction processes are summarized, which can be used for the laser sensing, with interaction cross-section and measurable physical and chemical quantities.

2.2.1. Mie Scattering

Mie scattering is the interaction process for small particles with diameter near or larger than the light wavelength. The scattering cross section of spherical particle targets is shown in Fig. **(2)** as a function of the diameter D normalized to the light wavelength λ. The scattering cross section is normalized to the geometrical cross section of the particle. Large increase in the scattering cross section is observed as compared to small particles like molecules, and highly sensitive detection can be realized. For a visible and near-infrared laser beam, various aerosol particles and cloud water and ice droplets with 0. 1~10 μm diameter size belong to this process.

Table 1: Optical interaction process and detectable quantities

Interaction process	Cross section (m^2)	Target	Detectable quantity: *Applications*
Mie scattering	$10^{-28} \sim 10^{-10}$	particle	density, size, shape, cloud, aerosol, wind, visibility: *Meteorology, pollution control*
Rayleigh scattering	$\sim 10^{-29}$	molecule	density, temperature, pressure, humidity, wind, extinction: *Meteorology*
Raman scattering	$10^{-35} \sim 10^{-30}$	molecule	composition, density, humidity, temperature, extinction: *Meteorology, industrial*
Fluorescence	$\sim 10^{-10}$ $10^{-28} \sim 10^{-10}$	atom molecule	composition, density, temperature: *Pollution control, industrial*
Absorption	$\sim 10^{-10}$ $10^{-24} \sim 10^{-14}$	atom molecule	composition, density, humidity: *industrial, pollution control*
Reflection	r/π*	solid	surface reflectance, roughness
Doppler effect	-	atom molecule	wind, particle velocity, temperature: *Meteorology, industrial*

The frequency of Mie scattering is equal to the laser frequency and this scattering belongs to the elastic process. Polarization of the scattering is depolarized for some particles. By measuring the degree of depolarization of the scattered signal using a polarizer, some insight on the shape of particles can be obtained. For instance, non-spherical aerosol particles and ice crystals can be distinguished from spherical water droplets using a polarization Mie lidar [14].

Mie scattering lidars have been developed and used for observation of aerosol and clouds [14] in the atmosphere. Various meteorological information have been found: the boundary layer and the stratosphere, dust particles [15], Asian dust. Space borne Mie scattering lidars have been used for global sensing [16].

In the fields of industrial application, measurements of smoke stack plumes were used for the control of air pollution emitted from industrial areas [17]. Mie lidars have also been used as ceilometers for measuring cloud height and slant angle visibility in many airports for traffic control [12].

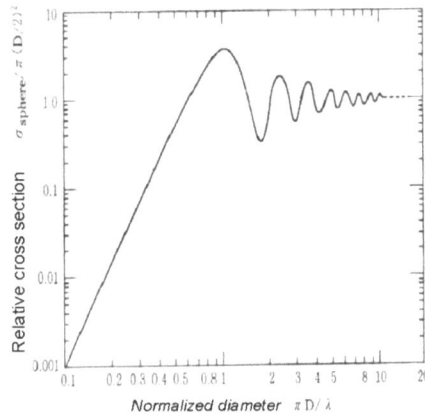

Fig. (2). Increase in the Mie scattering cross section as a function of normalized particle diameter to the wavelength.

2.2.2. Rayleigh Scattering

Light scattering by molecules is called Rayleigh scattering. The Rayleigh volume-backscattering coefficient β of a molecule is given by

$$\frac{\beta}{\beta_{ST}} = 1.47 \times 10^{-6} \left[\frac{550nm}{\lambda(nm)} \right]^4 m^{-1} sr^{-1} \qquad (8)$$

where λ (nm) is the laser wavelength and β_{ST} is the volume backscattering coefficient at standard temperature and pressure (STP), which are defined as 273 K and 1 atm (101. 3 kPa). From this relation, it is evident that the Rayleigh scattering cross section increases in the short ultraviolet wavelength and efficient measurement of molecules can be realized. Since the intensity of Rayleigh scattering is proportional to the molecular density, Rayleigh scattering can be used for atmospheric molecular density measurement.

The center frequency of Rayleigh scattering is equal to the laser frequency and the scattering belongs to the elastic process. The spectrum is broadened by the Doppler shift of the molecular thermal motion velocity.

Fig. (3) shows the superposition of the spectrum of Rayleigh scattering of the atmosphere for two different temperatures and the spectrum of Mie scattering by aerosol particles. The center frequency of the two scattering phenomena overlap. In this case, Rayleigh scattering measurement is not possible in the lower atmosphere because of interference from Mie scattering. Therefore, atmospheric molecular density measurement using Rayleigh scattering is limited to high altitude, above the troposphere, where aerosol distribution is limited [10].

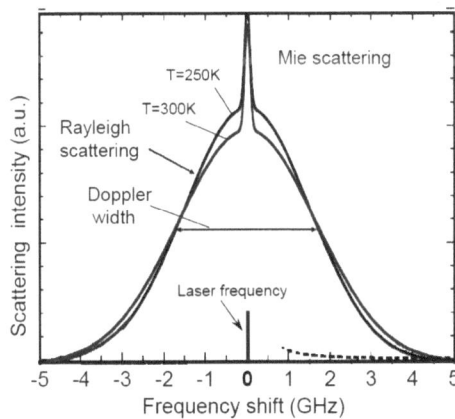

Fig. (3). Spectral distribution of Rayleigh scattering and Mie scattering of the atmosphere for laser wavelength 355 nm.

For observation of Rayleigh scattering in lower altitudes near the ground, an extremely narrow bandwidth spectral filtering device is necessary for blocking or separating the interference by intense Mie scattering. Atomic or molecular absorption filters or Fabry-Pérot interference filters are used for separating these two components, and this lidar system is known as High-Spectral Resolution Lidar (HSRL) [18, 19].

The spectral width of Rayleigh scattering is proportional to the square root of the temperature and can be used for measurement of molecular or atmospheric temperature. As meteorological and industrial applications, several HSR Rayleigh temperature lidars have been developed for the remote measurement of atmospheric temperature [20-22]. Three dimensional measurement of the temperature profile over urban areas is useful for observation and analysis for the environmental design of modern cities to reduce heat island phenomena and local environmental warming issues.

2.2.3. Raman Scattering

Raman scattering by molecules and chemical compounds is an inelastic scattering process and the scattering frequency is shifted by the amount of vibrational and/or rotational frequency of the molecule. Based on the Raman frequency shift, the species of molecules can be identified [4, 23, 24].

The density of an unknown molecule can be derived by comparing the molecular Raman intensity I_m and the nitrogen Raman intensity I_{N2} of the atmosphere

$$\frac{I_m}{I_{N2}} = \frac{N_m \sigma_m(\pi)}{N_{N2} \sigma_{N2}(\pi)} \tag{9}$$

where $\sigma_m(\pi)$ and $\sigma_{N2}(\pi)$ are the backscattering cross sections, N_m and N_{N2} are the densities of the unknown molecule and of nitrogen, respectively. In this case, all system parameters are assumed to be the same for the two Raman wavelengths in the lidar equation (2).

The Raman scattering cross sections have been measured for many molecules and are listed [4, 9]. The values are about 3 orders of magnitude smaller than the Rayleigh scattering cross section and thus the detection sensitivity of Raman lidar is limited mostly by the small signal power.

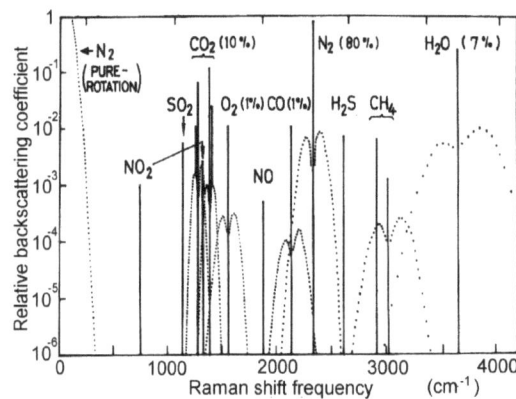

Fig. (4). Raman spectra of vibration-rotation Raman shift estimated for oil plume in the atmosphere [4].

Fig. **(4)** shows a spectrum of the vibration-rotation Raman spectra of the backscattering coefficient, calculated for oil smoke in the atmosphere [4]. The relative density of the composition molecules is indicated. The central lines of each spectrum is the Q-branch, and dotted points are rotational lines of the vibration-rotation Raman spectra.

The first raman lidar was developed to observe vibrational Raman scattering by nitrogen molecules in the atmosphere [3]. A mobile Raman lidar system with a 30 cm diameter receiver telescope was developed, and was used for the detection of SO_2 in the stack plume of industrial area. Because of the small Raman scattering cross section, the detection sensitivity was limited to the ~100 ppm level.

The detection sensitivity was improved by incorporating the UV laser and a large size telescope. High sensitive spectroscopic detecting was realized at 1 ppm level in 200 m range and was used for chemical analysis of jet engine exhaust gas [26].

An interesting and useful application of Raman lidar has been found in the remote measurement of atmospheric temperature by using vibration-rotation Raman spectra. The measurement is based on the temperature dependence of the rotational spectrum, which can be calculated assuming a Boltzmann distribution in the rotational energy levels [27, 28]. Another useful application of Raman lidar is the measurement of the extinction profile and transmittance of the laser beam by aerosols and clouds [29]. Because the height and horizontal distribution of N_2 density is predictable, the laser beam transmittance can be derived from the N_2 Raman scattering power in the atmosphere by using Eq. (7). The systems are also applicable for extinction measurement of dust aerosol in the atmosphere, which are effectively used for the analysis of the warming intensities in the local and global atmosphere.

Raman measurement is also useful for absolute intensity measurement of other chemical spectra by detecting water Raman spectrum for ocean and terrestrial lidar measurements [30].

The unique feature of the Raman scattering sensor is the ability of detecting diatomic molecules such as H_2 and N_2, which do not have absorption spectra from infrared, visible, and ultraviolet regions. Recently, a new industrial application was found in the Raman lidar sensor for the detection of hydrogen gas leaks, which can be used for safety surveillance of hydrogen handling facilities in the new hydrogen energy industry, as described in Chapter 4 of this book.

2.2.4. Fluorescence

In the fluorescence process, the substance is excited by the laser photons to the upper level and emit the photons after some time delay as fluorescence light. Because of the large cross-section of this process, highly sensitive detection of atoms and molecules can be realized in laser-induced fluorescence (LIF) spectroscopy [23].

In the case of atomic vapors, various fluorescence lidars have been developed for the detection of atomic vapor layers at high altitudes at 90-110 km of the atmosphere, called the mesosphere. Many metallic atomic vapors such as Na, K, Li, Fe, Ca, Ca^+ have been detected [24-26]. Extremely low vapor density of 1~10^3 atom/cm^3 could be detected. Temperature could be assessed by measuring the spectral shape and width of the fluorescence line with resolution less than 1 K [25]. This atomic fluorescence lidar requires a frequency tunable laser with narrow spectral width and an accurate frequency control technique.

Fluorescence lidars have been developed using pulsed uv lasers for the measurements of hydrocarbon oil slicks and chemical contaminants on the sea surface, and have realized highly sensitive detection and distribution mapping [30]. Application to the marine environment are described in Chapter 5. In addition, several studies have been performed for the utilization of fluorescence lidar for measurements of living terrestrial plants and vegetation, as discussed in Chapter 6.

2.2.5. Absorption

In the absorption process by an atom or gas molecule, the light intensity is attenuated in the optical path and transmission measurement is necessary. Absorption lidars are sensitive and practical schemes have been developed [31, 32], whose system arrangement is shown in Fig. **(1a)**. A topographic target is used as the reflector of the laser beam. The laser wavelength is alternately switched to the on-resonance absorption wavelength λ_{on} and the off-resonance wavelength λ_{off}, and the received power P_{on} and P_{off} are used to derive the integrated column concentration N_t by the relation

$$N_t = \frac{1}{2\Delta\sigma_a} \ln\left(\frac{P_{off}}{P_{on}}\right) \tag{10}$$

where $\Delta\sigma_a$ is the absorption cross section difference for the two wavelengths, and the laser power is assumed to be the same for the two wavelengths. The integrated column concentration is given by $N_t = \int_0^R n(z)dz$ (ppm-m) for the density of molecule $n(z)$ distributed in the optical path z. Spectral purity of the laser power is important for accurate measurement of gas concentration.

For the measurement of spatial distribution of trace gas, the system arrangement of the differential absorption lidar, DIAL, is applicable as shown in Fig. (1b) [33, 34]. In this scheme, the Mie backscattering by aerosol particles is detected as the distributed reflectors. By transmitting the two wavelength pulses, the range-resolved density is basically derived by [52]

$$N(R) = \frac{1}{2\Delta\sigma_a L} \ln\left[\frac{P_{off}(R+L)P_{on}(R)}{P_{off}(R)P_{on}(R+L)}\right], \tag{11}$$

where $P_{on}(R)$ and $P_{off}(R)$ are the received power for on-resonance and off-resonance wavelengths from range R, $P_{on}(R+L)$ and $P_{off}(R+L)$ are those from range $R+L$, respectively, and L is the range resolution of the measurement. The detection limit is given by the signal-to-noise ratio of the measured power and differential cross section of the gas and aerosols and interference gases.

Many DIAL systems have been developed for the measurement of industrial emissions. The systems employ ultraviolet and infrared tunable lasers for the measurement of molecules such as SO_2, NO_2, NH_3, HCl, Cl_2, CO, hydrazine, Hg, and O_3 [5, 33]. The mobile, shipborne and airborne systems have been used for the DIAL studies of pollution and industrial emissions [34].

Detection levels of 10 ppbv were demonstrated in the detection of SO_2 at range up to km [35]. A multiwavelength DIAL system has been developed using two dual-wavelength dye lasers and realized 0.5 ppbv detection accuracy for O_3 and SO_2 measurements at altitudes of about 1 km with a range resolution of 300 m [36].

2.2.6. Doppler Effect

Doppler frequency shift can be observed by moving particles in the atmosphere and the frequency is given by

$$f_d = \frac{2V_r}{c}\nu_0 \tag{12}$$

where V_r is the radial velocity of the target or radial wind velocity, for laser frequency ν_0.

The Doppler shift of the Mie scattering by aerosols is used to measure the wind velocity in Doppler lidar with coherent, heterodyne detection. High sensitivity detection can be realized in coherent Doppler lidar.

The first coherent Doppler lidar was reported by Huffaker *et al.* in 1970 [3] for wake vortex detection using a 10.6 μm CO_2 laser. Various meteorological and industrial applications have been found, such as clear air turbulence (CAT) sensors [37], meteorological wind monitoring for aviation safety, and wind survey of wind power generation. The current state-of-the art fiber laser based Doppler lidar system is described in Chapter 7.

2.2.7. Comparison of Detection Sensitivity of Spectroscopic Lidar Systems

Detection of specific low density atoms and molecules can be made by spectroscopic interaction schemes such as Raman, fluorescence, and absorption processes. These methods have been applied from *in situ* plasma diagnostics and combustion studies to remote sensing using lidar. The sensitivity depends on the received power, but it is basically limited by the cross section of the interaction process for the normalized system size defined by the laser power and receiver efficiency. It will be instructive to compare the minimum detectable density of atoms and molecules for the lidar system.

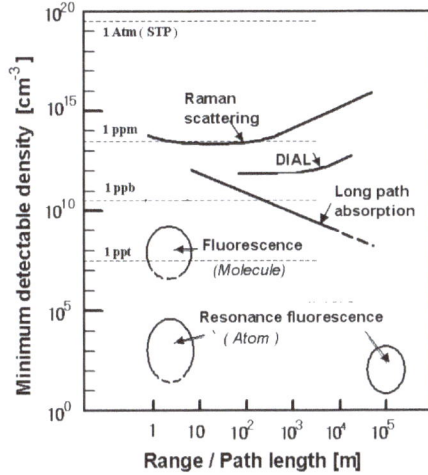

Fig. (5). Comparison of minimum detectable density and detection rage of the spectroscopic laser sensing system.

The minimum detectable density of atoms and molecules is plotted in Fig. **(5)** as a function of detection range for Raman, fluorescence and absorption lidar systems. Practical values for the laser power and receiver size are assumed. In the case of the long-path absorption scheme, the detection range corresponds to the path length. For the long range detection lidars, 100 m range resolution is assumed. The results show extremely high sensitivity of the fluorescence scheme for the detection of atoms and molecules and the absorption scheme for molecules. The Raman scheme is limited in sensitivity.

3. EYE-SAFE LIDARS

3.1. Eye Safe Laser Spectrum

For practical use of lidars in the open atmosphere, eye safety property of the laser beam is crucially important. In Fig. **(6)**, the maximum permissible exposure (MPE) of laser pulse to the human eye, and the laser wavelength relation is shown. The infrared wavelength longer than 1. 4 μm exhibits a MPE of more than 5 orders of magnitude higher than the visible and near infrared shorter than 1. 4 μm [38]. This spectral region is thus called the eye-safe spectral region. Ultraviolet wavelengths shorter than 400 nm wavelength also belong to the eye-safe region.

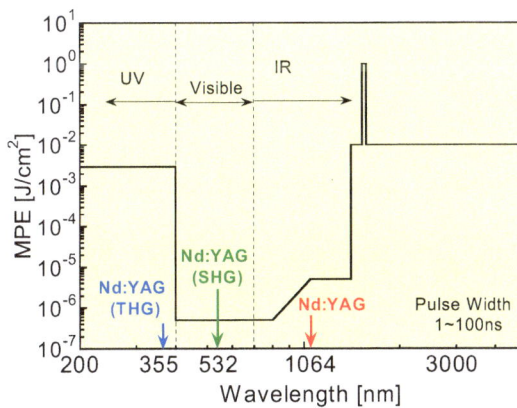

Fig. (6). Maximum permissible exposure and the wavelength of the laser pulse beam [38].

3.2. Compact 1.57 μm Mie Lidar

A compact Mie lidar with a 1. 57μm, eye-safe wavelength laser was developed by our group for observation of aerosols and plumes [39]. A block diagram of the compact Mie lidar system is shown in Fig. (7). A compact flashtube-pumped, Q-switched Nd:YAG laser with 120 mJ pulse energy at 1. 064 μm wavelength was used to pump an optical parametric oscillator (OPO) to generate the 1. 57 μm output wavelength. A KTP crystal of 20 mm length and 90 degree phase matching angle was used as the nonlinear crystal. The 20 cm diameter telescope and two InGaAs APD detectors were used for depolarization measurement of the aerosol backscatter. A photograph of the compact Mie lidar is shown in Fig. (8).

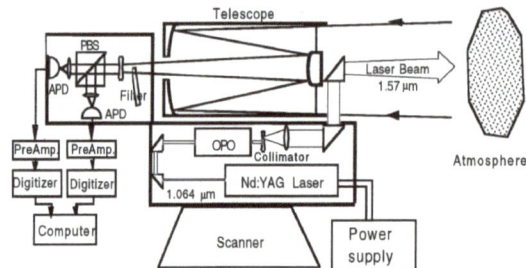

Fig. (7). System arrangement of the compact 1.57 μm near infrared Mie lidar.

Fig. (8). Photograph of the compact beam scanning Mie lidar in the eye safe wavelength.

The upper part of the optics is the receiving telescope with the OPO transmitter attached in the lower box. The output beam is automatically scanned.

The lidar was effectively used for the detection of aerosol and clouds in the atmosphere. In Fig. (9), the plan position indication (PPI) display is shown for the horizontal distribution of a dust plume from a stack located at a range of 700 m. The relative volume backscattering coefficient β, which is approximately proportional to the plume mass concentration, is indicated by color scale. The results indicate the detection sensitivity in clear air up to a maximum range of 1. 5 km in single shot measurement, and up to 4km in 100 shot average measurement.

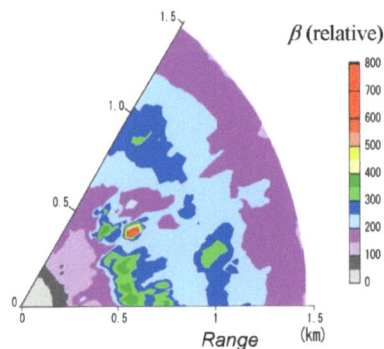

Fig. (9). A plan position indication (PPI) display of dust concentration distribution near the stack located at 700m range obtained by the compact beam scanning Mie lidar.

3.3 355 nm UV Mie Lidar

Ultraviolet lasers are used also in Mie lidars and used for eye-safety operation in industrial applications. The most reliable and practical UV solid-state laser is the third harmonic Nd:YAG laser at 354. 7 nm wavelength. The UV lidar has several other advantages over the visible and near lidars: a considerable increase in the Mie, Rayleigh and Raman scattering cross sections as indicated by Eqs. (8) and (9), and a high quantum efficiency of the detectors. On the other hand, the disadvantage of the UV lidar is the relatively large extinction of the laser beam by aerosols and the limitation in the maximum measurement range in the horizontal direction.

As an example of UV Mie lidar measurement, a PPI display of a horizontal distribution of the aerosol volume backscatter coefficient is shown in Fig. **(10)** [40]. Plumes from the two incineration towers are monitored clearly over the background aerosol concentration level. In this lidar, a 6 mJ energy output Nd:YAG laser at 355 nm wavelength was used with a 25 cm diameter telescope. The system has been installed for dust monitoring in steel industry firms.

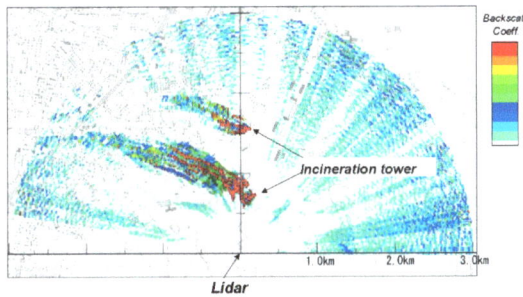

Fig. (10). An UV Mie lidar measurement of dust plume concentration from incineration towers and background level [40].

4. CONCLUSION AND OUTLOOK

In this chapter, basic principles and performance of laser remote sensing techniques and systems were briefly introduced. Optical interaction processes used in the laser remote sensors were classified. The detection sensitivity of the spectroscopic lidar systems was compared.

Laser remote sensing systems have been developed over almost a half century and have reached technical maturity. Various meteorological and industrial applications have been developed.

Further development in laser sensing techniques are still in progress by the emergence of compact, efficient and reliable high-power tunable laser sources such as fiber lasers, diode lasers, diode-pumped solid-state lasers, and quantum cascade lasers, as well as highly sensitive detectors in the infrared spectrum. Remote detection and imaging systems based on absorption lidar and other spectroscopic sensing techniques will have further potential of practical and wide applications for gas leak detection, gas emission and pollution monitoring, and environmental sensing and regulation.

REFERENCES

[1]   G. Fiocco, L. Smullin, "Detection of Scattering Layers in the Upper Atmosphere (60–140 km) by Optical Radar", *Nature*, Vol. 199, pp. 1275-1276, 1963.

[2]   M. Ligda, in *Proceedings 1st Conference on Laser Technology*, G. Adelman, T. Dowd (eds.), Vol. 1, p. 63, 1963.

[3]   D. Leonard, "Observation of Raman Scattering from the Atmosphere using a Pulsed Nitrogen Ultraviolet Laser", *Nature*, Vol. 216, pp. 142-143, 1967.

[4]   H. Inaba, T. Kobayashi, "Laser Raman Radar- Laser Raman scattering methods for remote detection and analysis of atmospheric pollution", *Opto-electronics*, Vol. 4, pp. 101-123, 1972.

[5]   G. Gimmerstad, "Differential-Absorption Lidar for Ozone and Industrial Emissions", in *Lidar, Range resolved optical remote sensing of the atmosphere*, C. Weitkamp (ed.), Springer, pp. 187-212, 2004.

[6]   R. Huffaker, A. Jelalian, J. Thompson, "Laser-Doppler system for detection of aircraft trailing vortices", *Proc. IEEE*, Vol. 58, pp. 322-326, 1970.

[7]   E. Hinkley (ed.), *Laser Monitoring of the Atmosphere*, Springer-Verlag, 1976.

[8]   D. Killinger, A. Mooradian (eds.), *Optical and Laser Remote Sensing*, Springer-Verlag, 1983.

[9]   R. Measures, *Laser Remote Sensing, Fundamentals and Applications*, John Wiley Sons, 1984.

[10]   R. Measures (ed.), *Laser Remote Chemical Analysis*, Wiley Interscience, 1987.

[11]   W. Grant, E. Browell, C. She (eds.), *Selected Papers on Laser Applications in Remote Sensing*, SPIE Milestone Series, MS 141, 1997.

[12]   C. Weitkamp (ed.), *Lidar, Range resolved optical remote sensing of the atmosphere*, Springer, 2004.

[13]   T. Kobayashi, "Techniques for laser remote sensing of the environment", *Remote Sensing Reviews*, Vol. 3, pp. 1-56, 1987.

[14]   K. Sassen, "Polarization in Lidar", in *Lidar, Range resolved optical remote sensing of the atmosphere*, C. Weitkamp (ed.), Springer, pp. 19-42, 2004.

[15]   J. Reagan, M. McCormick, J. Spinhirne, "Lidar sensing of aerosols and clouds in the troposphere and stratosphere", *Proc. IEEE*, Vol. 77, pp. 433-448, 1989.

[16]   M. McCormick, "Airborne and Spaceborne Lidar", in *Lidar, Range resolved optical remote sensing of the atmosphere*, C. Weitkamp (ed.), Springer pp. 355-398, 2004.

[17]   E. Uthe, W. Viezee, B. Morley, J. Ching, "Airborne lidar tracking of fluorescent tracers for atmospheric transport and diffusion studies", *Bulletin of the American Meteorological Society*, Vol. 66, pp. 1255-1262, 1985.

[18]   S. Shipley, D. Tracy, E. Eloranta *et al.*, "High spectral resolution lidar to measure optical scattering properties of atmospheric aerosols. 1: Theory and instrumentation", *Applied Optics*, Vol. 22, pp. 3716-3724, 1983.

[19]   E. Eloranta, "High Spectral Resolution Lidar", in *Lidar, Range resolved optical remote sensing of the atmosphere*, C. Weitkamp (ed.), Springer, pp. 143-164, 2004.

[20]   J. Hair, L. Caldwell, D. Krueger, C. She, "High-Spectral-Resolution Lidar with Iodine-Vapor Filters: Measurement of Atmospheric-State and Aerosol Profiles", *Applied Optics*, Vol. 40, pp. 5280-5294, 2001.

[21]   D. Hua, M. Uchida, T. Kobayashi, "UV high-spectral-resolution Rayleigh-Mie lidar with a dual-pass Fabry-Pérot etalon for measuring atmospheric temperature profiles of the troposphere", *Optics Letters*, Vol. 29, pp. 1063-1065, 2004.

[22]   M. Imaki, Y. Takegoshi, T. Kobayashi, "Ultraviolet high-spectral-resolution lidar using Fabry-Pérot filter for the accurate measurement of extinction and lidar ratio", *Japan Journal of Applied Physics*, Vol. 44. 3063-3067, 2005.

[23]   H. Inaba, "Detection of Atoms and Molecules by Raman Scattering and Resonance Fluorescence", in *Laser Remote Sensing of the Atmosphere*, E. D. Hinkley (ed.), Springer-Verlag, Berlin, 1976.

[24]   U. Wandinger, "Raman lidar", in *Lidar, Range resolved optical remote sensing of the atmosphere*, C. Weitkamp (ed.), Springer, pp. 241-272, 2004.

[25]   S. Nakahara, K. Ito, S. Ito, *et al.*, "Detection of sulphur dioxide in stack plume by laser Raman radar", *Opto-electronics*, Vol. 4, pp. 169-177, 1972.

[26]   T. Hirschfeld, E. Schildkraut, H. Tannenbaum, D. Tanenbaum, "Remote spectroscopic analysis of ppm-level air pollutants by Raman spectroscopy", *Applied Physics Letters*, Vol. 22, pp. 38-40 (1973).

[27]   A. Behrendt, T. Nakamura, T. Tsuda, "Combined Temperature Lidar for Measurements in the Troposphere, Stratosphere, and Mesosphere", *Applied Optics*, Vol. 43, pp. 2930-2939, 2004.

[28]   D. Hua, J. Liu, K. Uchida, T. Kobayashi, "Daytime Temperature Profiling of Planetary Boundary Layer with Ultraviolet Rotational Raman Lidar", *Japan Journal of Applied Physics*, Vol. 46, pp. 5849-5852, 2007.

[29]   A. Ansmann, M. Reibeselle, C. Weitkamp, "Measurement of atmospheric aerosol extinction profiles with Raman lidar", *Optics Letters*, Vol. 15, pp. 746-748, 1990.

[30]   F. E. Hoge, "Oceanic and terrestrial lidar measurements", in *Laser Remote Chemical Analysis*, R. Measures (ed.), Wiley Interscience, pp. 409-503, 1987.

[31]   T. Kobayashi, N. Sugimoto, H. Kuze, "Laser Remote Sensing Techniques of Leak Gases", *Review of Laser Engineering*, Vol. 33, pp. 295-299, 2005. (in Japanese)

[32]   C. Webster, R. Menies, E. Hinkley, "Infrared Laser Absorption: Theory and Applications", in *Laser Remote Chemical Analysis*, R. Measures (ed.), Wiley Interscience, pp. 163-272, 1987.

[33]   W. Staehr, W. Lahmenn, C. Weitkamp, "Range-resolved differential absorption lidar: optimization of range and sensitivity", *Applied Optics*, Vol. 24, pp. 1950-1956, 1985.

[34] K. Fredriksson, "Differential Absorption Lidar for Pollution Mapping", in *Laser Remote Chemical Analysis*, R. Measures (ed.), Wiley Interscience, pp. 273-332, 1987.

[35] T. Fukuchi, T. Fujii, N. Goto, K. Nemoto, N. Takeuchi, "Evaluation of differential absorption lidar DIAL measurement error by simultaneous DIAL and null profiling", *Optical Engineering*, Vol. 40, pp. 392-397, 2001.

[36] T. Fujii, T. Fukuchi, N. Cao, K. Nemoto, N. Takeuchi, "Trace Atmospheric SO_2 Measurement by Multiwavelength Curve-Fitting and Wavelength-Optimized Dual Differential Absorption Lidar", *Applied Optics*, Vol. 41, pp. 524-531, 2002.

[37] M. Imaki, T. Kobayashi, "Ultraviolet HSL Doppler lidar for wind field and aerosol properties", *Applied Optics*, Vol. 44, pp. 6023-6030, 2005.

[38] ANSI standard Z136. 1, 2007

[39] T. Kobayashi Y. Enomoto, D. Hua, C. Galvez, T. Taira, "A compact, eye-safe lidar based on optical parametric oscillators for remote aerosol sensing", in *Advances in Atmospheric Remote Sensing with Lidar,* A. Ansmann, R. Neuber, P. Rairoux, U. Wandinger (eds.), Springer, pp. 11-14, 1997.

[40] T. Yokozawa, M. Ruike, T. Azuma, Y. Ohmura, T. Kobayashi, "Development of UV eye-safe lidar for industrial applications", *Abstracts of 21st Japanese Laser Sensing Symposium*, D3, 46, 2001.

CHAPTER 2

Optical Design for Near Range Lidar

Tatsuo Shiina[*]

Graduate School of Advanced Integration Science, Chiba University, 1-33 Yayoi-cho, Inage-ku, Chiba-shi, Chiba 263-8522, Japan

Abstract: Nowadays, measurement of air flow and certain gas species in near range are needed for safety and environmental monitoring. Lidar is the appropriate tool for these applications. However, conventional lidar optics has a blind area because of the distance necessary to overlap the transmitted beam and the receiver's field of view. To detect the near range lidar echo with a narrow field of view, the optical design should be compact and simple. The near range from zero to a few hundred meters (< a few km) is the target distance. In this chapter, the theoretical calculation of the lidar echo for the near range measurement lidar is presented. The inline optics, which has common optics for the transmitting and receiving optics, is introduced. The signal-to-noise ratio is also estimated from the viewpoint of lowering the transmitted beam power for eye-safety. The actual near range lidar setup, which is based on the analysis, is also presented. The calculated results are compared and evaluated with the optical specification of the lidar. Some studies using the near range lidar, and several types of near range lidars are introduced.

Keywords: Compact lidar, near field lidar, inline lidar, Raman lidar, micropulse lidar, polarization lidar, LED lidar, hydrogen gas, lidar optics, lidar design.

1. INTRODUCTION

The demand for a compact lidar, which can detect hazardous gases and monitor the atmosphere in the near range, is high. For example, the former capability can be applied to leak gas detection inside a factory or in outdoor conditions (there have been numerous examples of gas poisoning due to stagnation of volcanic gases in bowl-shaped depressions), and the latter capability can be applied to map the air flow of certain suspended particulate matter (SPM) in a large hall, or around buildings. Conventional biaxial lidar, however, has a blind area, or nondetectable area of lidar echo, in near range, because separate optics are used for transmitting the laser beam and for receiving the backscattered light. In the blind area, the transmitted beam does not enter the receiver's field of view (FOV). The blind area also occurs in the case of the coaxial lidar, in which the laser beam is transmitted along the optical axis of the receiver, usually *via* a reflector located behind the secondary mirror of the reflecting telescope. By broadening the receiver's FOV, the blind area can be shortened, but this results in an increase of the background light and deterioration of the signal-to-noise ratio of the lidar echo. Furthermore, making such arrangement to cover the near range will lower the detection efficiency of the lidar echo in far range. Due to these difficulties, there has not been much progress in the "near range" lidar. Many of the lidars commercialized for "near range" are downsized versions of conventional lidars, with the inherent problem of the blind area unsolved.

This chapter presents "inline" type compact lidar optics for near range detection and sensing. The term "inline" used here represents the use of common optics for the transmitter and receiver. By this design, the transmitted beam is always in the receiver's field of view (FOV), and the "inline" typed lidar system has no blind area. The lidar which has common optics for the transmitter and receiver has been proposed [1], but the fabrication of the actual system had been difficult because of the direct reflection from the optics entering the detector. Efficient separation of the received echo from the transmitted beam is technically difficult.

The first accomplishment of an "inline" lidar is Spinhirne's micro pulse lidar (MPL) [2-4]. The MPL has common optics for the transmitter and receiver, while there is no consideration for near range detection.

---

*Address correspondence to Tatsuo Shiina: Graduate School of Advanced Integration Science, Chiba University, 1-33 Yayoi-cho, Inage-ku, Chiba-shi, Chiba 263-8522, Japan; Tel: +81-43-290-3470; Fax: +81-43-290-3039; E-mail: shiina@faculty.chiba-u.jp

Tetsuo Fukuchi and Tatsuo Shiina (Eds)

The author started development of the "inline" MPL, which enables detection at near range for the prediction of local disasters such as heavy rain and lightning strikes [5-7]. The near range lidar is accomplished by installing additional optics to separate the received echoes from the transmitted beam. The inline optics allows detection of near range echoes with the narrow FOV. The author has improved these designs to fabricate a compact Raman lidar for detection of hydrogen gas leaks [8].

In this chapter, the design and specification of the in-line lidar optics for near range detection is presented in section 2. The lidar echo intensity and signal-to-noise ratio using the laser and detector performances are estimated. Applications for near range detection are presented in section 3. As an example, the development of a compact Raman lidar system for the detection of hydrogen leak gas is presented. Several kinds of inline lidars for the other purposes are introduced in section 4. Section 5 concludes the chapter.

2. METHOD

2.1. Lidar Measurement at Near Range

As the fundamental structure of the lidar optics, there are three types as shown in Fig. **(1)**. The biaxial type optics shown in Fig. **(1a)** has separate optics for the optical transmitter and optical echo receiver. The coaxial type optics shown in Fig. **(1b)** transmits a pulsed laser beam from a portion of the receiver's aperture, along the optical axis of the receiver. The "inline" type optics shown in Fig. **(1c)** has common optics for the transmitter and the receiver.

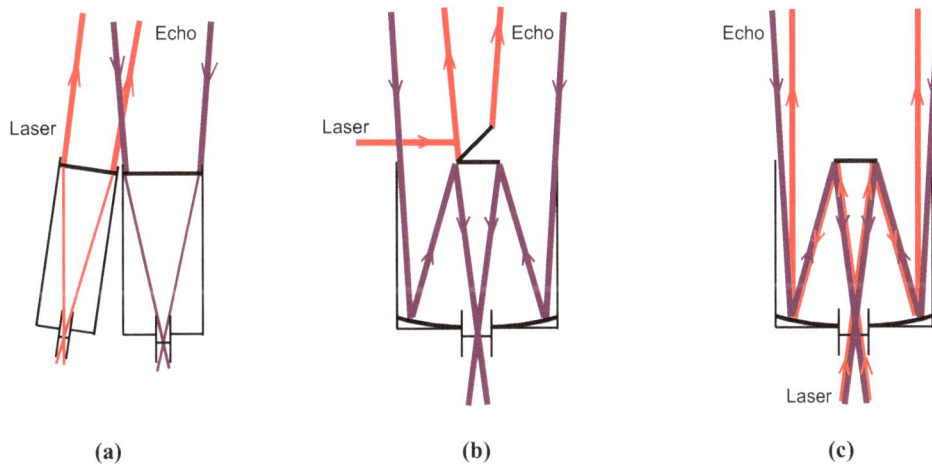

(a)   **(b)**   **(c)**

Fig. (1). Optical setup for three types of transmitting and receiving optics. (a) biaxial type optics, (b) coaxial type optics, (c) inline type optics.

In the case of the biaxial type optics, the easy way to shorten the blind area at near range is to broaden the receiver's FOV, at the expense of the increase in the background light decrease in the dynamic range of lidar echo. As the path of the transmitted laser beam is set at a slant angle against the receiver's FOV, the blind area of near range can be controlled with the fixed FOV. The slant path, however, will deviate from the receiver's field at a certain distance and beyond. To avoid this problem, the laser path is slanted at a small angle with respect to the optical axis of the receiver, which results in a large blind area.

In the case of the coaxial typed optics, the laser path is parallel to the optical axis of the receiver's FOV, which is set by reflecting the laser beam by a reflector located on the back side of the secondary mirror located at the center of the receiver's aperture. The coaxial type optics also includes the case that the laser beam is transmitted from a certain point in the telescope aperture by a small mirror or a prism. To narrow the receiver's FOV to eliminate background light, a pinhole (Field Stop Aperture [FSA]) is installed at the focal point of the telescope. In this case, the near range lidar echo is out of focus and cannot pass through the pinhole. As a result, the coaxial lidar also has a blind area [9-11].

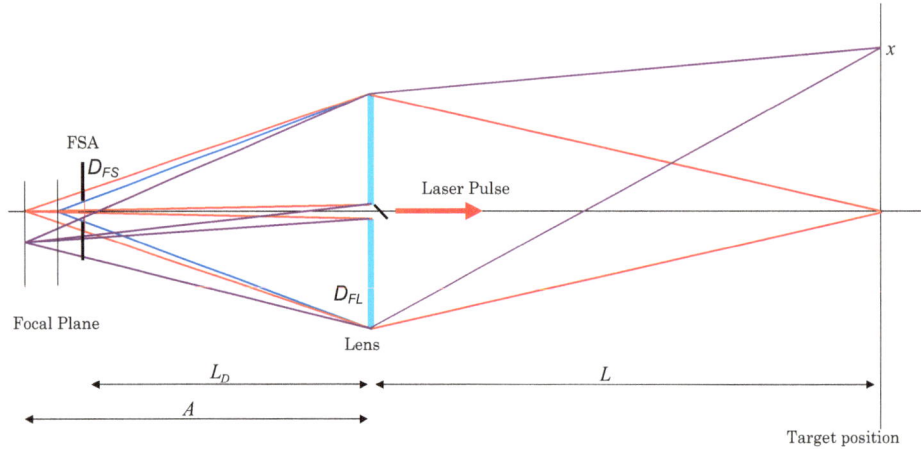

Fig. (2). Geometric arrangement of lidar optics.

In the case of inline type optics, the transmitted laser beam and the receiver's FOV have the same path in the lidar optics and in open air. That is, the laser path always overlaps with the receiver's FOV, even if the FOV is narrow, of the order of 0.1 mrad. As the intensity distribution of the transmitted laser beam can be enlarged up to the telescope aperture, there is no blind area in near range [7]. On the other hand, the lidar echoes should be separated from the transmitting path in the lidar optics. In general, it will make the optics complex.

2.2. Optical Design

By modeling the lidar optics as shown in Fig. (1), the optical design is performed. Fig. (2) shows the geometric arrangement of the coaxial and inline optics. In this illustration, the lens is a model of the primary mirror of the telescope. The field stop aperture (FSA) is arranged at the focal plane of the primary mirror. The lens and FSA decide the field of view by the law.

$$\tan\theta = -D_{FS} / 2f_{FL} \tag{1}$$

where θ is the angular spread of the FOV, D_{FS} is the diameter of the FSA, and f_{FL} is the focal length of the lens. The laser pulse can be located at a certain position of the lens aperture with a small diameter as the coaxial optics or the inline optics. The backscattered light (echo light) from the scattering target (such as suspended particles) is focused by the imaging effect of the lens. As the echo light diffuses, a certain part of the echo light cannot pass through the FSA because of defocusing, which is especially conspicuous in the near range. The transparent quantity of the echo light is estimated by the geometric boundary condition as shown in Fig. (3).

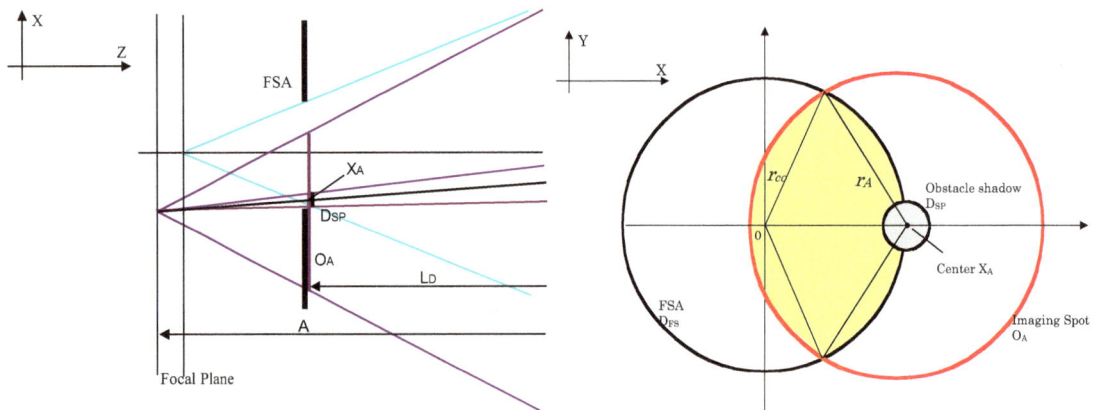

Fig. (3). Boundary condition of FSA: side view (above) on-axis view (below).

Table 1: Design layout of lidar optics

Main Lens	Aperture : D_{FL} Focal length : f_{FL}
Obstacle	Diameter : D_{ss}
FSA	Diameter : D_{FS}
Target	Distance : L *Position : X*
Focal Distance	$A = L \cdot f_{FL} / (L - f_{FL})$
Image Size at FSA	Diameter: $O_A = (A - L_D)D_{FL} / A$ Center Position: $X_A = L_D X / L$
Obstacle at FSA	Diameter: $D_{sp} = (A - L_D)D_{ss} / A$ Center Position: $X_A = L_D X / L$

The waveform of the laser beam is designed as an arbitrary shape, while in this chapter it has been assumed to be Gaussian. An obstacle such as a secondary mirror of the reflecting telescope can also be taken into account. Its shadow is also imaged on the focal plane. The parameters of the analysis are summarized in Table **1**.

The lidar echo characteristics were evaluated by calculating the lidar echo power $P(L)$ and its signal-to-noise ratio $SNR(L)$ [1,12].

$$P(L) = P_0 KY(L)Arl\beta(L)T(L)^2 / L^2 + Pb \quad T(L) = \exp[-\int_0^L \alpha(x)]dx \qquad (2)$$

$$SNR(L) = \frac{\sqrt{M}\sqrt{\eta\Delta t / h\nu}P(L)}{\sqrt{\mu}\sqrt{P(L) + Pb + Pd}} \qquad (3)$$

where P_0 is the transmitted power, K is the system optical efficiency, A_r is the receiver's area, c is the speed of light, l is the pulse length of light, $\beta(L)$ is the backscattering cross section, $Y(L)$ is the geometrical form factor, $T(L)$ is the transmittance, P_b is the background light power, $\alpha(L)$ is the atmospheric extinction coefficient, M is the number of signal summations, η is the detector's quantum efficiency, μ is the detector's noise factor, h is Planck's constant, ν is the light frequency, P_d is the equivalent dark current power. The geometrical form factor $Y(L)$ is the overlap function between the receiver's FOV and the transmitted beam, which is determined by the telescope specification, the beam divergence, and the size of the FSA.

2.3. Calculation

2.3.1. Coaxial Lidar

As the first approach, the coaxial lidar optics is explained to examine the near range echo characteristics and to compare to those of the "inline" optics. In this design, the laser beam was transmitted from a corner prism at the center of the lens aperture as shown in Fig. **(3)**. As this prism makes its shadow on the focal plane of the lens, the transparent echo intensity at FSA is estimated by considering the shadow effect. The diameter of the transmitted beam is small, and its divergence can be ignored in the near range up to 50m. The parameters are summarized in Table **2**.

The lidar system is based on a flashlamp pumped Nd:YAG laser (3rd harmonic) as the laser source and a photomultiplier (PMT) as the detector. The diameter of the output beam and the receiver's lens are 6 mm and 200 mm, respectively. In the optical design, the concave lens is inserted as the FSA. The transparent echo light is collimated and enters the detector. The detector also has a convex lens and a second FSA to control the receiver's FOV. Here, the Receiver's FOV was fixed at 8 mrad.

Table 2: Lidar and Atmospheric specifications for coaxial optics

Lidar Parameters	
Beam Intensity	P_0=1.25 MW (10 mJ/ 8ns)
Optical Efficiency	K=0.3
Receiver's Aperture	D_{FL}=0.2 m
Receiver's Area	A_r=$10^{-2}\pi$
Summation	M=10.0
Sampling Time	Dt=1 ns
Plank Constant	h=6.63x10^{-34}
Laser Wavelength	355 nm
Optical Frequency	v_0=c/λ
Quantum Efficiency	η=0.2
Optical Pulse Length	L_0=2.4 m
Noise Factor	μ=1
Atmospheric Parameters	
Backscattering Coefficient	β=3.4x10^{-9}
Extinction Coefficient	α=50.0β
Background Light	P_b=6.4x10^{-10}
Electric Noise	P_d=0

Fig. (**4**) shows change of the receiver's FOV in near range. In comparison to the receiver's aperture, the size of the obstacle (prism or mirror) to transmit the laser beam at the center of the aperture is sufficiently small, and the shape of FOV is always the same as shown in Fig. (**4a**). The receiving efficiency at the center rises gradually along to the distance.

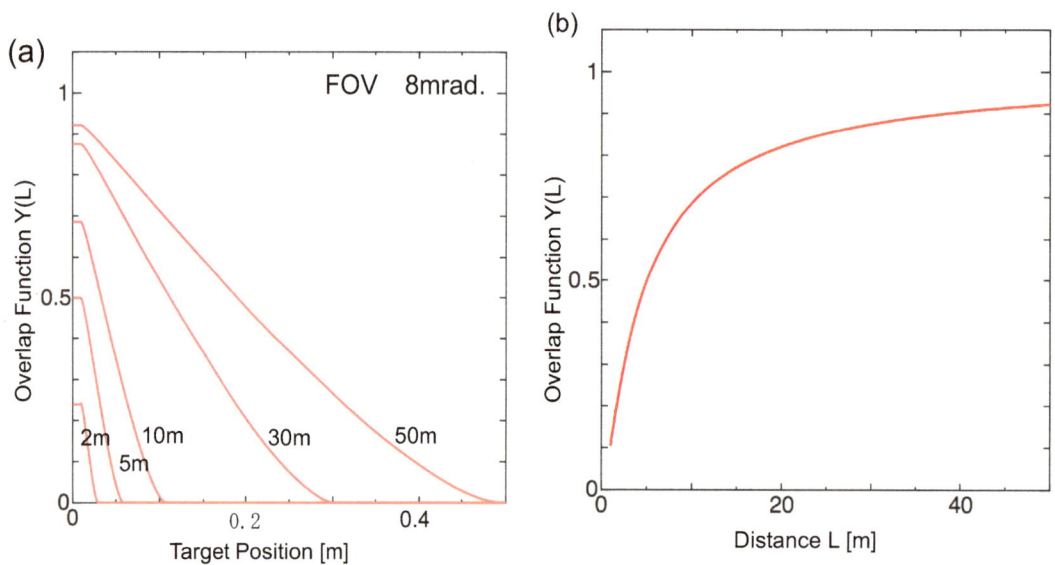

Fig. (4). Field of view and overlap Function in the case of coaxial type optics. (a) Field of view, (b) Overlap Function.

The overlap function was calculated by estimating the transparent intensity of echo light at the FSA against received intensity, which passes through the receiver's lens. The overlap function rises from the immediate vicinity (nearly 0 m) and approaches unity, as shown in Fig. (**4b**).

Fig. (**5**) shows the signal-to-noise ratio. The estimation was performed assuming a summation time of 1 second (10 pulses transmitted 10 Hz). The atmospheric attenuation is the value at the Raman scattering wavelength by atmospheric N_2 (387 nm). The observation conditions with the background light ($P_b > 0$) and without the background light ($P_b = 0$) were simulated. The ideal overlap function ($Y(L) = 1$) means that the echo passes completely through the FSA throughout the measurement range. Because of the near range measurement, the signal-to-noise ratio does not show a large difference between the cases of without background light and with background light.

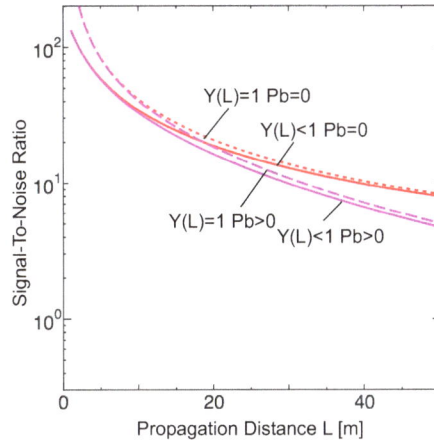

Fig. (5). Signal-to-noise ratio of the coaxial lidar.

On the other hand, each curve includes not only the defocusing effect of the lens, but also the shadow effect of the obstacle. As a result, the blind area still remains at the nearest range (in front of the lidar system). Even if the lidar optics is of a coaxial type, the obstacle such as a secondary mirror of a reflecting telescope, a prism or a mirror to transmit the laser pulse, is sufficiently small in comparison to the receiver's FOV, the blind area can be controlled or shortened as much as possible.

2.3.2. Inline Lidar

The parameters for the inline type lidar were set as shown in Table **3** to estimate the lidar echo power and signal-to-noise ratio. They were designed for the compact Raman lidar for detection of hydrogen gas leak. The system is compact so that it can be carried on a compact vehicle. The detection range is 0-50 m to observe the gas leak from a safe distance. It is the same range as the former design. Furthermore, the system should be also designed to be eye-safe because of the scanning observation of the horizontal direction.

Table 3: Lidar and atmospheric parameters for inline optics

Lidar Parameters	
Beam Intensity	P_0=40 kW(0.12 mJ/3 ns)
Optical Efficiency	K=0.5
Receiver's Aperture	D_{FL}=0.05 m
Receiver's Area	A_r=2.5x10^{-4}□
Summation	M=1000.0
Sampling Time	D_i=1 ns
Plank Constant	h=6.63x10^{-34}

Table 3: cont....

Laser Wavelength	349 nm
Optical Frequency	$v_0 = c/\lambda$
Quantum Efficiency	$\eta = 0.2$
Optical Pulse Length	$L_0 = 0.45$ m
Noise Factor	$u = 1$
Atmospheric Parameters	
Backscattering Coefficient	$\beta = 3.4 \times 10^{-9}$
Extinction Coefficient	$\alpha = 50.0\beta$
Background Light	$P_b = 6.4 \times 10^{-10}$
Electric Noise	$P_d = 0$

In the in-line optics, the lidar echo should be separated from the transmitting beam. A half mirror is an easy way to separate the echo from the outgoing beam for Mie scattering detection, but only half of the laser power will be transmitted to the air and also the half of the lidar echo can be detected. For Raman scattering detection, a dichroic mirror is suitable to reflect the outgoing beam and to pass the Raman scattering echoes with high efficiency. Fig. (6) shows the simple inline type lidar optics. It consists of a laser head, detector, beam expander and beam separator (half or dichroic mirror). The FSA is also installed into the expander to restrict the receiver's FOV. The expander is functionally equivalent to the telescope in lidar optics. Here, it consists of transmissive lenses, and there is no consideration of the secondary mirror and its effect in this simulation. Calculations were conducted for the measurement range of 0-50 m. The transmitted laser beam was assumed to have a Gaussian distribution with a diameter of 30 mm. When the FSA diameter was 400 μm, the spread angle of the receiver's FOV became 2 mrad as shown in Fig. (7a). Fig. (7b) shows the overlap function $Y(L)$ of the lidar equation, which indicates the degree of overlap between the transmitted beam and the receiver's FOV. The overlap functions are calculated with FSA diameters of 400 μm and 150 μm, respectively. When the echo from near range is received, a certain part of the echo is cut off, if the focal point of the lidar echo lies behind the focal point of the objective lens. The rise in the overlap functions correspond to such off-focus conditions.

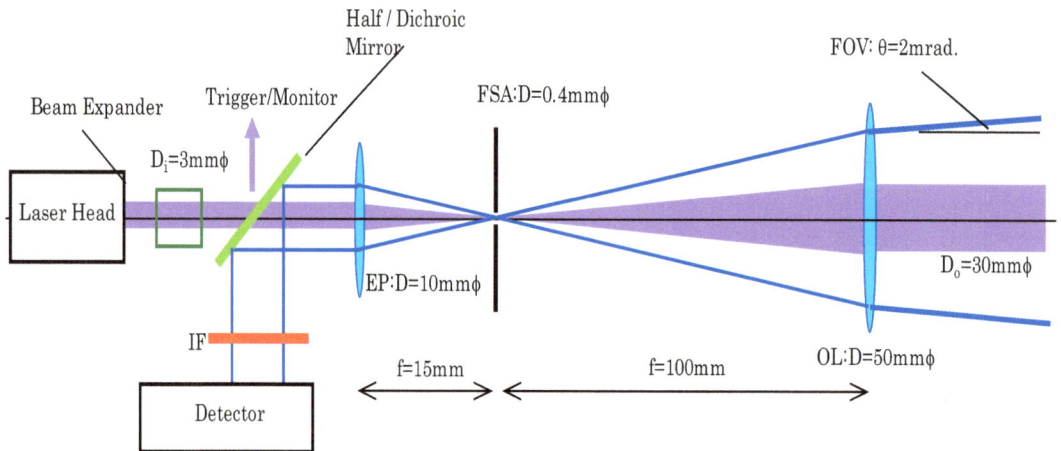

Fig. (6). Inline type compact lidar.

The signal-to-noise ratios were also estimated by using eq. (3), and the results are shown in Fig. (8). Fig. (8a) shows the case of the receiver's FOV of 2 mrad. The estimation was performed at the summation of 1 second. The atmospheric attenuation is the value at the Raman scattering wavelength by atmospheric N_2 (387 nm). The observation conditions with the background light ($P_b > 0$) and without the background light ($P_b = 0$) were simulated. The ideal overlap function ($Y(L) = 1$) means that the echo passes completely through FSA throughout the measurement range.

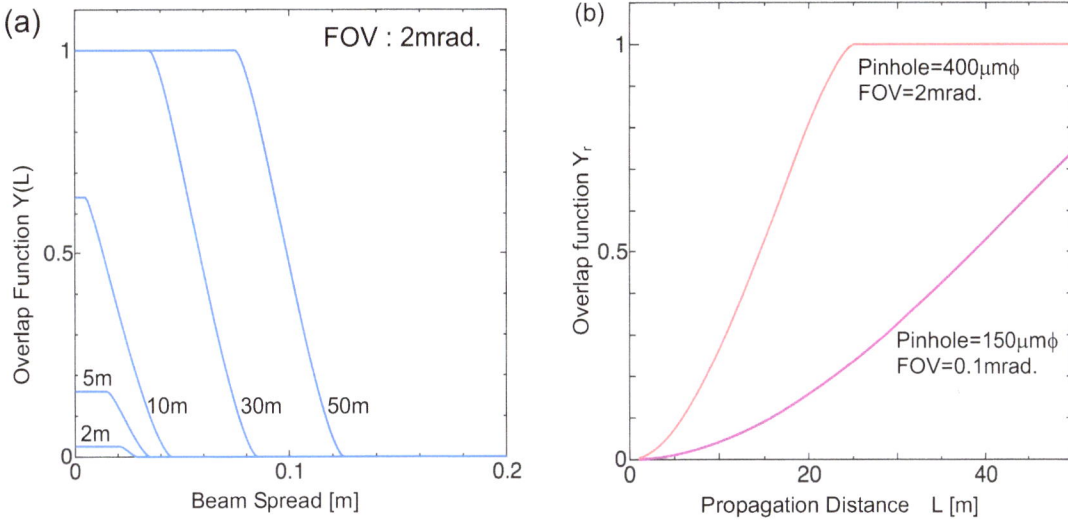

Fig. (7). Receiver's FOV and overlap function of compact Raman lidar. (a) Field of View, (b) Overlap function.

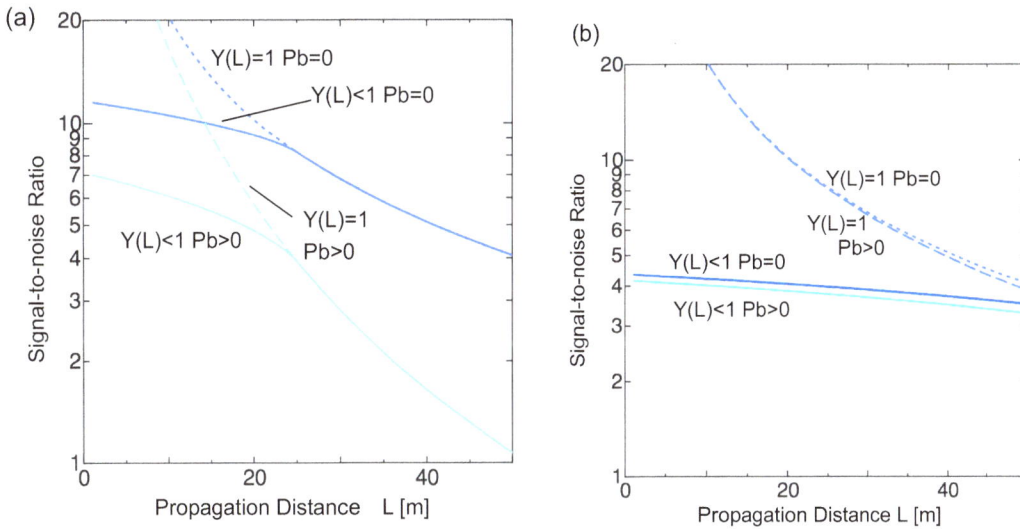

Fig. (8). Estimations of Signal-to-noise ratio of lidar echoes. (a) Receiver's FOV = 2 mrad, (b) Receiver's FOV = 0.1 mrad.

The signal-to-noise ratio *SNR* in the case of the ideal overlap function *Y(L)*=1 decreases monotonously, while that in the case of the calculated overlap function *Y(L)*<1 of Fig. **(7b)** is lower at closer range because of the initial rise of the overlap function.

When the FSA aperture becomes small of 150 μm as shown in Fig. **(7b)**, the overlap function rises more slowly. The decrease in the signal-to-noise ratio, however, will be smaller than the former case, as shown in Fig. **(8b)**. The estimation with diameter 150 μm is equivalent to the receiver's FOV of 0.1 mrad. In the measurement range of 0-50 m, the signal-to-noise ratio changes little.

Influence of the background light is also negligibly small. This performance is suited for airflow mapping or quantitative determination of the concentration of gas leakage. The inline typed compact lidar has such a unique performance for near range atmospheric sensing or gas detection.

3. APPLICATIONS

3.1. Inline Typed Raman Lidar

The lidar system was developed based on the concept of the lidar design of inline optics. The schematic is shown in Fig. **(9a)**, the overview is shown in Fig. **(9b)**, and the specifications are summarized in Table **4**. The system is compact as the size is 580 mm x 520 mm x 230 mm. The outgoing laser energy is 120 μJ, which is eye-safe when the beam diameter is broadened to >10 mm. The receiver's FOV is 2 mrad. A narrower FOV is selectable by using FSA of smaller aperture. The optical stability, however, will be sensitive when the smaller FSA is used.

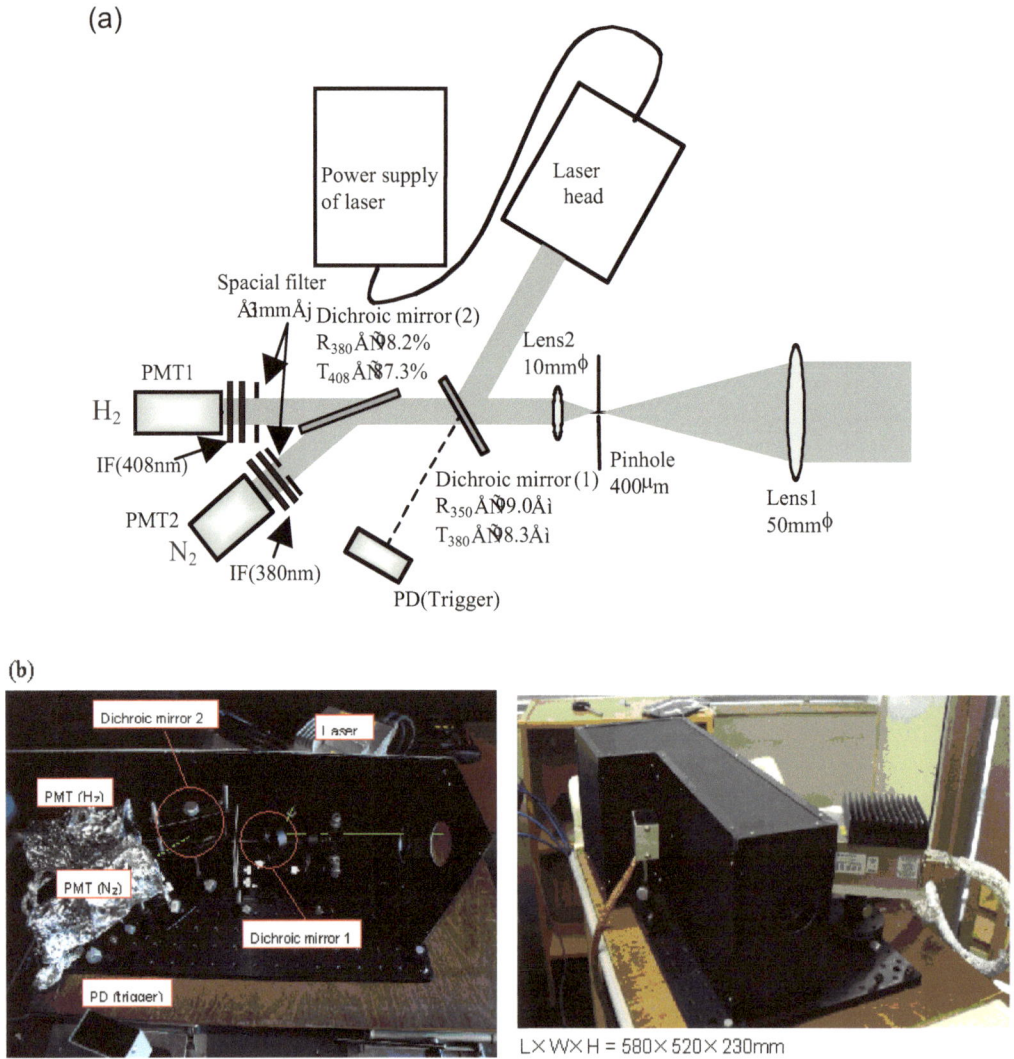

(a)

(b)

Fig. (9). Inline typed compact Raman lidar. (a) Lidar optics, (b) Overview of lidar setup.

Here, FSA of the diameter of 400 μm is selected in a hands-on form. The laser pulse reflected by dichroic filter (1) and goes to the air. The backscattered Raman signals goes back through the filter (1), and separated by dichroic filter (2) due to the wavelength. The detection of atmospheric N_2 is for the quantitative estimation of H_2 concentration.

Fig. **(10)** shows an example of the measured N_2 Raman signal. The result was estimated by subtracting the directly reflected light from the optics from the observed data. The N_2 Raman echo is observed from 0 m to

50 m, which corresponded to the system design. At the near distance of 2-3 m, there was a large peak in the observed data, which was caused by the direct reflection from the optics. The system calibration is conducted by subtracting the direct reflection light from the observation echo. The rise in the signal in Fig. **(10a)** is caused by the calibration. It can be further suppressed by decreasing the direct reflected light. The decay curve of the observation echo was inversely proportional to the square of the distance. When the atmosphere was calm, the overlap function was estimated by the observation data. The echo signal of N_2 was corrected by multiplying square of distance. By using the lidar equation, the distance corrected data was divided by the round-trip transmittance.

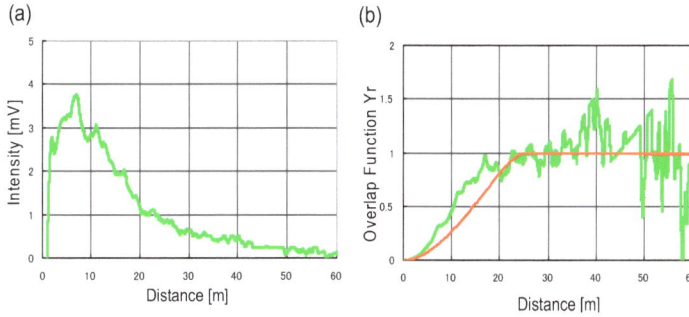

Fig. (10). Near range Raman lidar echo. (a) Near range lidar echo of N_2 in air. (b) Range corrected N_2 echo.

Table 4: Inline typed Raman Lidar

Laser	
Manufacturer	Spectra Physics
Product	Explorer
Power	12 µJ@5 ns
Wavelength	349 nm
RPF	<5 kHz
Detector	
Manufacturer	Hamamatsu Photonics
Product	H6780-03
Sensitivity	4.3×10^4 A/W
Dichroic Filter	350 nm/380 nm for laser /echo 380 nm/408 nm for N_2 / H_2
Interference Filter	Band width 4nm

$$Y(L) = \frac{P(L)L^2}{C\beta(L)T(L)^2} \qquad\qquad (4)$$

Here, $\beta(L)$ is constant in the near range and is derived from the slope method [1]. The result represents the overlap function. Inserting the pinhole diameter of 400 µm as the FSA, the actual receiver's FOV is 2.5 mrad. The overlap function rises from the 0 m and increases up to 20 m as shown in Fig. **(10b)**. It becomes unity from that distance. It means that the echo is defocused and stopped partially at the FSA up to 20 m, while at further distance, it passes through the FSA. The characteristics of the overlap function affects the receiving characteristics of signal-to-noise ratio. The characteristics of Fig. **(10b)** coincides with the calculation results of Fig. **(7b)**. The difference is caused by the difference of the receiver's FOV between the calculation (2 mrad) and the actual setup (2.5 mrad). As the rise in the overlap function is estimated, the nearest echo of 0-8 m signal is also correct. It means that the gas detection can be performed from the nearest distance of 0 m. It is also identical to the theoretical decay curve. The H_2 Raman signal was also confirmed to be detectable in the same observation range.

The H_2 gas concentration can be estimated by measuring the ratio between the intensities of the Raman scattering signals from atmospheric N_2 and H_2 gas. Fig. **(11)** shows the current result of H_2 gas detection.

An experiment was conducted to measure the H_2 gas filled in a cell. The cell was positioned at a distance of 7-8 m from the lidar. By changing the concentration of H_2 gas, the echo intensity was monitored with the developed lidar system. The result, shown in Fig. **(11)**, shows that the low concentration gas of about 10% was detected successfully.

However, the fluctuation of echo intensity was large. It came from the background light and thermal noise of the detector. The measurement was performed by a oscilloscope with the photomultiplier operating in analog mode. Since the explosion limit of H_2 in air is 4% [13], the measurement should be executed with a accuracy of less than 1%. To conduct the stable measurement, the echo should be fully summed with a photon counting mode. The performance of the compact Raman lidar can be further improved.

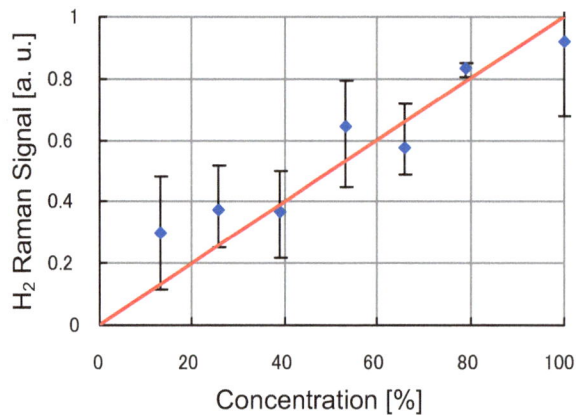

Fig. (11). Relation between H_2 Raman signal and H_2 gas concentration.

Typical lidar configurations, such as the biaxial or the coaxial type lidar, have a long blind area because a certain distance is required before the transmitted beam enters into the receiver's FOV. For near range measurement, a wide FOV could be used, but this results in higher background noise and suffers from the effects of the multiple scattering. The inline type lidar has the solution to observe the near range with the narrow FOV. The Raman lidar is a suitable application for the inline type optics because the dichroic mirror separates the lidar echo from the outgoing laser beam with high efficiency. The lidar optics is simple. The optical design of the inline optics allows estimation and control of the blind area.

When the inline optics is applied to the Mie lidar, additional optics is needed to separate the receiving echo from the outgoing beam. As a solution, an optical circulator and a pair of Axicon prisms can be installed into the lidar optics to realize the inline optics. This is described in the following section.

4. OTHER APPROACHES

In this section, several kinds of the inline and coaxial type Mie lidars for near range measurement are introduced. Additional optics is needed to separate the receiving echo from the outgoing laser beam. The optical circulator, a pair of Axicon prisms, and the other optical techniques were installed into the lidar optics to realize the near range measurement. All optical components were selected to realize a high polarization extinction ratio and to withstand the high power of the light source. All optical components have are slightly tilted with respect to the optical axis to eliminate the directly reflected light.

4.1. Inline typed Micro Pulse Lidar

Fig. **(12)** shows the inline typed MPL (Micro Pulse Lidar) for local disaster prediction. This lidar can estimate the depolarization due to scattering by particles by receiving the orthogonally polarized echoes. In other words, the system can distinguish the ice-crystals and the spherical water droplets. Rough weather which may lead to local disasters can be predicted by monitoring the ice-crystal's flow at low altitudes.

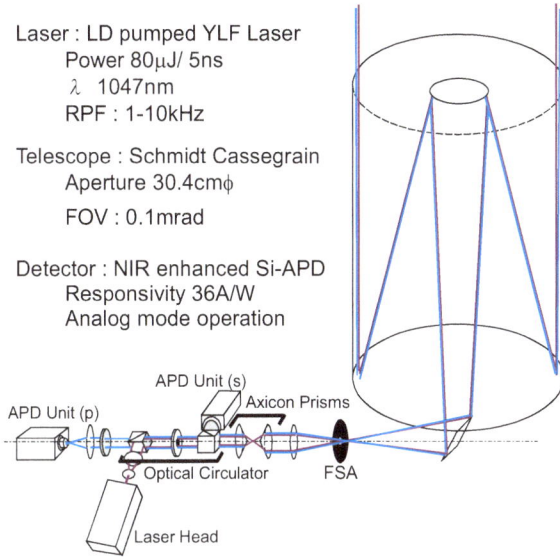

Fig. (12). Inline typed micro pulse lidar.

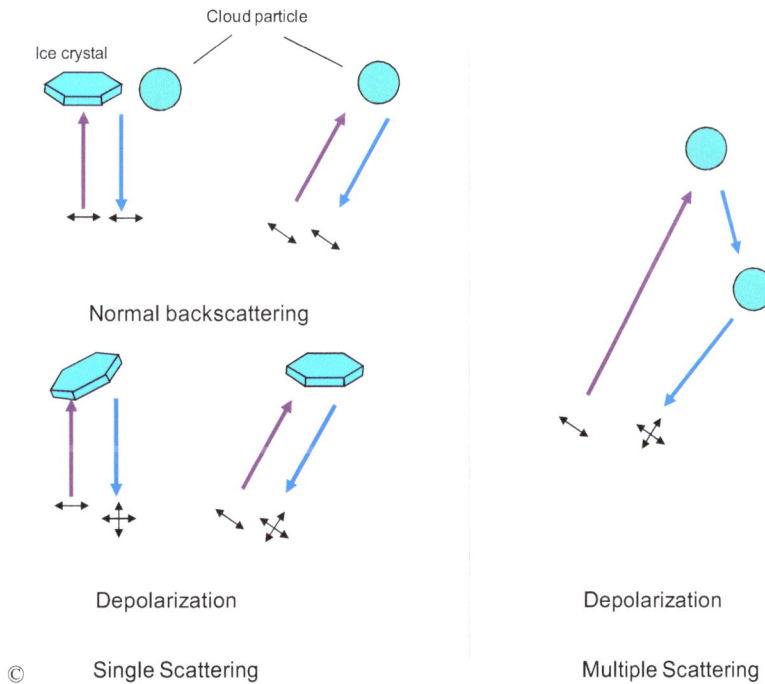

Fig. (13). Normal backscattering and depolarization by ice-crystals and cloud particles.

Cloud particles, which we intend to measure, have a diameter of about 5 µm, and cause Mie scattering [14-18]. The polarization plane of incident bream and backscattering echo against a cloud particle and an ice-crystal are illustrated in Fig. **(13)**. In the scattering, the backscattering echoes maintain the incident polarization plane when the beam hits the cloud particles (spherical particles). On the contrary, non-spherical particles such as ice-crystals change the incident polarization. This is known as the depolarization effect. It is caused by the difference of Fresnel's refraction coefficient between the parallel and the orthogonal polarization and multiple reflection inside a crystal. When the beam is normally incident on the surface of an ice-crystal, depolarization never occurs. When the beam hits the crystal at an angle, one can distinguish the ice-crystals from the spherical particles such as cloud particles by examining the orthogonal polarization components of the lidar echoes.

The spherical particles, however, cause depolarization in the case of multiple scattering due to multiple particles [19-21]. To reduce the contribution of multiple scattering, it is effective to narrow the receiver's FOV. The biaxial and coaxial lidars with narrow FOV have a broad blind area, because a certain distance is needed for the transmitted beam to enter into the receiver's FOV (the overlap function has a slow rise with respect to distance). In contrast, the inline type lidar can have a narrow FOV with no blind area.

In the inline typed MPL, the optical circulator and the pair of Axicon prisms were installed into the lidar optics. The laser source is a LD pumped YLF laser, of wavelength 1.047 μm, pulse width 5 ns, output energy 80 μJ, and maximum pulse repetition frequency 50 kHz.

The transmitted beam is collimated by the beam expander just behind the laser head and passes though the optical circulator. The optical circulator allows measurement of the orthogonally polarized echoes with two detectors, that is, the same polarized component (p-component) as the transmitting beam and the orthogonal one (s-component) [5]. The s-component echo is reflected at the polarized beam splitter and enters the detector APD(s), while the p-component echo passes through the circulator optics and goes to the detector APD(p). The insertion loss of the optical circulator as the transmitter is 2 dB and the isolation between the orthogonal polarized echoes is about 20 dB. As the insertion loss of the Faraday rotator was large, the sensitivity of the s-component echo became twice of that of the p-component. It is convenient to estimate the depolarization effect by comparing the weak s-component echo with the p-component. All optical elements have small tilts (1.5 degrees) against the optical axis. The laser head also has a tilt of 3 degrees against the optical axis. As the inline MPL could not reject the directly reflected light from the lidar optics perfectly, the reflected light influenced the lidar echo signal (ringing or saturation of the lidar receiver's electrical circuit, *etc.*). Pinholes were inserted in front of the detectors (APDs) in order to prevent the light directly reflected from the inline optics from entering the detectors. Its aperture was determined in combination with the tilts of the inline optics.

When the transmitted beam passes through the telescope, part of the beam is reflected by the secondary mirror, obstructing the transmitted beam and causing a large noise signal upon returning to the detector. Therefore, a pair of Axicon prisms was installed to prevent this reflection. The prism pair creates an annular beam, whose hole overlaps the secondary mirror of the reflecting telescope. The insertion loss of the prisms was 0.932 dB. The insertion loss of the whole lidar optics was about 3 dB. The total energy of the output beam was 8 kW. The eyepiece focuses the transmitting annular beam at the field stop aperture (FSA). The FSA is a pinhole of 650 μm diameter. It was put at the focal point of the telescope. At the focal point, the annular beam changes its beam shape into the nearly non-diffractive beam, of which spot size is less than 4 μm diameter [22-30]. The beam then passes though the FSA without any obstruction. The non-diffractive beam retransformed its beam shape into the annular beam again through the enlargement in the telescope. The annular beam finally goes to the telescope.

A Schmidt-Cassegrain reflecting telescope was used for the transmitter and receiver. The annular beam was expanded up to the telescope aperture (304 mm diameter). The quality of the transmitting beam was verified by the identification between the annular beam enlarged by the telescope and that in front of the eyepiece. The light directly reflected from the in-line optics was suppressed to below 0.1 mW in front of the detectors. The FOV of the telescope is 0.1 mrad, determined by the focal length of the telescope and the aperture size of the FSA. The divergence of the transmitting beam can be controlled by the beam expander. Though the divergence does not depend on the telescope's FOV, it is basically less than the FOV.

The detectors are NIR enhanced Si-APDs (PerkinElmer 30954E). They have a responsivity of 36 A/W at wavelength 1064 nm and at the operating voltage. The responsivity rises up over 500 A/W near the breakdown voltage. The voltage over the breakdown is in the region of the Geiger mode. In our system, the APDs were used in the analog mode and Geiger mode on a case-by-case basis. All optics including the laser head and the detectors were mounted on the telescope tube. It allows arbitrary direction measurement of lidar echoes from the near distance with the narrow FOV. The incident beam energy into an human eye was estimated as 5.34 J/s. It is lower than the maximum permissible exposure (MPE) value of 11.4 J/s in the same condition [31]. The inline MPL ensures eye safety even in front of the system.

The actual disaster prediction of the local weather change will be accomplished by the lidar mounted on a vehicle. The illustration is shown in Fig. **(14)**. The inline MPL is compact, and the compact car is available for the on-site observation. The prediction time is 5-10 minutes before the heavy rain. The MPL is eye safe, and it is useful to observe in locations such as urban canyons, near an airport, near a port.

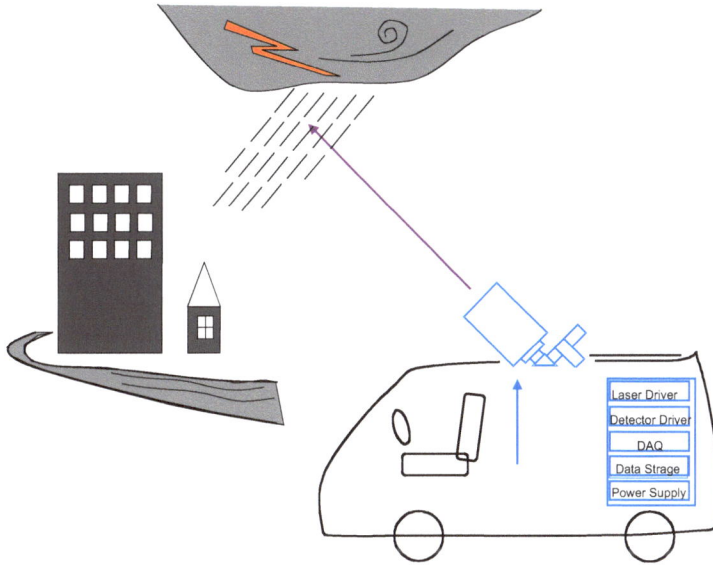

Fig. (14). Lidar vehicle measurement for local weather disaster prediction.

At first, the authors would have grasped not only the omen of the local weather change, which lead to heavy rain, but also the sign of the lightning attack. The omen of heavy rain was successfully detected before the fact, but it is hard to distinguish the sign of the lightning discharge from the observation data of the orthogonal polarization.

4.2. High Precision Polarization Lidar

After the development of the inline MPL, a high precision polarization lidar was designed to detect lightning strikes.

The interaction between the light and the atmosphere is caused by scattering (Mie, Rayleigh, Raman), and also by the magneto-optical effect and electro-optical effect. The magneto-optical effect (Faraday effect) is associated with the lightning discharge. It has been reported as an optical measurement method for magnetic confinement fusion reactor [32]. The polarization plane of a beam propagating parallel to the magnetic flux is rotated in a partially ionized atmosphere (plasma) as shown in Fig. **(15)**.

The rotation angle is proportional to the product of the ionization electron density n_e and the magnetic flux density B along the beam propagation path. The linearly polarized beam can be regarded as a combination of the clockwise and the counterclockwise circularly polarized beams. The refractive indices of the ionized atmosphere for each circularly polarized beam are as follows.

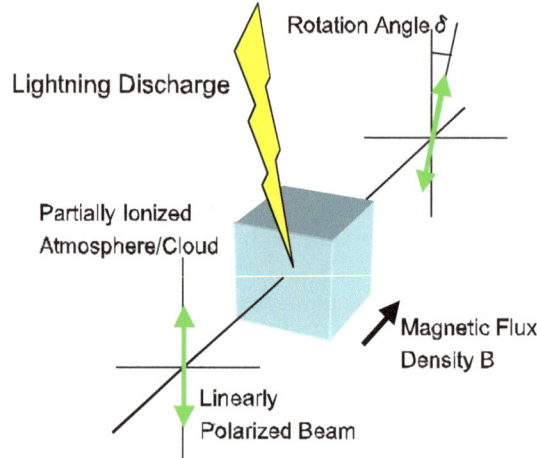

Fig. (15). Faraday effect.

$$n_{\pm} = \left(1 - \frac{\omega_{pe}^2}{\omega^2} \frac{\omega}{\omega \pm \omega_{ce}}\right)^{1/2}$$

$$\omega_{pe} = \sqrt{\frac{e^2 n_e}{\varepsilon_0 m_e}} \qquad \omega_{ce} = \frac{eB}{m_e}$$

(5)

where ω_{pe}, ω_{ce} are the plasma and electron cyclotron frequencies, respectively, e is the fundamental charge, m_e is the electron mass, and ε_0 is the permittivity of free space. Therefore, the rotation angle of polarization of the beam propagated at distance L $(=L_2-L_1)$ is obtained as follows.

$$\delta = \frac{\pi}{\lambda} \int_{L_1}^{L_2} (n_+ - n_-) dl$$

$$= 2.62 \times 10^{-13} \lambda^2 \int_{L_1}^{L_2} n_e B dl$$

(6)

where λ is the wavelength of the propagating beam. Since δ is proportional to λ^2, the rotation angle for visible light is small. Therefore, the polarization angle rotation must be measured with high accuracy in order to detect lightning discharge.

When the Faraday effect is applied to lightning measurement, the atmosphere needs to be partially ionized, and the magnetic flux due to the lightning discharge must exist. Cloud-to-cloud discharge, which causes 20-30 times continuous discharge, satisfies those conditions [33-36].

The analysis and experimental results have shown that the rotation angle of polarization plane of the propagating beam is less than 1 degree, so that the mutually perpendicular polarization components must be measured with a sensitivity and accuracy of >30 dB in order to detect lightning discharges.

The rotation of the polarization plane only occurs in a nearly perfectly ionized atmosphere, so the signal cannot be detected unless the transmitted beam intersects the discharge path. On the other hand, the shock wave (variation in the neutral gas density) generated by the discharge can be detected over a broader range, while it causes no rotation of the polarization plane [37]. Both of the orthogonal polarization echo was forced to fluctuate with the same law. This was confirmed by high voltage discharge experiment [38].

Therefore, the scenario of the lightning detection using the lidar system is designed as follows. At first the system roughly scans the sky in the direction in which the occurrence of a cloud-to-cloud lightning discharge (or thunderclap) is likely. If a shock wave is detected, the 3-dimensional lightning position is estimated. Next, by scanning the neighborhood of the lightning position with higher precision to intersect the propagating beam and the lightning discharge path, the rotation angle of the polarization plane will be measured. In general, lower area (bottom) of clouds will be scanned, as the beam penetrates only a few hundred meters in clouds.

The distribution of the discharge location and its change will lead to the prediction of lightning strike. The lidar system must be capable of measurement at near range with a narrow field of view in order to eliminate the effects of multiple scattering. The use of in-line optics is effective in meeting this requirement. The system must also have scanning capability to search the discharge. The concept of the lidar system for lightning detection is shown in Fig. **(16)**. For the detection of the small rotation angle, differential detection should be used.

The optical design is shown in Fig. **(17)** [39,40]. This lidar is also installed the optical circulator and a pair of Axicon prisms to accomplish the inline lidar optics. They makes possible to capture the near range echo. This lidar has the high power laser of 200 mJ at wavelength 532nm. The outgoing laser beam is in balanced with the orthogonal polarization components, and the polarization independent optical circulator is originally developed.

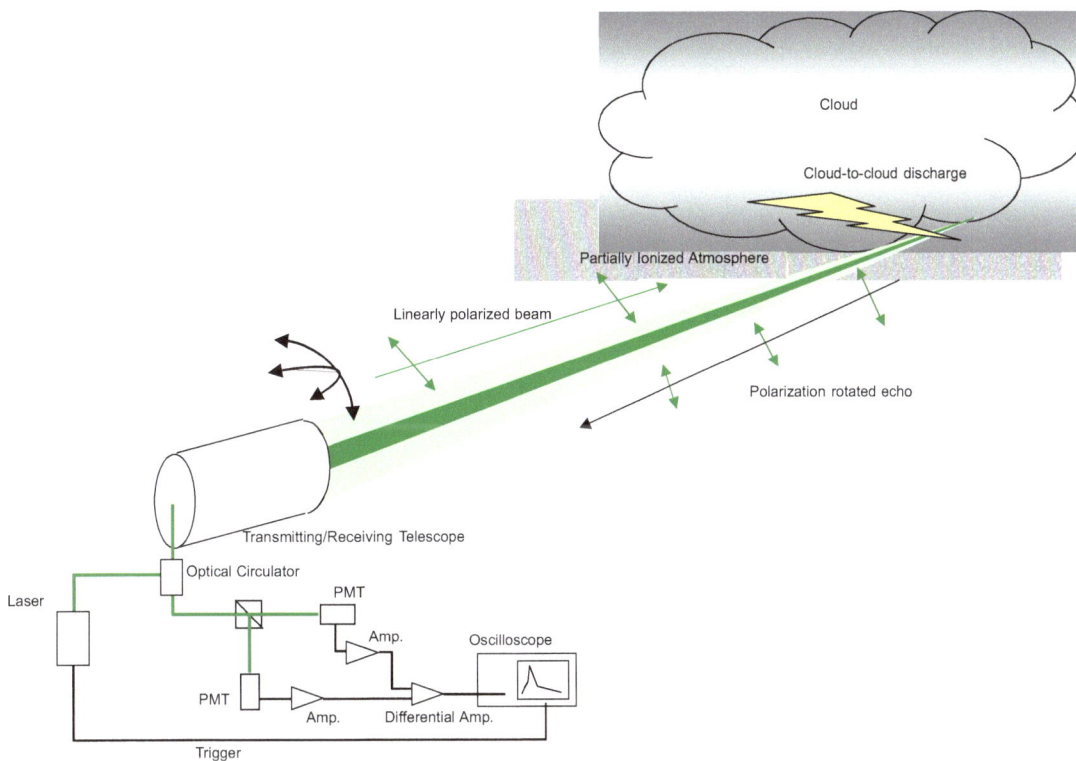

Fig. (16). Concept of Lightning discharge measurement by lidar.

The detection range is from the nearest of 50m to the far range of 20 km, while the receiver's FOV is 0.177 mrad. The main feature is the high precision polarization detection. By using the Glan laser prisms and air-gaped Faraday rotators, the accuracy of the orthogonal polarization detection is more than 30 dB. It is for the detection of the small rotation angle of the polarization plane, which is caused by the lightning discharge (Faraday effect). As results of the numerical analysis, the grand-based experiment and high voltage discharge experiment, the rotation angle of the polarization plane of the propagating beam is estimated as less than 1 degree [38,41,42]. To obtain such a small rotation angle, the lidar has the differential detection in balanced with the orthogonal polarization beams and the high precision polarization independent optical circulator. The system has been accomplished successfully and started continuous measurement through a year.

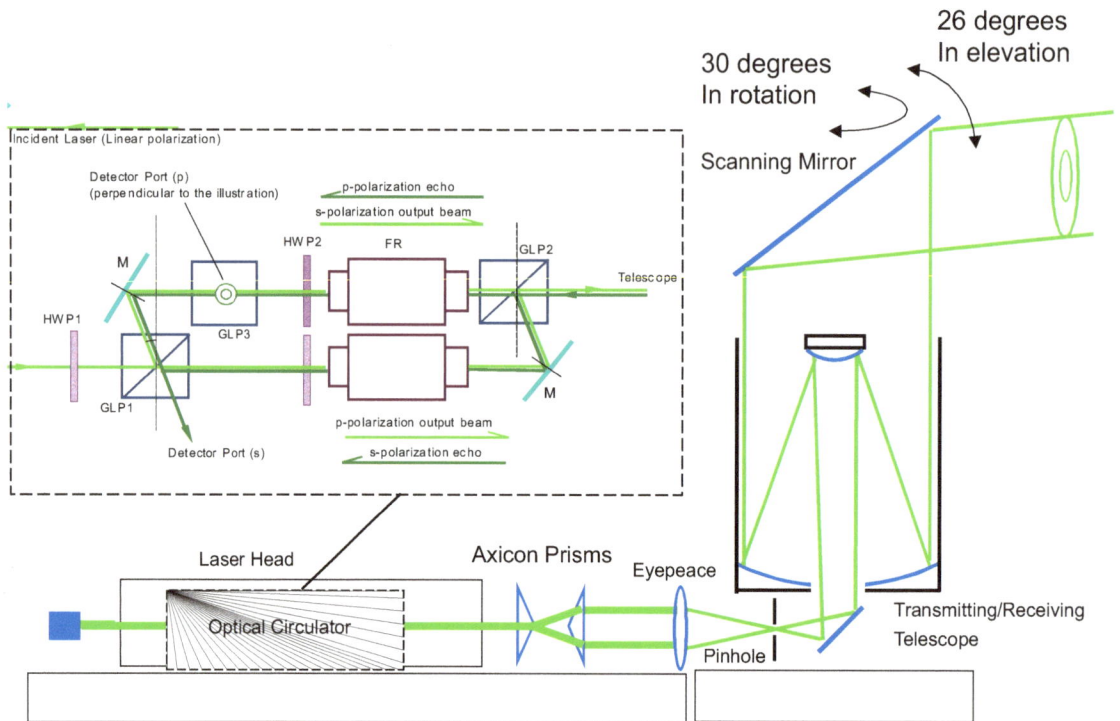

Fig. (17). Optical design of high precision polarization lidar.

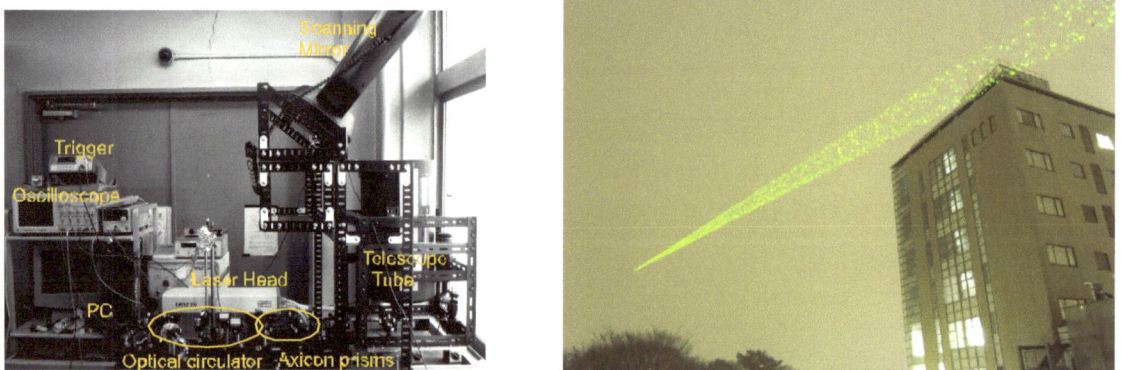

Fig. (18). Photograph of high precision polarization lidar and transmitted laser beam.

Photographs of the high precision polarization lidar and transmitted laser beam are shown in Fig. **(18)**. In the photograph, the laser beam propagates from the upper right to the lower left. The speckle patterns are due to the scattering by particles in the air.

4.3. Compact LED Lidar

The unique lidar with a Light-Emitting Diode (LED) module as a light source is under development. The optical layout of the system is shown in Fig. **(19a)**, and a photograph of the system is shown in Fig. **(19b)** [43].

The LED module is very compact, cheap, and comes in a variety of wavelengths in comparison with laser diode modules and withstands rough treatment. It is developed for a certain gas detection in a near range, and air flow in a closed space, and so on. This lidar is a coaxial type. The usual coaxial lidar, which transmits the outgoing beam from the secondary mirror of the telescope, has the large blind area. The LED mini lidar, however, has no blind area because of the consideration of the arrangement of the outgoing beam.

(a)

Fresnel Lens
190mm

É",f=140mm

Mirror

Power
Supply

Drive
circuit

FSA:3mmÉ"
+ IF

PMT

Photon Counter

Outgoing Beam
60mmÉ"

(b)

23cm x 23cm x 21cm

Mirror

LED Unit

PMT Unit

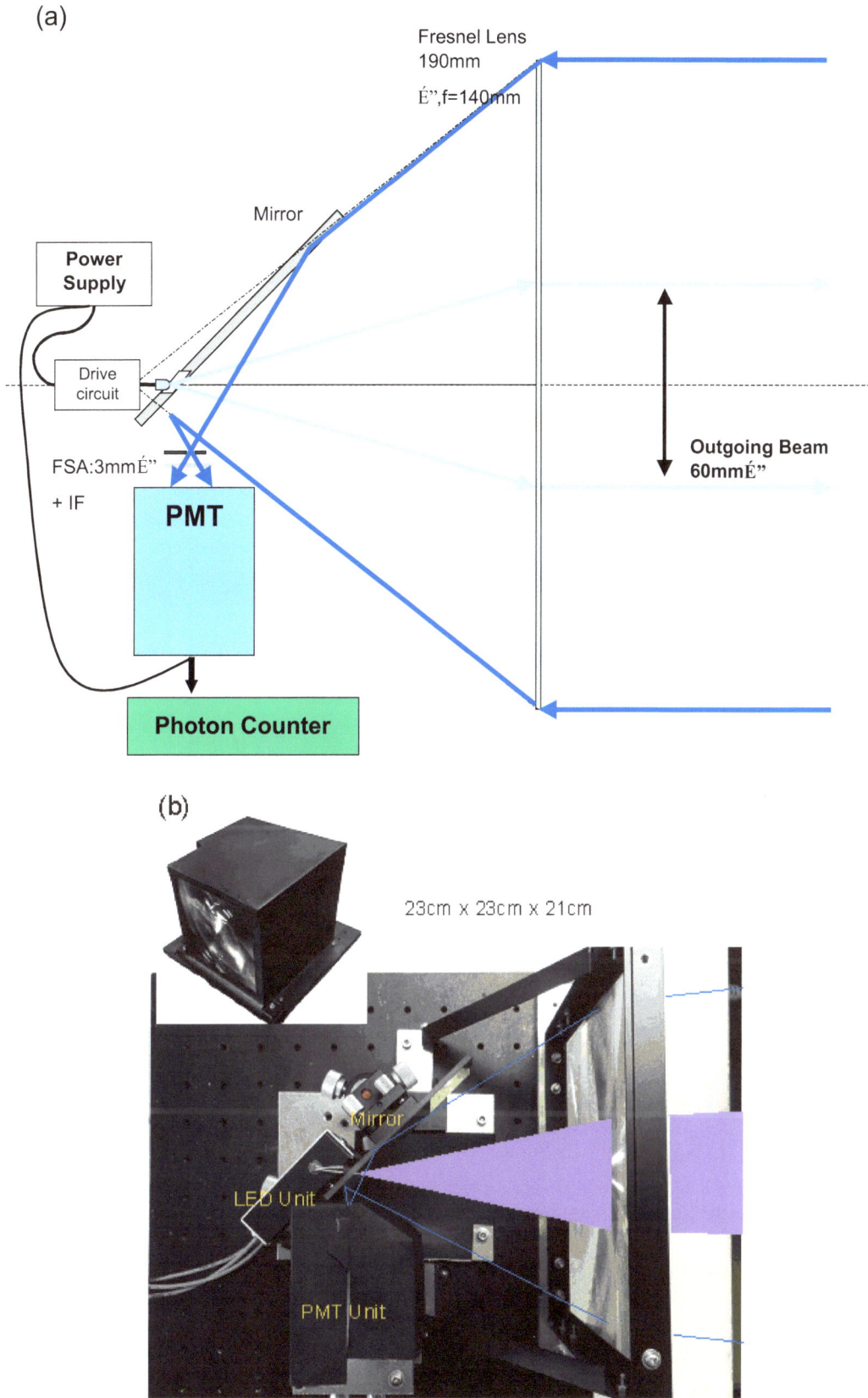

Fig. (19). Coaxial typed LED mini lidar. (a) Optical layout of LED mini lidar, (b) Photograph of LED mini lidar.

The handmade LED module consists of a LED lamp with a dome-lens of plastic mold and its driver circuits. It is fixed in front of the focal point of the Fresnel lens (Objective lens), while the detector (PMT unit with FSA and interference filter) is arranged at the focal point. The LED lamp with the lens has a light emitting aperture of the order of mm.To collimate the outgoing beam, its size should be expanded. In this setup, the beam aperture is 60 mm with a beam divergence of 9.5 mrad. The receiver's FOV is 10 mrad. The optical setup is accomplished by a mirror with a light emitting hole (patent pending). That is, LED module and detector unit were arranged by the mirror. The output power of the pulsed LED is 200 mW at blue and red color LEDs. The detection range is from 0 m to a few hundred meters. The pulse width is 10 ns at NUV-LED of 493 nm. Pulse repetition rate is >100 kHz. The lidar system is compact and is mounted on a breadboard of about 20 cm square. LED module, detector unit and the power supply were set in the box.

Monitoring of plant vegetation and near range gas and atmosphere measurement were conducted in the range of a few m to 100 m. Fig. **(20)** shows a possible application for the LED mini lidar. The system is so compact that handy system will be also selectable by installing a buttery unit. Data acquisition is accomplished by an oscilloscope for hard target detection and multi-channel scaler for the atmosphere measurement.

The LED mini lidar is thought to be applicable to air quality monitoring in large closed spaces, such as exhibit halls, factory buildings, auditoria, and gymnasia. Since the LED can be pulsed at high repetition rates, flow visualization of particles by particle imaging velocimetry should be possible by using a highly sensitive, high-speed imaging device such as a CCD camera with an image intensifier.

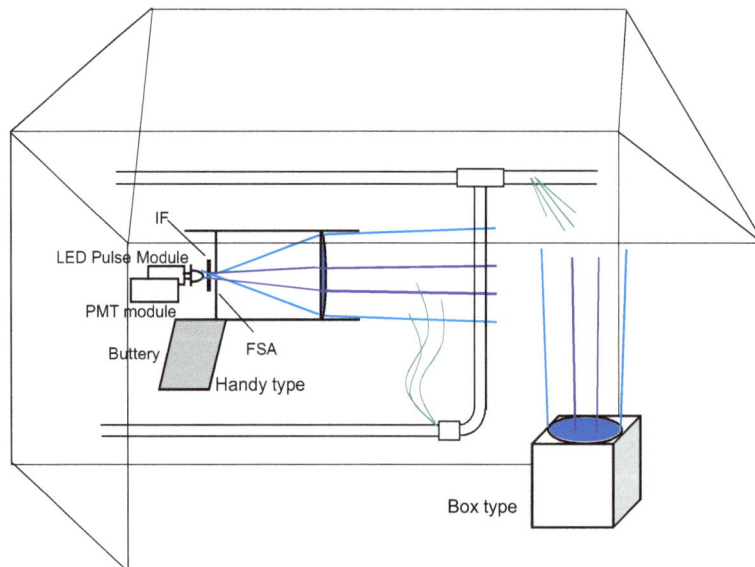

Fig. (20). Conceptual diagram of indoor measurement using LED mini lidar.

5. CONCLUSION

The near range lidar makes possible to detect certain kinds of dangerous gas and dusts, and to monitor an airflow in a closed space. The detection range can design from zero meters to the adequate distance. The inline optics is a key technique. The inline optics may make complex the system alignment and the direct reflection from the optics should be cared not to insert into the detector module. The near range measurement does not only have the merit to detect the atmosphere or certain gas in front of the system, but also has a function of hazard avoidance. The near range lidar often throws the outgoing beam in horizontal direction or to the low altitude. So the beam power should be eye-safe. Otherwise, the system should be stopped to throw the beam when an obstacle including a person is in the monitoring area. The blind area will cloud its judgment. No blind area is identical.

Now there are several varieties of optical design for near range measurement. Not only the inline optics, but also the coaxial optics can eliminate the blind area. The optical design is selectable due to the system structure and application.

ACKNOWLEDGEMENTS

The part of this study is funded by the Grant-in-Aid for Science Research Japan. The author expresses his gratitude to Japan Society for the Promotion Science.

REFERENCES

[1]   R. M. Measures, *Laser Remote Sensing; Fundamentals and Applications*, John Wiley & Sons, New York, 1984.

[2]   J. D. Spinhirne, "Micro Pulse Lidar", *IEEE Transactions on Geoscience and Remote Sensing*, Vol.31, 48-55, 1993.

[3]   J. D. Spinhirne, J. A. R. Rall, and V. S. Scott, "Compact Eye Safe Lidar System", *Review of Laser Engineering*, Vol.23, 112-118, 1995.

[4]   H. S. Lee, I. H. Hwang, J. D. Spinhirne, V. S. Scott, "Micro Pulse Lidar for Aerosol and Cloud Measurement", *Advances In Atmospheric Remote Sensing with Lidar*, Springer, 7-10, 1997.

[5]   T. Shiina, E. Minami, M. Ito, and Y. Okamura, "Optical Circulator for In-line Type Compact Lidar", *Applied Optics*, Vol. 41, No. 19, 3900-3905, 2002.

[6]   T. Shiina, K. Yoshida, M. Ito, and Y. Okamura, "In-line type micro pulse lidar with annular beam -Experiment-", *Applied Optics*, Vol.44, No.34, 7407-7413, 2005.

[7]   T. Shiina, K. Yoshida, M. Ito, and Y. Okamura, "In-line type micro pulse lidar with annular beam -Theoretical Approach-", *Applied Optics*, Vol.44, No.34, 7467-7473, 2005.

[8]   H. Miya, T. Shiina, T. Kato, K. Noguchi, T. Fukuchi, I. Asahi, S. Sugimoto, H. Ninomiya and Y. Shimamoto, "Compact Raman Lidar for Hydrogen Gas Leak Detection", *Proceedings of CLEO Pacific Rim 2009*, 1-2, 2009.

[9]   T. Halldorsson and J. Langerholc, "Geometrical form factors for the lidar function", *Applied Optics*, Vol. 17, No. 2 pp. 240-244, 1978.

[10] J. Harms, "Lidar return signals for coaxial and noncoaxial systems with central obstruction", *Applied Optics* Vol. 18, No. 10, pp.1559-1566, 1979.

[11] N. Sugimoto, I. Matsui, and Y. Sasano, "Design of lidar transmitter-receiver optics for lower atmospheric observations: geometrical form factor in lidar equation", *Japan Journal of Optics*, Vol. 19, pp. 687-693, 1990.

[12] T. Fujii and T. Fukuchi Eds., *Laser Remote Sensing*, Chapter 3, Taylor&Francis, 2005.

[13] H. Ninomiya, S. Yaeshima, K. Ichikawa, and T. Fukuchi, "Raman lidar system for hydrogen gas detection", Optical Engineering, Vol.46, 094301, 2007

[14] S. R. Pal and A. I. Carswell, "Polarization properties of lidar backscattering from clouds", *Applied Optics*, Vol. 12, No. 7, pp.1530-1535, 1973.

[15] J. S. Ryan, S. R. Pal, and A. I. Carswell, "Laser backscattering from dense water-droplet clouds", *Journal of the Optical Society of America*, Vol. 69, No. 1, pp.60-67, 1979.

[16] K. Sassen, "The polarization lidar technique for cloud research: A review and current assessment", *Bulletin of the American Meteorological Society*, Vol.72, No.12, pp.1848-1866, 1991.

[17] G. Zaccanti, P. Bruscaglioni, M. Gurioli, and P. Sansoni, "Laboratory simulations of lidar returns from clouds : experimental and numerical results", *Applied Optics*, Vol. 32, No. 9, pp.1590-1597, 1993.

[18] M. Gai, M. Gurioli, P. Bruscaglioni, A. Ismaelli, and G. Zaccanti, "Laboratory simulations of lidar returns from clouds", *Applied Optics*, Vol. 35, No. 27, pp.5435-5442, 1996.

[19] K. Sassen and R. L. Petrilla, "Lidar depolarization from multiple scattering in marine stratus clouds", *Applied Optics*, Vol. 25, No. 9, pp.1450-1458, 1986.

[20] M. Kerscher, W. Krichbaumer, M. Noormohammadian, and U. G. Oppel, "Polarized Multiply Scattered LIDAR Signals", *Optical Review*, Vol. 2, No. 4, pp.304-307, 1995.

[21] P. Bruscaglioni, A. Ismaelli, G. Zaccanti, M. Gai, and M. Gurioli, "Polarization of Lidar Returns from Water Clouds : Calculations and Laboratory Scaled Measurement", *Optical Review*, Vol. 2, No. 4, pp.312-318, 1995.

[22] P. Belanger and M. Rioux, "Ring Pattern of a Lens-Axicon Doublet Illuminated by a Gaussian Beam", *Applied Optics*, Vol. 17, No. 7, pp.1080-1086, 1978.

[23] J. Durnin and J. J. Miceli, Jr., "Diffraction-Free Beams", *Physical Review Letters*, Vol. 58, No. 15, pp.1499-1501, 1987.

[24] G. Indebetouw, "Nondiffracting Optical Fields: Some Remarks on their Analysis and Synthesis", *Journal of the Optical Society of America A*, Vol. 6, No. 1, pp.150-152, 1989.

[25] G. Scott and N. McArdle, "Efficient Generation of Nearly Diffraction-Free Beams using an Axicon", *Optical Engineering*, Vol. 31, No. 12, pp.2640-2643, 1992.

[26] R. Arimoto, C. Saloma, T. Tanaka, and S. Kawata, "Imaging Properties of Axicon in a scanning optical system", *Applied Optics*, Vol. 31, No. 31, pp.6653-6657, 1992.

[27] L. L. Doskolovich, S. N. Khonina, V. V. Kotlyar, I. V. Nikolsky, V. A. Soifer, and G. V. Uspleniev, "Focusators into a Ring", *Optics and Quantum Electronics*, Vol. 25, pp.801-814, 1993.

[28] K. Kono, Y. Mitarai, and T. Minemoto, "New Super-Resolution Optics with Double-Concave-Cone Lens for Optical Disk Memories", *Journal of Optical Memory and Neural Networks*, Vol. 5, No. 4, pp.279-285, 1996.

[29] K. Kono, M. Irie, and T. Minemonto, "Generation of Nearly Diffraction-Free Beam Using a New Optical System", *Optical Review*, Vol. 4, No. 3, pp.423-428, 1997.

[30] V. Soifer, V. Kotlyar, and L. Doskolovich, *Iterative Methods for Diffractive Optical Elements Computation*, Taylor & Francis, 1997.

[31] American National Standards Institute, "American national standard for the safe use of lasers", ANSI Z136.1-1986, pp.1-96, 1986.

[32] K. Kawahata and S. Okajima, "Interferometry and Polarimetry –Principle of Interferometry and Polarimetry-", *Japan Society of Plasma Science and Nuclear Fusion Research*, Vol.76, No.9, pp.845-847, 2000 (in Japanese).

[33] E. Franzblau and C. J. Popp, "Nitrogen oxides produced from lightning", *Journal of Geophysical Research*, Vol.94, No.D8, pp.11.089-11.104, 1989.

[34] E. Franzblau, "Electrical discharges involving the formation of NO, NO_2, HNO_3, and O_3" , *Journal of Geophysical Research*, Vol.96, No.D12, pp.22.337-22.345, 1991.

[35] J. Stith, J. Dye, B. Ridley, P. Laroche, E. Defer, K. Baumann, G. Hubler, R. Zerr, and M. Venticinque, "NO signature from lightning flashes", *Journal of Geophysical Research*, Vol.104, No.D13, pp.16.081-16.089, 1999.

[36] Society of Atmospheric Electricity of Japan, *Atmospheric Electricity*, Chapter 2, Corona publishing Tokyo, 2003 (in Japanese).

[37] T. Fukuchi, K. Nemoto, K. Matsumoto, and Y. Hosono, "Visualization of High-speed Phenomena using an Acousto-optic Laser Deflector", *IEEJ Transactions on Fundamentals and Materials*, Vol.125-A, No.2, pp.113-118, 2005 (in Japanese).

[38] T. Fukuchi and T. Shiina, "Measurement of rotation of polarization plane of laser radiation propagating though impulse discharge in air", *IEEJ Transactions on Electrical and Electric Engineering*, in press.

[39] T. Shiina, T. Honda, and T. Fukuchi, "High-precision polarization lidar - lidar in-line optics -", *CLEO Pacific Rim 2007 proceedings*, pp.1499-1500, 2007.

[40] T. Shiina, M. Miyamoto, D. Umaki, K. Noguchi and T. Fukuchi, "Fundamental Measurement by In-line typed high-precision polarization lidar", *SPIE Asia-Pacific Remote Sensing 2008 proceedings*, Vol.7153, pp.71530B-1 - 71530B-8, 2008.

[41] T. Shiina, T. Honda, and T. Fukuchi, "Measurement of polarization plane rotation of propagating light in a partially ionized atmosphere under discharge conditions", *Electrical Engineering in Japan*, Vol.171, No.3, 1-6, 2010.

[42] T. Shiina, T. Honda, and T. Fukuchi, "Evaluation of polarization angle rotation of propagating light in a partially ionized atmosphere under discharge conditions", *Electrical Engineering in Japan,* Vol.163, No.4, 1-7, 2008.

[43] M. Koyama and T. Shiina, "Light Source Module for LED Lidar", *proceedings of 25th International Laser Radar Conference*, S01P-32-1 – 4, 2010.

CHAPTER 3

Gas Sensing Using Laser Absorption Spectroscopy

Tetsuo Fukuchi[*]

Electric Power Engineering Research Laboratory, Central Research Institute of Electric Power Industry, 2-6-1 Nagasaka, Yokosuka-shi, Kanagawa 240-0196, Japan

Abstract: Gas detection is important for industrial safety, environmental protection, and environmental monitoring. Absorption spectroscopy is a widely used technique for gas detection. By applying laser remote sensing techniques, gas sensing with large standoff distances becomes possible. The industry standard method uses the double frequency technique, in which the laser wavelength is modulated about a center wavelength corresponding to the absorption peak of the target gas species, and the optical signal is detected at the second harmonic of the modulation frequency. Alternatively, by selecting two wavelengths which correspond to strong and weak absorption of the target gas species, two-dimensional imaging or visualization of gas leaks becomes possible. These techniques have been applied to detection and imaging of natural gas (methane gas) leaks. Laser sensing is also used for *in situ* gas analysis in power plants and incineration plants. Recent progress in quantum cascade lasers has enabled detection of minor species in harsh environments such as flue gas.

Keywords: Absorption spectroscopy, 2f method, differential absorption method, methane gas, visualization, flue gas, stack exhaust gas, differential absorption lidar, atomic laser absorption spectroscopy, quantum cascade laser.

1. INTRODUCTION

Various gases are used in the energy and chemical industries, which include flammable species such as methane (natural gas) and hydrogen, and toxic species such as hydrogen sulfide and cyanide compounds. In addition, combustion of fossil fuels result in production of nitrogen oxides and sulfur oxides. These gas species are subject to environmental regulation and industrial safety regulation, so detection of leaks and emission monitoring are of major interest.

Most gas species can be detected using chemical, physical, electronic sensors, or standard analytic techniques such as mass spectrometry and Fourier Transform InfraRed (FTIR) spectroscopy. Recent progress in Cavity RingDown Spectroscopy (CRDS) has enabled extremely sensitive detection of minor species. However, these techniques require contact with the gas sample, which necessitates the placement of a large number of sensors to cover an installation, or manual sensing using a portable sensor with gas sampling. The former is not realistic when the potential leak location extends over a large area (such as a gas pipeline), and the latter may be hazardous to the operator, especially if the gas species is toxic.

By using laser remote sensing techniques, gas species can be detected at a safe standoff distance. There are several laser detection techniques which can be employed for gas sensing, including Tunable Diode Laser Absorption Spectroscopy (TDLAS) [1], Laser-Induced Fluorescence (LIF) [2], Raman scattering and resonant Raman scattering [2], and Coherent Anti-Stokes Raman Scattering (CARS) [3]. TDLAS, Raman scattering, and LIF have been employed in lidar (Light Detection And Ranging) studies of the atmosphere, and all methods have been applied to plasma diagnostics and combustion studies. In this chapter, the application of TDLAS to gas sensing is described. A brief description of the methodology, and application to methane gas detection are presented. In addition, recent progress in gas detection using quantum cascade lasers is presented. Gas sensing using differential absorption lidar (DIAL) is also briefly described.

*****Address correspondence to Tetsuo Fukuchi:** Electric Power Engineering Research Laboratory, Central Research Institute of Electric Power Industry, 2-6-1 Nagasaka, Yokosuka-shi, Kanagawa 240-0196, Japan; Tel: +81-46-856-2121, Fax: +81-46-856-3540, E-mail: fukuchi@criepi.denken.or.jp

Tetsuo Fukuchi and Tatsuo Shiina (Eds)

2. METHODOLOGY

2.1. Measurement Principle

Gas detection by absorption spectroscopy is based on the Beer-Lambert law, which describes the attenuation of propagating light due to absorption,

$$I(\lambda) = I_0(\lambda)\exp[-\alpha(\lambda)L] \tag{1}$$

where $I_0(\lambda)$ and $I(\lambda)$ are the transmitted light intensities at wavelength λ in the absence and in the presence of the target gas species, respectively, $\alpha(\lambda)$ is the absorption coefficient of the target gas species, and L is the optical path length. The absorption coefficient α is proportional to the density n (number of target gas molecules per unit volume) or the concentration c (proportion of the density of target gas molecules relative to ambient air density n_a),

$$\alpha(\lambda) = n\sigma(\lambda) = cn_a\sigma(\lambda) \tag{2}$$

where $\sigma(\lambda)$ is the absorption cross section of the target gas species molecule, which can be found in the literature. The absorption cross section usually depends on the temperature and pressure. The former is mostly due to changes in the rotational or vibrational level populations with respect to temperature, and the latter is mostly due to collisional broadening of the absorption lines. This is described in section 2.4.

Open path gas detection (meaning that the laser light propagates through open air containing the gas species of interest) can only measure the line-integrated gas concentration, not the concentration itself. The measurable quantities are described in section 2.3.

The transmitted light intensity decreases due to factors other than absorption by the gas species of interest. These may include attenuation due to scattering by the atmosphere, geometrical spread of the beam, and absorption by other species. Methods to eliminate these effects are presented in sections 2.5 and 2.6.

2.2. Instrumental Configurations

Laser sensing of gas species using absorption spectroscopy can be categorized into two modes: transmission mode and reflection mode. In transmission mode, as shown in Fig. (1a), the laser source and detector are placed on opposite sides of the sensing region. This configuration is useful for atmospheric monitoring over a fixed, long optical path, or for smokestack or flue gas monitoring at a fixed location. In reflection mode, as shown in Fig. (1b), the detector detects laser light which is reflected by a hard target (topographic target) behind the sensing region. This configuration is useful for gas leak detection in closed areas or from pipelines and joints, in which some light is always reflected from an object in the optical path. Reflection mode can also be used for fixed, long optical paths by using a retroreflector instead of a hard target.

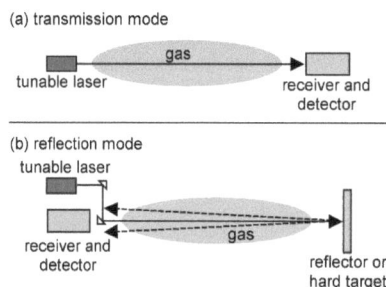

Fig. (1). Instrumental configurations for laser remote sensing of gas species in (a) transmission mode and (b) reflection mode.

For applications such as stack exhaust gas sensing, in which neither mode can be used, lidar, which is described in section 5, is appropriate.

2.3. Measurable Quantities

The signal power received by the detector can be written as

$$P = P_0 Y \eta RT \exp(-\alpha L) \tag{3}$$

where P_0 is the emitted power, Y is the geometrical factor which indicates the fraction of the emitted light received by the detector, η is the detector efficiency, T is the atmospheric transmission over the entire optical path, R is the reflectivity of the hard target in reflection mode ($R=1$ in transmission mode), α is the absorption coefficient of the target gas species, and L is the optical path length. P_0, Y, η are instrumental parameters, and α, R, T are environmental parameters which depend on the site and atmospheric condition.

In eq.(3), the parameters η, R, T generally have wavelength dependence. However, in gas sensing, a single spectral line of the target gas species is selected and the laser wavelength is scanned in a narrow wavelength region about the line center. The variation of η, R, T within this wavelength region can be considered to be negligible, so these parameters can be assumed to be constant throughout the wavelength scan. Therefore, the wavelength dependence mainly arises from the term exp(-αL). Wavelength scanning is described in section 2.5. Alternatively, two wavelengths corresponding to the line center and off-line can be used. This is described in section 2.6.

An important notion in laser gas sensing is the "column density". As evident in Fig. **(1)**, the amount of absorption measured by laser sensing is a line-integrated value. Therefore, the measured quantity is not the density of the target gas species n, but the line-integrated density $\bar{n}L$, which is often called the column density, given by eq. (4):

$$\bar{n}L = \int_0^L n(z)dz \tag{4}$$

Here L is the total optical path length, which is the distance between the laser and detector in transmission mode, or twice the distance from the laser (and detector) to the hard target. The quantity $\bar{n}$ corresponds to the average density over the entire path, which is obtained by dividing the column density by the total path length. The path length can be measured using, for example, a distance meter or range finder. The column density divided by the atmospheric density (which is equal to the line-integrated concentration) is often called the "column concentration"

$$\bar{c}L = \frac{\bar{n}L}{n_a} = \int_0^L c(z)dz \tag{5}$$

A consequence of line-integrated measurement is that high gas concentration localized in a narrow region produces the same results as low gas concentration distributed over a wide region. Consider these two cases: (a) 1000 ppm of gas existing only within a region of depth 1 cm, and (b) 1 ppm of gas existing uniformly over a path length of 10 m. Both will result in the same column concentration of 10 ppm-m. These two cases will be indistinguishable.

The minimum column concentration which can be measured by a laser gas sensing device is frequently used as the sensitivity, or detection limit, of the device. The sensitivity for gas leak detection is generally in the order of ppm-m.

2.4. Linewidth Considerations

Generally, gas species possess characteristic absorption lines which can be exploited for selective sensing or detection of the target gas species. The linewidth of the absorption line is broadened due to the following two effects: Doppler broadening, which is due to the thermal motion of the gas molecules, and pressure broadening, which is due to the collision with other molecules. Doppler broadening follows a Gaussian distribution,

$$g_D(v) = \frac{c}{v_0}\sqrt{\frac{m}{2\pi kT}}\exp\left[-\frac{(v-v_0)^2}{\Delta v_D^2}\right]$$ (6)

where $v = c/\lambda$ is the light frequency, v_0 is the light frequency at the center of the absorption line, c is the light speed, m is the mass of the gas molecule, k is the Boltzmann constant, T is the temperature, and Δv_D is the Doppler linewidth, which is defined as the full width at half maximum (FWHM) of the Gaussian distribution [3],

$$\Delta v_D = \frac{2v_0}{c}\sqrt{\frac{2\ln 2kT}{m}}$$ (7)

On the other hand, pressure broadening follows a Lorentzian distribution [3],

$$g_L(v) = \frac{\Delta v_L}{2\pi}\frac{1}{(v-v_0)^2 + (\Delta v_L/2)^2}$$ (8)

where Δv_L is the collisional linewidth, which is equal to the FWHM of the Lorentzian distribution. Δv_L increases with increasing pressure. The factor 2π is a normalization constant, so that the integral of $g_L(v)$ over all values of v is equal to 1.

Pressure broadening is usually dominant for atmospheric pressure and room temperature conditions. At elevated temperatures (around 2000K), Doppler and pressure broadening become comparable. In this case, the spectral shape of the absorption line is given by a Voigt profile, which includes both broadening effects [3].

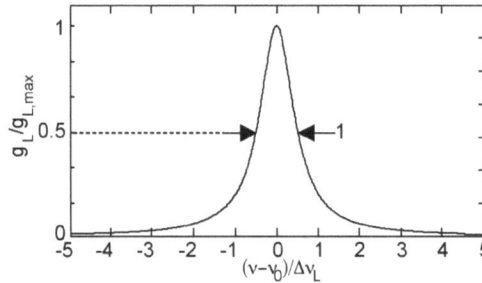

Fig. (2). Typical absorption line profile.

A typical absorption line profile following a Lorentzian distribution is shown in Fig. (2). The horizontal axis is expressed as the difference in frequency from the line center v_0 in units of Δv_L.

In laser gas sensing, the effective absorption is given by the convolution of the spectral profile of the laser and the absorption profile of the gas. Therefore, the laser linewidth must be narrowed to a fraction of the broadened linewidth. This is because, if the laser linewidth is too wide, a significant proportion of the light is unabsorbed even if the center laser wavelength coincides with the center wavelength of the absorption line, resulting in an effective decrease in the sensitivity.

2.5. The 2*f* Method

The standard method for gas sensing using a tunable laser is the double frequency (2*f*) method. In this case, the laser wavelength is modulated in such a way that the center wavelength corresponds to the absorption peak of the target spectral line. In diode lasers, the wavelength is modulated by modulation of the injection current.

Suppose that the spectrum of the absorption line of the target gas species has the form shown in Fig. (**3a**), and the laser wavelength is sinusoidally modulated about the peak wavelength λ_0 in the order A, B, C, D, E with a period of T_m (frequency $f=1/T_m$), as shown in Fig. (**3b**). At B and D, the absorption is maximal, so the detected signal becomes minimal. On the other hand, at A, C, and E, the absorption is minimal, so the detected signal becomes maximal. Therefore, the detected signal has a time dependence shown in Fig. (**3c**).

In Fig. (**3c**), a signal of period $T_s=T_m/2$ (*i.e.*, a signal with frequency component of $2f$) clearly appears in the detected signal (higher harmonics also exist). The $2f$ signal can be selectively detected by an instrumental configuration shown in Fig. (**3d**), in which the detector output is mixed with a reference signal of frequency $2f$ and the DC component is detected after passing through a low-pass filter (LPF).

The $2f$ signal is proportional to the absorption by the target gas species. Therefore, by calibrating the instrument at known column concentrations, the proportionality constant between the $2f$ signal amplitude and column concentration can be obtained. This value can then be used to convert the $2f$ signal amplitude to column concentration when gas sensing is performed (*i.e.* unknown column concentrations are measured).

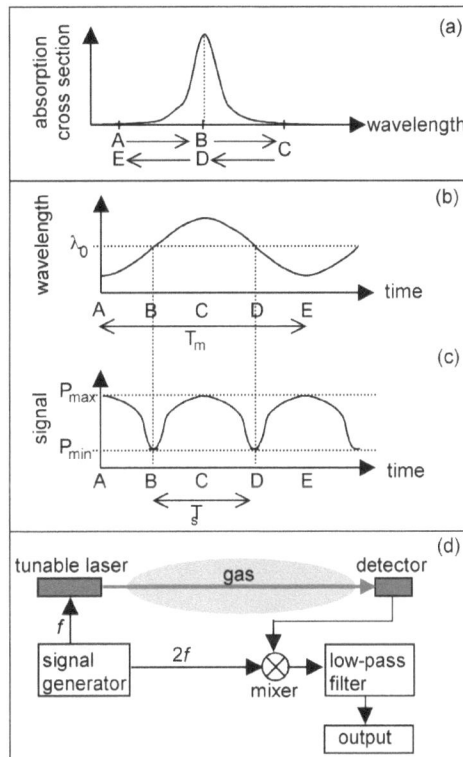

Fig. (3). Schematic diagram illustrating the $2f$ detection method: (a) absorption line profile, (b) time dependence of laser wavelength, (c) time dependence of detected signal, (d) example of instrumental configuration.

2.6. The Differential Absorption Method

In the $2f$ method, the laser wavelength is modulated about the center of the absorption line. Alternatively, the laser wavelength can be fixed to the absorption line center, and the attenuation due to absorption by the target gas species can be measured. In this case, another reference wavelength is needed to account for the attenuation due to scattering or the reflectivity of the hard target in reflection mode.

Suppose that the laser wavelength is limited to two wavelengths corresponding to A and B in Fig. (**3a**). In this case, wavelength A is usually called the "*off*" wavelength λ_{off} and B is called the "*on*" wavelength λ_{on}, as shown in Fig. (**4a**). If the laser wavelength is alternated between λ_{on} and λ_{off}, as shown in Fig. (**4b**), the detected signal will have a response as shown in Fig. (**4c**).

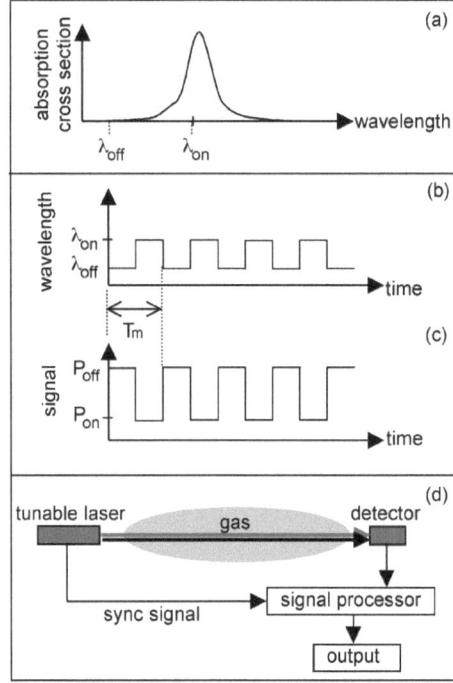

Fig. (4). Schematic diagram of the differential absorption method: (a) absorption line profile and wavelengths used, (b) time dependence of laser wavelength and detected signal, (c) example of instrumental configuration.

An example of the instrumental configuration for gas sensing using differential absorption is shown in Fig. (4d). The detector output signals at the two wavelengths $P_{on}=P_r(\lambda_{on})$ and $P_{off}=P_r(\lambda_{off})$ are accumulated based on a sync signal from the laser, which indicates whether the emitted wavelength is λ_{on} or λ_{off}.

Assuming that λ_{on} and λ_{off} are sufficiently close, so that the parameters η, R, T can be assumed to be the same, the ratio of $P_{on}=P_r(\lambda_{on})$ to $P_{off}=P_r(\lambda_{off})$ becomes

$$\frac{P_{on}}{P_{off}} = C\exp[-(\alpha_{on}-\alpha_{off})L] \tag{9}$$

where C is an instrumental constant. By testing the device in a controlled environment in which the target gas species does not exist ($\alpha_{on}=\alpha_{off}$), C can be determined from the signal ratio. Therefore, the column density of the target gas species can be obtained by

$$\bar{n}L = \frac{1}{\sigma_{on}-\sigma_{off}}\ln\left(C\frac{P_{off}}{P_{on}}\right) \tag{10}$$

The main advantage of the differential absorption method is the measurement rate speed. In the $2f$ method, the time response of the gas sensor is limited by the modulation frequency, but in the differential absorption method there is no such limitation. The differential absorption method is especially suited for two-dimensional imaging of the gas (measurement of the column density in the plane perpendicular to the laser beam direction, and displayed as an image).

The differential absorption method requires λ_{on} to be tuned exactly on the center of the absorption line. Therefore, this method requires far more wavelength stability than the $2f$ method. This method is essentially the same as differential absorption lidar (DIAL), in which two wavelengths are emitted into the atmosphere to measure the concentration of a trace species. DIAL uses atmospheric backscatter, whereas gas sensing in reflection mode uses a hard target. DIAL is briefly described in Section 5.

3. METHANE GAS SENSING

3.1. Absorption Characteristics of Methane

Methane (CH_4) is the major component of natural gas, which is used in industrial plants as well as in households. Therefore, a large number of pipes for its transportation have been installed throughout the world. Methane gas leaks from pipes, pipe joints, and facilities, or natural emission of methane gas may lead to serious incendiary accidents, so their detection is of high importance for industrial and residential safety. Since methane is flammable (the explosion limit is 5.3 vol%), its detection from a safe distance is preferable. Laser gas sensing is especially suited for this purpose.

The methane molecule exhibits absorption in the near infrared to mid-infrared region, which can be exploited for sensing. The absorption cross section of methane in the infrared region is shown in the left figure of Fig. **(5)** [4]. The absorption bands consist of a large number of single absorption lines, each of which is broadened by the effects described in section 2.4.

The strongest absorption bands exist at 3.3 μm and 7.7 μm, which ensure high sensitivity. However, as these bands are overlapped by absorption bands of water vapor, the absorption line used for methane detection must be carefully chosen, to avoid overlap with neighboring water vapor absorption lines. Weaker absorption bands exist at 2.3 μm and 1.65 μm, which are free of overlap by absorption bands of water vapor. The 1.65 μm band is shown in the right figure of Fig. **(5)** [4], and is commonly used for methane gas sensing because semiconductor lasers at this wavelength are readily available at low cost.

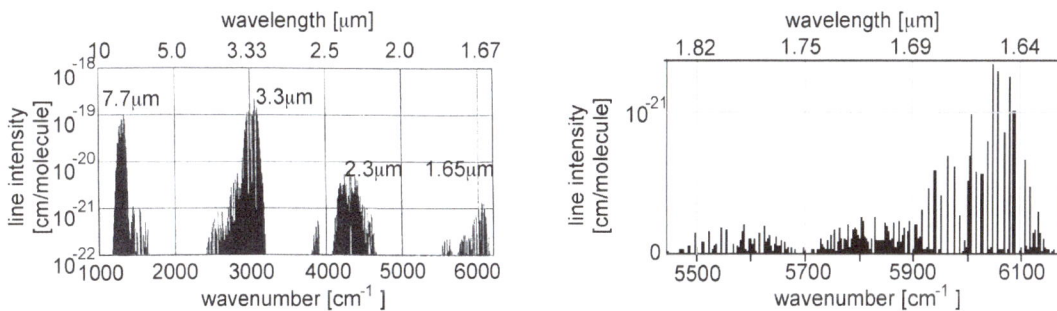

Fig. (5). Absorption characteristics of methane in the infrared region (left figure) and expanded view of the 1.65 μm absorption band (right figure).

3.2. Sensing of Methane Gas Using Semiconductor Laser

Methane gas sensors using a semiconductor laser have been developed and are commercially available. An example of the Laser Methane detector developed by Tokyo Gas Co. and Tokyo Gas Engineering Co. and commercialized by Anritsu Corp. is shown in Fig. **(6)**. This device measures the column concentration and display the time variation on a LCD screen. It is battery operated, and hand-held operation is possible [5,6].

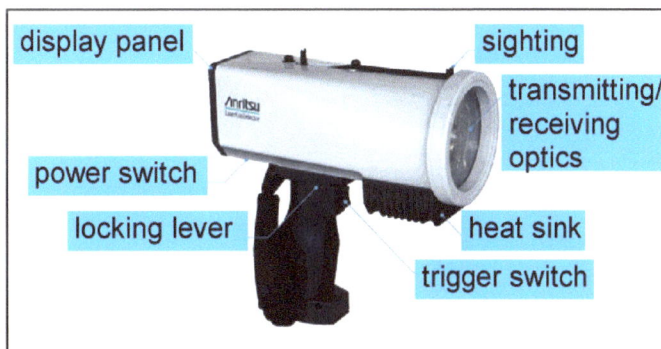

Fig. (6). Photograph of a methane sensor using a semiconductor laser [6].

This device modulates the laser wavelength at a frequency of 10 kHz, and detects the 20 kHz component based on the $2f$ method. The laser light is illuminated on gas pipes or ground surface, and reflected or scattered light is received by the optical receiver. Column concentrations of 0-10,000 ppm-m can be measured with a precision of +/-10% [6]. Recently, a smaller model, Laser Methane Mini, has also been developed [7].

Similar devices have been developed by other manufacturers. For example, the Remote Methane Leak Detector (RMLD[TM]) was developed by Physical Sciences Inc. and Heath Consultants Incorporated [8,9]. The device can measure methane plumes at 30 m (100 ft) range by collecting the scattering from topographic targets behind the gas.

Similar devices for detection of other gas species are also commercially available. For example, the GasFinder series from Boreal Laser can measure hydrogen fluoride (HF), hydrogen chloride (HCl), hydrogen sulfide (H_2S), ammonia (NH_3), carbon dioxide (CO_2), hydrogen cyanide (HCN), ethylene (C_2H_4), acetylene (C_2H_2) in addition to methane [10]. The Enhanced Laser Diode Spectroscopy (ELDS[TM]) open pass gas detection (OPGD) series from Senscient is available in models to measure H_2S, CH_4, H_2S+CH_4, propane, butane, NH_3, CO_2, ethylene, HCl, HF [11].

It is safe to say that the technology of gas sensing using semiconductor lasers has reached technical maturity, as a large variety of gas sensors have become commercially available.

3.3. Visualization of Methane Gas Using Semiconductor Laser

In some cases, it is advantageous to measure the two-dimensional distribution of methane gas (in the plane perpendicular to the line of sight). This can visualize the positions of high methane gas content, which is effective in locating the leak in the event of a gas leak.

A methane gas imaging device using an InGaAs semiconductor laser has been developed by the research group at Glasgow University [12]. A photograph of the device is shown in Fig. **(7a)**. This device can obtain a two-dimensional image of methane gas by scanning the laser light within the field of view of the optical receiver, as shown in Fig. **(7b)**. The optical receiver uses a Fresnel lens of 150 mm diameter. The $2f$ method is used for measuring the column concentration, whose lower detection limit is 10 ppm-m.

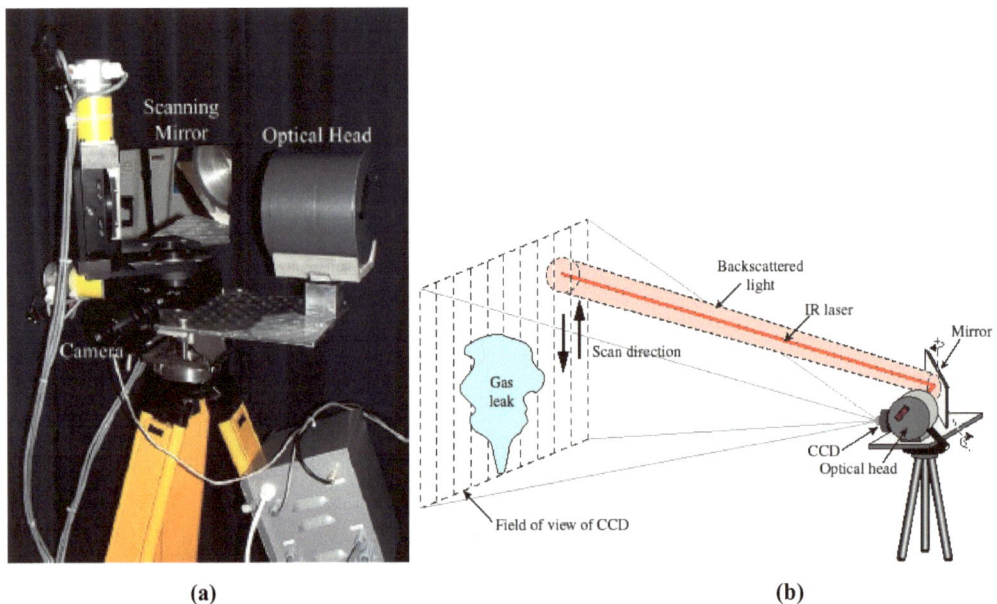

(a)   (b)

Fig. (7). Visualization of methane gas using a semiconductor laser: (a) photograph of device, (b) principle of visualization using beam scanning [12].

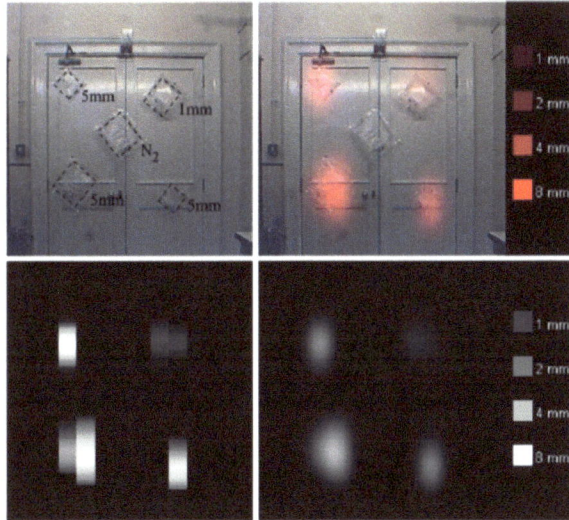

Fig. (8). Example of methane gas visualization: location of plastic bags containing known concentrations of methane (upper left), intensity map of raw signal (lower left), result of smoothing (lower right), superposition onto the visual image (upper right) [12].

An example of methane gas visualization is shown in Fig. **(8)**. Plastic bags with different concentrations of methane gas were attached onto the wall, as shown in the upper left. The numerical figures represent the column concentration in ppm-m. The bag at the center only contained N_2. The laser beam over the area including the bags. The intensity map of the raw signal is shown in the lower left. Spatial smoothing was applied to the intensity map, and the result is shown in the lower right. The superposition of the smoothed image and the visual image is shown in the upper right, in which the intensity of the smoothed image is colored in red. There is a clear correlation between the intensity and the column concentration.

Simulated methane gas leaks in indoor conditions and outdoor conditions in the presence of foliage was successfully visualized. For indoor conditions, the image for bags containing about 5 ppm-m of methane gas is shown in the upper left. The image for a simulated gas leak, in which methane gas was released at a rate of 1 lit/min, is shown in the upper right. For outdoor conditions, the image of a bag containing 5-10 ppm-m of methane gas located in foliage is shown in the lower right. The image for a simulated gas leak, in which methane gas was released at a rate of 1 lit/min inside the foliage, is shown in the lower right.

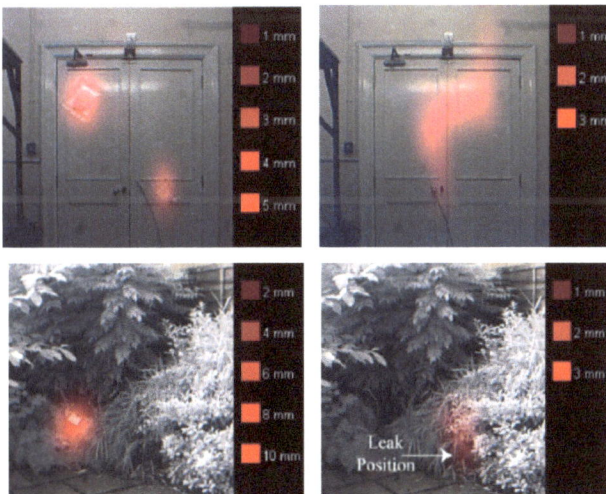

Fig. (9). Example of methane gas visualization: bags with known methane concentration placed on the wall (upper left), simulated methane leak (upper right), bag with known methane concentration placed on foliage (lower left), simulated methane leak behind foliage (lower right) [12].

The results showed that methane gas can be visualized in outdoor conditions in the presence of obstructions such as foliage [12].

3.4. Visualization of Methane Gas Using All Solid State Laser

A methane gas visualization device based on solid state laser was developed by the Japan Gas Association under the project "Development of gas pipe leak countermeasure technology" by the Ministry of Economy, Trade, and Industry (Japan) from 1998 to 2004. A portable model for indoor gas leak visualization, a vehicle-mounted model for detection of gas leaks from buried pipelines, and a helicopter-mounted model for large scale leak caused by natural disaster were developed [13].

The portable model, which is mounted on a backpack, uses an optical parametric oscillator of wavelength 3.27 μm, and can visualize methane gas up to a distance of 5 m. A photograph of the device is shown in Fig. **(10a)** [14].

The device emits the laser as a horizontal linear beam and detects the reflected or scattered light with a linear sensor array of 256 elements. The beam is then scanned in the vertical direction, providing a two-dimensional image. An example of the device in use is shown in Fig. **(10b)**.

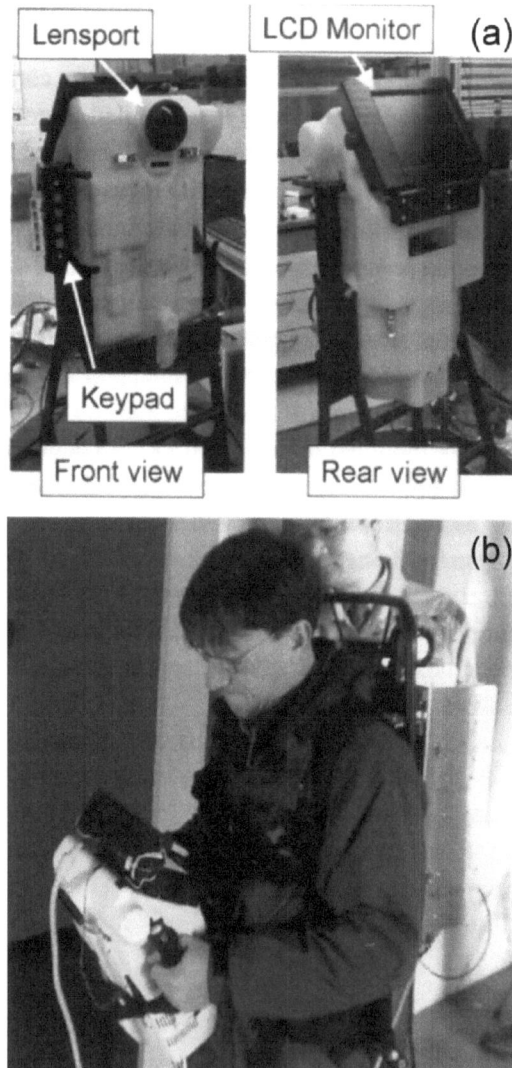

Fig. (10). Portable methane gas visualization device: (a) photograph of device, (b) example of device in use [14]

The vehicle-mounted model has the optics mounted on the rooftop the power supplies and electronics inside the vehicle, as shown in Fig. **(11a)** and **(11b)**. It can visualize methane gas up to a distance of 30 m in front of the vehicle at a cruising speed of <40 km/h. The visualization result is displayed on a panel mounted on the passenger side of the vehicle, as shown in Fig. **(11c)**. Both the portable and vehicle-mounted models can detect column concentrations of >10 ppm-m [15]. A vehicle-mounted model has also been developed by SRI International [16].

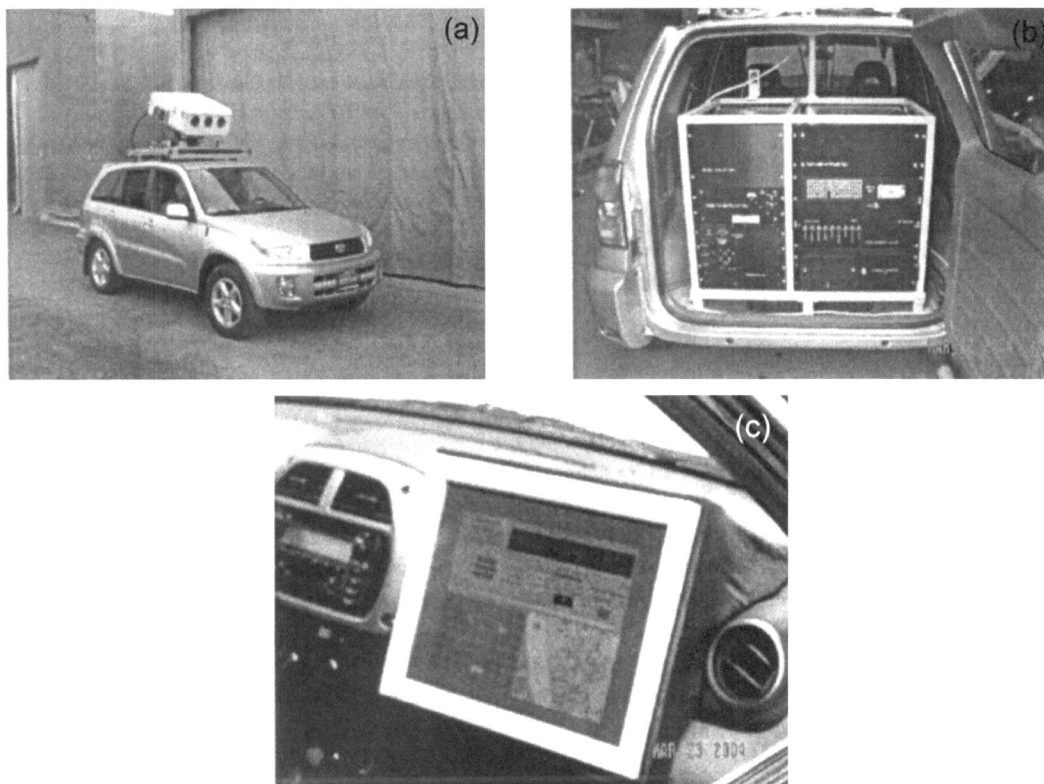

Fig. (11). Vehicle-mounted methane gas visualization device: (a) exterior view, (b) view inside the vehicle, (c) display panel mounted on passenger side [15].

The helicopter-mounted model which was developed in this project is based on passive spectral imaging at 7.7 μm and does not use a laser. Therefore, its sensitivity is lower, and minimum detectable column concentration is about 1000 ppm-m (0.1 %-m). Passive imaging was selected because helicopters operate mainly during daylight (in the presence of natural light), and a low sensitivity will suffice for visualization of large scale leaks caused by a major disaster (*e.g.* underground gas pipeline rupture due to an earthquake) [13].

A helicopter-borne methane leak detector has been developed by the Natural Sciences Center of A. M. Prokhorov General Physics Institute, Russia [17]. This system is based on a diode laser operating in the 1.65 μm region, which is transmitted downward from the helicopter, and uses a telescope of diameter 25 cm to collect the reflection from the ground, at a distance of 50-200 m below.

Since the spatial distribution of methane in gas leaks changes dynamically because of local winds, visualization without beam scanning is preferable. This may be a subject of future development.

4. FLUE GAS MONITORING

4.1. Measurement of Sulfur Oxides in Combustor Exhaust

Flue gas results from combustion of fossil fuels, biomass, garbage, *etc.* Since the temperature of the gas is generally high (150 to 350 degrees C), and the gas often contains sulfur or chlorine compounds which are reactive and corrosive, the harsh environment is often unsuited for conventional sensors which require contact with the gas. Since laser gas sensing is a non-contact technique, it is suited for harsh environments.

Sulfur oxides are always present in the flue gas from combustion of oil or coal. The main constituent is sulfur dioxide (SO_2), but some of it is oxidized to sulfur trioxide (SO_3) or sulfuric acid (H_2SO_4). Since SO_3 and H_2SO_4 are very reactive and corrosive, ammonia (NH_3) injection is frequently used in thermal power plants in order to neutralize them. The reaction of sulfur oxides and ammonia results in ammonium salts, mainly ammonium sulfate, which is eliminated from the flue gas by the electric precipitator. A schematic diagram of the flue gas flow in a thermal power plant is shown in Fig. **(12)**.

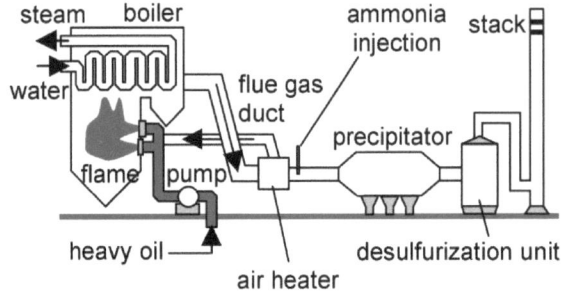

Fig. (12). Flue gas flow in a thermal power plant.

Since SO_2 and SO_3 have absorption bands in the ultraviolet region, ultraviolet absorption spectroscopy is commonly used for their measurement. However, the large abundance of particulate matter (ash) in the flue gas makes open path measurement difficult, as ultraviolet light is strongly scattered and attenuated.

SO_2 and SO_3 also have absorption bands in the infrared region (~7 μm), which can be used for their measurement by TDLAS. Until recently, the poor availability of tunable lasers in the 7 μm region had hampered its implementation. The only available laser sources were lead salt lasers and different frequency mixing. The former has been applied to SO_2 and SO_3 measurement in the exhaust plume of an aircraft engine by Aerodyne Research [18]. However, the need for liquid nitrogen cooling has prevented its use in industrial environments such as power plants.

Recent advances in quantum cascade laser (QCL) technology has made tunable laser sources available in the mid-infrared to far-infrared regions [19]. QCLs have been applied to measurement of various gases, such as CO_2, CO, NO, NH_3, CH_4, N_2O, O_3, SO_2 [19-21]. A QCL gas sensor for SO_2 and SO_3 for application to combustor exhaust streams has been developed by Physical Sciences Inc. and has achieved sensitivities of 1 ppm-m [22].

A QCL gas sensor for measurement of SO_2 and SO_3 in flue gas in oil-fired thermal power plants has been developed by the Central Research Institute of Electric Power Industry [23]. This sensor operates in reflection mode, and consists of a probe, optical unit, and measurement and control instrumentation. A schematic diagram of the device is shown in Fig. **(13)**.

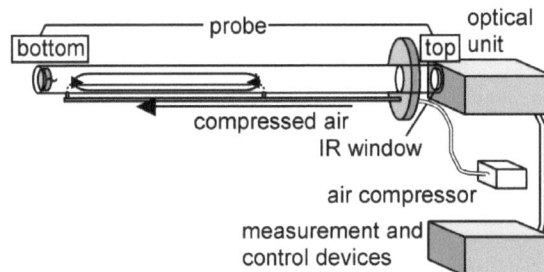

Fig. (13). Schematic diagram of QCL gas sensor for measurement of sulfur oxides in flue gas [23].

The probe is a stainless steel tube with a reflector at one end and two apertures of length 1 m located at diametrically opposite sides along the tube. The probe is inserted directly into the flue gas. The flue gas passes through the probe, entering from one aperture and exiting from the other. This is shown schematically in Fig. **(14)**.

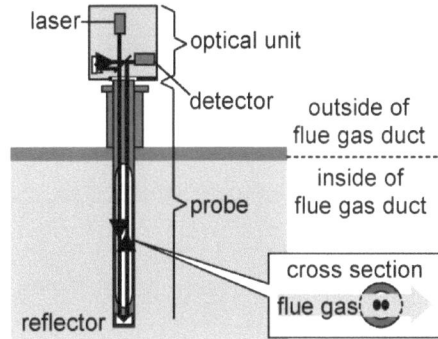

Fig. (14). Flue gas measurement using QCL gas sensor [23].

Both ends of the probe are maintained at positive air pressure using an air compressor, so that the flue gas is confined in the region between the two apertures.

The optical unit contains a quantum cascade laser, an optical chopper, infrared detector, and optics. The laser is tunable from 6.9 to 7.4 μm and operates at a repetition rate of 100 kHz. The laser light is sequentially split into two paths, a reference path and a measurement path, by the chopper. The measurement light is directed into the probe, reflected from the far end of the probe, and directed to the infrared detector. The reference light is reflected inside the optical unit and enters the same detector.

A photograph of the QCL gas sensor installed on the flue gas duct of a thermal power plant (between the boiler exit and air heater) is shown in Fig. **(15)**.

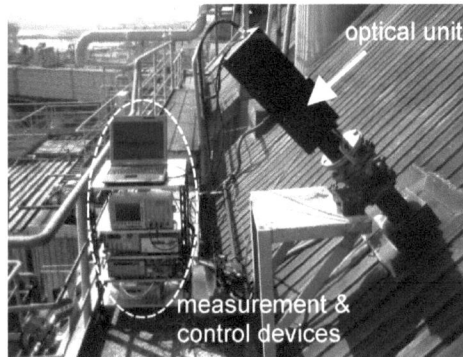

Fig. (15). Photograph of QCL gas sensor installed on the flue gas duct of a thermal power plant [23].

4.2. Measurement of Flue Gas in Waste Incineration Plants

Flue gas of waste incineration plants contain environmentally toxic species, such as HCl, HF, CO, NO, NO_2, and SO_2, as well as heavy metals, dioxins, and particulate matter. These substances are subject to strict exhaust regulations and the operator is required to eliminate them from the flue gas before releasing into the atmosphere. Although most gaseous species can be measured by gas sampling and FT-IR, *in situ* measurement is also possible by TDLAS.

A typical configuration of a TDLAS device for measurement of flue gas in waste incineration plants is shown in Fig. **(16)**. The device operates in transmission mode, and often several devices are installed at different locations between the furnace and stack for process monitoring.

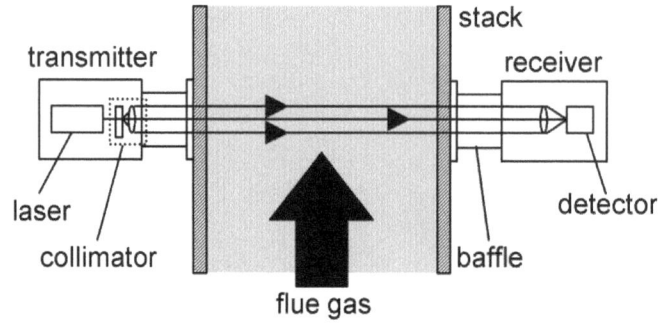

Fig. (16). Configuration of typical TDLAS device for measurement of flue gas in waste incineration plants.

A TDLAS monitor for measurement of HCl, HF, CO has been developed and has proven performance in a waste incinerator plant [24]. Presently, TDLAS sensors for HCl and HF are available from several manufacturers.

Metal atoms in flue gas can also be measured by TDLAS. Atomic laser absorption spectroscopy is briefly presented in Section 6.2.

5. STACK EXHAUST GAS MONITORING

5.1. Measurement Method

Laser remote sensing is especially suited for measurement of gas species in exhaust gas or plumes from stacks, since the gas exit is high above the ground.

Assuming that the measurement is done from the ground, at a certain horizontal distance from the stack, the beam is transmitted so that it intercepts the gas near the gas exit, as shown in Fig. (17).

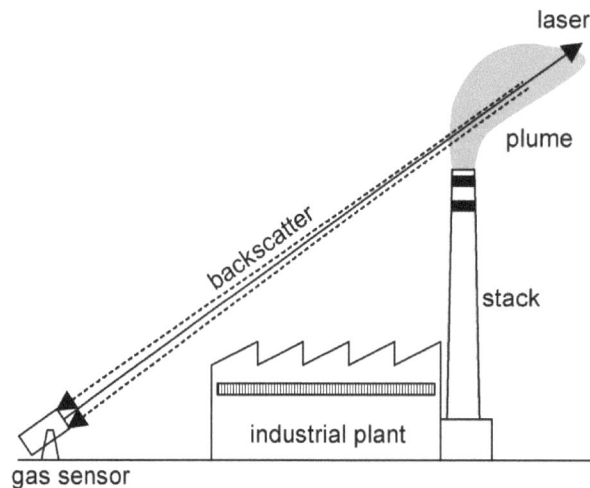

Fig. (17). Schematic diagram of gas sensing in stack exhaust gas.

In Fig. (17), a location to install a hard target or receiver behind the gas exit is obviously unavailable, so the configurations shown in Fig. (1) are inapplicable. In this case, the applicable method is differential absorption lidar (DIAL), which is similar to the differential absorption method described in section 2.6, except for the fact that it uses atmospheric backscatter instead of a hard target. A schematic diagram of a DIAL system is shown in Fig. (18), in which a Newtonian telescope is used for the receiver (Cassegrain telescopes can also be used). Since the DIAL technique is thoroughly described in various texts [25-27], only a brief description is given here. DIAL systems have mostly been used for atmospheric studies rather than industrial applications.

5.2. Measurement Principle of DIAL

A DIAL system emits laser pulses at two wavelengths λ_{on} and λ_{off}, which correspond to large and small absorption by the target gas species, and collects atmospheric backscatter (Rayleigh scattering due to molecules or Mie scattering due to aerosols). The pulses at are usually generated by two independent lasers, or by a single laser with a rapid tuning mechanism. The two wavelengths λ_{on} and λ_{off} are usually sufficient close (~1 nm) so that a single interference filter and detector can be used to detect the backscatter at the two wavelengths, as shown in Fig. (18).

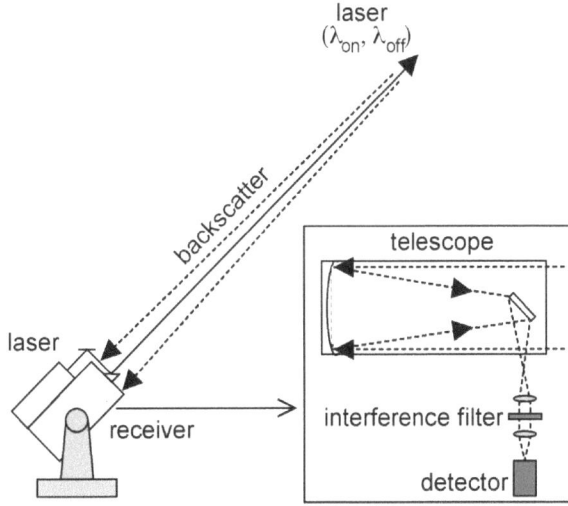

Fig. (18). Schematic diagram of a DIAL system.

The power received at wavelength λ_{on} and λ_{off} from the range between r and $r+\Delta r$ can be written as

$$P_{on}(r) = P_{0,on} Y(r) \eta_{on} A \beta_{on} \frac{\Delta r}{r^2} \exp(-2\alpha_{on} r) \tag{11}$$

$$P_{off}(r) = P_{0,off} Y(r) \eta_{off} A \beta_{off} \frac{\Delta r}{r^2} \exp(-2\alpha_{off} r) \tag{12}$$

where $P_{0,on}$ and $P_{0,off}$ are the laser powers emitted at λ_{on} and λ_{off}, $Y(r)$ is the overlap function between the emitted laser beam and the receiver field of view, η_{on} and η_{off} are the detection efficiencies at λ_{on} and λ_{off}, A is the receiver area, Δr is the range resolution, r is the distance from the DIAL system, α_{on} and α_{off} are the extinction coefficients and β_{on} and β_{off} are the atmospheric backscatter coefficients at λ_{on} and λ_{off}.

Writing the equivalent of eqs.(11) and (12) for range $r+\Delta r$ (*i.e.* replacing r in eqs.(11) and (12) by $r+\Delta r$), one obtains

$$\frac{P_{on}(r+\Delta r)}{P_{off}(r+\Delta r)} \frac{P_{off}(r)}{P_{on}(r)} = \exp[-2(\alpha_{on} - \alpha_{off})\Delta r] \tag{13}$$

The extinction coefficient is given by

$$\alpha = \sum_i n_i \sigma_i + \alpha_{atm} \tag{14}$$

where n_i and σ_i are the concentration and absorption cross section of gas species i, and α_{atm} is the extinction due to atmospheric absorption and backscatter.

Suppose one is aiming at measuring gas species j. The wavelengths λ_{on} and λ_{off} are chosen so that σ_j differs largely at λ_{on} and λ_{off}, and σ_i ($i \neq j$) differs little. Also, if λ_{on} and λ_{off} are within ~1 nm, the difference in α_{atm} is negligible (unless one is trying to measure very low concentrations), so that

$$\alpha_{on} - \alpha_{off} \cong n_j(\sigma_{j,on} - \sigma_{j,off}) \tag{15}$$

Substitution into eq.(13) yields

$$n_j(r) = \frac{-1}{2\Delta\sigma_j \Delta r} \ln\left[\frac{P_{on}(r+\Delta r)}{P_{off}(r+\Delta r)} \frac{P_{off}(r)}{P_{on}(r)}\right] \tag{16}$$

where $\Delta\sigma_j = \sigma_{j,on} - \sigma_{j,off}$. This is the basis of the DIAL equation, which gives the concentration of the target gas species n_j as a function of the distance r, based on the return signals P_{on} and P_{off}.

Although eq. (16) is somewhat similar to eq. (10), it is important to note that the concentration is obtained as a function of the distance r. This can be easily understood in the following manner. In Fig. (1b), the entire beam is reflected by the hard target, so the range is fixed by the distance to the target. On the other hand, in Fig. (18), the atmosphere acts as a diffuse target, so signals from different depths (distances) return to the receiver. The range can be calculated by $r=ct/2$, where c is the speed of light and t is the time relative to the firing of the laser. Obviously, the return signal due to atmospheric backscattering is much smaller compared to a hard target, so a large receiver area is required to collect sufficient backscattered light to obtain a return signal. DIAL systems for atmospheric studies in the lower troposphere (where most pollutants exist) ordinarily have receiving telescopes of aperture 20-50 cm. Systems for studies of stratospheric ozone have larger aperture, because the distance is larger.

In eq. (16), the atmospheric conditions corresponding to the return signals at λ_{on} and λ_{off} are assumed to be identical. Therefore, the temporal separation between the pulses at λ_{on} and λ_{off} should be smaller than characteristic times of atmospheric fluctuations, which are in the order of 1 ms.

DIAL systems, as well as all lidar systems in general, have a minimum and maximum distance for measurement. The minimum distance is determined by the overlap function between the laser beam and the telescope field of view, which is dependent on the system design. For example, for the system configuration shown in Fig. (18), measurement is not possible at the immediate near-field because the backscatter is obstructed by the secondary mirror. The maximum distance is usually determined by the signal-to-noise ratio of the backscatter, and is dependent on the laser power and telescope area. For near-field applications, lidar designs such as those presented in Chapter 2 can be used.

For stack exhaust gas sensing and gas emissions measurement, the minimum and maximum measurement distances are in the order of 10 m and 100 m, respectively. For these distances, a laser pulse energy of ~mJ and a telescope diameter of 20-30 cm would be appropriate. Larger telescopes will result in substantially higher cost, larger system size and weight, and more difficulty in handling.

5.3. Gas Measurement Using DIAL

Gas species which have been measured by DIAL are shown in Table 1 [26,27]. These are mostly atmospheric studies, and the measured concentrations are in the ppb-ppm range. Most DIAL measurements are performed in the UV (wavelengths below 300 nm) or in the infrared, with the exception of NO_2, which can be measured in the visible.

As an example, the absorption cross sections of SO_2 and NO_2 are shown in Fig. (19) with the commonly used wavelengths for DIAL measurement (λ_{on}=300.1 nm, λ_{off}=299.3 nm for SO_2, λ_{on}=448.1 nm, λ_{off}=446.8 nm for NO_2). The structure in the absorption cross section shows that the linewidth of the lasers to be used should be much less than 0.1 nm.

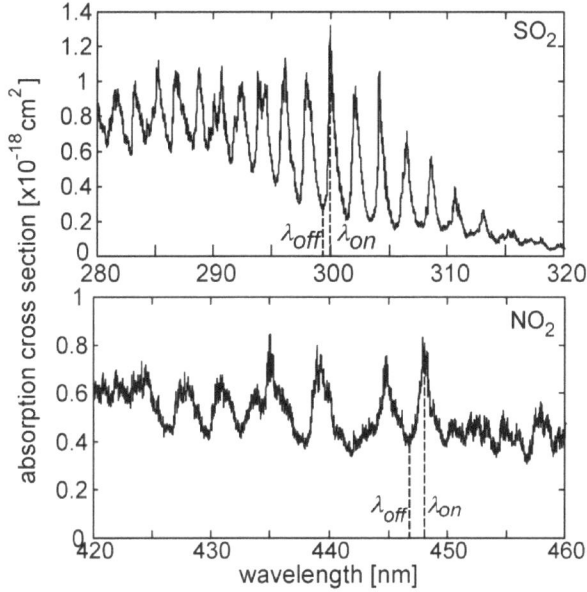

Fig. (19). Absorption cross sections of SO_2 and NO_2 indicating wavelengths commonly used for DIAL measurement.

Table 1: Gas species measured by DIAL

Species	Wavelength[*1]
O_3	266-300 nm[*2]
SO_2	300.1 nm
NO_2	448.1 nm
NO	226.8 nm
benzene (C_6H_6)	253 nm
toluene (C_7H_8)	267 nm
Hg	253.65 nm
Cl_2	330 nm
CH_4	1.67, 3.39 μm
HCl	3.64 μm
N_2O	3.89 μm
Freon-12 (CCl_2F_2)	10.72 μm
ethylene (C_2H_4)	10.53 μm
NH_3	9.22, 10.71 μm
SF_6	10.55 μm

[*1] λ_{on} unless otherwise noted.

[*2] The absorption spectrum of ozone does not show fine structure in this region, so two arbitrary wavelengths, usually separated by 5-10 nm, can be used.

Ozone (O_3) is a special case in which the absorption spectrum varies gradually in the region 266-300 nm, with no appreciable structure. Therefore, two arbitrary wavelengths without tunable capability can be used. Excimer lasers or Nd:YAG lasers (fourth harmonic) coupled with Raman shifters have been used for this purpose.

As for tunable lasers used in DIAL systems, UV lasers are mainly limited to second harmonic generation of a dye laser or OPO (Optical Parametric Oscillator). Sum frequency mixing (SFM) is also occasionally used. Commonly used infrared lasers are OPO, difference frequency mixing (DFM), and CO_2 lasers. CO_2 lasers are only linewise (not continuously) tunable, which limits the measurable species to those whose absorption line coincides with an emission line of the CO_2 laser. Tunable diode lasers have not been used for DIAL applications in atmospheric studies because of their limited power and large divergence, which makes them unsuitable for lidar applications with ranges in the order of 100 m to km.

For gas species of very low concentration, eq. (15) is not necessarily valid, so one must account for the difference in α_{atm} and/or the absorption by other species. One method to improve the measurement sensitivity is the use of more than two wavelengths to measure a single species (multiwavelength DIAL) [27].

If the flux of exhaust gas is known, the mass flux of the target gas species can be calculated, and this can be compared against local environmental regulations regarding emission of pollutants into the atmosphere.

An example of stack emissions monitoring by DIAL is the measurement of atomic mercury emissions from industrial plants . Atomic mercury has a very strong absorption line at 253.65 nm, which can be exploited for DIAL measurement. A DIAL system for atomic mercury measurement has been developed by the research group at Lund Institute of Technology in Sweden [26,28-29]. The system has been used to monitor atomic mercury emissions from chlor-alkali plants and mercury mines. The system could measure the spatial distribution of atomic mercury in the atmosphere, whose peak concentration was in the order of $\mu g/m^3$, and the obtained flux was in the order of g/h [28,29].

Other examples of stack emission monitoring include monitoring of SO_2 from power plants [30] and emission of HCl from incineration ships [31].

These research results show that DIAL is a powerful tool in monitoring stack exhaust gases and their dispersion in the atmosphere.

6. OTHER APPLICATIONS

6.1. Remote Detection of Hazardous Gases

The techniques described in section 3 for sensing and visualization of methane gas can be applied to other gases which have absorption bands that can be accessed by tunable diode lasers. Remote detection of hazardous or poisonous gases is useful for safety considerations. If a contact sensor or gas sampling device is used, the operator must approach the danger area, so protective gear such as masks must be used. Laser remote sensing would allow detection from a safe distance.

Some possible applications are detection of SO_2 and H_2S in areas with volcanic activity, and monitoring of CO concentration in closed areas with combustion devices (such as indoor heaters).

6.2. Metal Vapor Monitoring Using Atomic Laser Absorption Spectroscopy

As noted in the extremely high sensitivity of DIAL measurement of atomic mercury, laser absorption spectroscopy can measure trace amounts of metal atoms. This is because the spectral shape of atomic absorption is much sharper and higher in peak value than that of molecular absorption.

In any industrial process which uses atomic vapors, laser absorption spectroscopy offers a highly sensitive sensing method. Since most metals are solids at room temperature (mercury being an exception), atomic vapors are generated by heating metals to very high temperature in ovens or furnaces, electron or ion beam bombardment, of plasma generation by an electrical discharge or laser-induced breakdown (LIBS). LIBS is covered in Chapter 10 of this book.

Since monitoring of metal vapors is not within the scope of this chapter, only a brief explanation is given. The interested reader is requested to consult the references. Refs. [32-34] provide a good review.

TDLAS has been applied to detection of various metals. Metals which can be measured using commercially available diode lasers (or by frequency doubling) are listed in Table 2 [32]. This is by no means a complete list, and more elements should be added as diode lasers in previously unavailable wavelengths are developed.

Specific industrial applications of atomic laser absorption spectroscopy include, to name a few, control of uranium vaporization rates in the atomic vapor laser isotope separation (AVLIS) program [35,36], measurement of titanium vaporization rates in electron beam physical vapor deposition for manufacture of high-performance aircraft alloys [37], measurement of Yb vapor for optimization of isotope selection process [38]. The isotopic shift is usually resolvable in laser atomic absorption spectroscopy.

It is also possible to apply atomic laser absorption spectroscopy to detection of trace metals in flue gas or combustor exhaust. However, as the metals usually exist as oxides, it is necessary to atomize them prior to detection, which can be efficiently be done using laser-induced breakdown.

Table 2: Metals which can be measured using commercially available diode lasers [32]

Element	Wavelength [nm]	Element	Wavelength [nm]
Al	396.15	Mo	390.30
Ag	328.07*	Na	330.23*
B	208.95*	Nb	410.09
Ba	350.12*	Nd	405.89
Be	234.86*	Ni	341.48*
Bi	223.06*	Os	426.09
Ca	422.67	Pd	340.46*
Co	390.99	Rb	780.03
Cr	425.44	Re	346.05*
Cs	852.14	Rh	343.49*
Cu	324.75*	Ru	392.59
Dy	421.17	Sb	217.58*
Er	400.80	Sc	391.18
Eu	459.40	Se	203.99*
Fe	385.99	Sm	476.03
Ga	403.31	Sr	460.73
Gd	405.82	Tb	431.89
Hf	377.76	Ti	398.98
Ho	410.38	Tl	377.57*
In	410.18	Tm	420.37
Ir	208.88*	U	404.27
K	766.49	V	318.39*
La	403.72	W	407.44
Li	670.78	Y	410.24
Lu	335.96*	Yb	398.80
	331.21*	Zr	468.78
Mn	403.08		

* accessible by frequency doubling

7. FUTURE DEVELOPMENTS

7.1. Gas Sensing Using Quantum Cascade Lasers

The recent development in QCLs has greatly increased the availability of tunable lasers in the mid infrared range. Since QCLs are robust, compact, and simple to operate (do not require cryogenic cooling), they are suited for harsh environments in industrial applications, as described in the example for flue gas sensing in 3.2. In the future, laser gas sensing could extend to gas species which were previously difficult to measure outside of controlled laboratory environments.

QCLs for the terahertz region are also being developed [39]. The terahertz region, which lies between the infrared region and the microwave region, contains vibrational transitions of complex molecules.

7.2. White Light Gas Sensing

A new type of lidar using "white light" is also under development and may become a useful gas sensing device.

When extremely high intensity laser pulses, usually ultrashort pulses of widths in the order of 10 fs generated by a Ti:Sapphire laser of center wavelength 780-800 nm, are emitted into the air, continuum radiation with a very broad frequency spectrum is generated. This continuum has a spectrum which extends from the UV to the mid-IR, and is frequently called "white light" [40]. The continuum also is efficiently generated using gas cells filled by using noble gases at high pressure [41]. The white light continuum can be used for remote sensing of various gases in the atmosphere, in the same manner as Differential Optical Absorption Spectroscopy (DOAS) using conventional continuum light sources such as halogen lamps. A schematic diagram of the "white light lidar" is shown in Fig. **(20)**.

The main advantage of the "white light lidar" over conventional DOAS instruments is the fact that lidar provides range resolution, enabling spatial profiling of gas species. The advantage over DIAL is that multiple gas species can be simultaneously measured, without the need to use tunable laser sources for each species. So far, measurement of atmospheric O_2, water vapor, CO_2 using "white light lidar" have been reported [40,41]. Although this technology is still in the research stage, it has the potential for on-site applications.

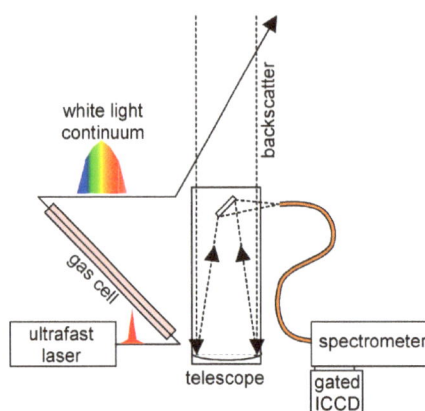

Fig. (20). Schematic diagram of laser remote sensing using white light continuum generated using an ultrafast laser and gas cell.

7.3. Gas Sensing Using Light-Emitting Diodes

In addition to lasers, Light Emitting Diodes (LEDs) may become widely used for remote gas sensing. The principal advantage of using LEDs is their low cost. So far, blue light and white light LEDs have been applied to measurement of NO_2 over an open path of about 50 m [42]. With the growing availability of LEDs in the ultraviolet range, low-cost ultraviolet light sources are becoming available. For example, with currently commercially available ultraviolet LEDs in the 280-300 nm range, measurement of SO_2 should be possible. The output spectra of a commercially available ultraviolet LED (UVTOP295 from Sensor Electronic Technology, Inc.) [43], and the relative transmitted intensity after absorption by a SO_2 column concentration of 300 ppm-m (*e.g.* concentration 100 ppm, path length 3 m) is shown in Fig. **(21)**. The change in the spectrum should be easily detectable using a compact spectrometer.

In conventional ultraviolet DOAS, lamp sources such as deuterium lamps are used. The replacement by LEDs may provide a lower cost, more robust device for measurement of gases, which may be better suited to the industrial environment. The low power consumption of LEDs should allow battery-operated gas sensors. The use of LEDs in the infrared region also offers opportunity for low-cost sensing of gases.

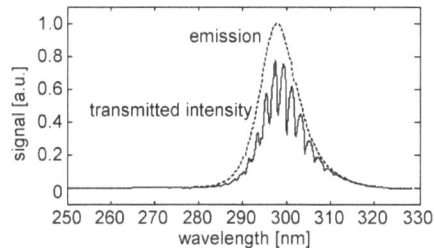

Fig. (21). Typical output spectrum of LED (broken line) and simulated spectrum after absorption by 300 ppm-m of SO_2 gas (solid line).

8. CONCLUSION

In this chapter, laser gas sensing and its industrial applications were presented. Detection of methane gas using the *2f* method seems to have achieved technical maturity, as a variety of laser gas sensors have become commercially available. Laser measurement of flue gas in power plants, waste incineration plants, and other industrial facilities using burners or combustors has also achieved technical levels for practical use. Gas species such as SO_2, NO, NO_2, NH_3, CO can be measured by commercially available laser gas sensors. For measurement of gases in stack exhaust and plumes, differential absorption lidar systems can be used.

With the growing industrial capacity of developing countries and worldwide concern for the atmospheric environment, the market for laser remote gas sensors is expected to expand.

Furthermore, the recent advances in quantum cascade lasers should result in a large increase of gases species measurable by laser remote sensing. On the other hand, application of light-emitting diodes should provide robust, low-cost gas sensors which are applicable to the industrial environment.

ACKNOWLEDGEMENT

The author expresses his gratitude to Dr. Takao Kobayashi for help and advice.

REFERENCES

[1]   H. Schiff, G. Mackay, J. Bechara, "The use of tunable diode laser absorption spectroscopy for atmospheric measurements", in *Air Monitoring by Spectroscopic Technique*s, M. Sigrist, ed., Wiley, New York, 1994.

[2]   H. Inaba: "Detection of Atoms and Molecules by Raman Scattering and Resonance Fluorescence", in *Laser Remote Sensing of the Atmosphere*, E. D. Hinkley, ed., Springer-Verlag, Berlin, 1976.

[3]   A. Eckbreth, *Laser diagnostics for combustion temperature and species*, Taylor & Francis, New York, 1996.

[4]   L.S. Rothman, R. Gamache, R. Tipping, C. Rinsland, M. Smith, D. Benner, V. Devi, J.-M. Flaud, C. Camy-Peyret, A. Perrin, , A. Goldman, S.T. Massie, L. Brown, R. Toth, "The HITRAN Molecular Database: Editions of 1991 and 1992," *Journal of Quantitative Spectroscopy and Radiative Transfer*, Vol. 48, pp. 469-507, 1992.

[5]   Anritsu Corporation, "SA3C15A Laser Gas Detector"; T. Iseki, "Techniques for methane gas leak detection using a near-infrared diode laser", *Review of Laser Engineering*, Vol. 33, No. 5, pp. 300-305, 2005 (in Japanese)

[6]   Crowcon Detection Instruments Ltd., http:// www.afcintl.com/product/tabid/93/productid/131/sename/lasermethane-detector-from-crowcon/ default.aspx (accessed Dec. 8, 2010)

[7]   Crowcon Detection Instruments Ltd., http:// www.afcintl.com/product/tabid/93/productid/418/sename/laser-methane-mini-methane-leak-detector-from-crowcon/default.aspx (accessed Dec. 8, 2010).

[8]   M. Frish, R. Wainner, J. Stafford-Evans, B. Green, M. Allen, "Standoff Sensing of Natural Gas Leaks: Evolution of Remote Methane Leak Detector", PSI-SR-1204, Physical Sciences Inc., 2004.

[9]   Heath Consultants Incorporated, "RMLD^{TM}", http://www.heathus.com/_hc/index.cfm/products/gas/rmld/ (accessed Dec. 8, 2010).

[10]   Boreal Laser Incorporated, "GasFinder Open Path Monitor - Specifications", http://www. boreal-laser.com/gasfinder-specifications (accessed Dec. 8, 2010).

[11]   Senscient Ltd., "Senscient ELDS™ Open Path Gas Detectors (OPGD)", http://www.senscient. com/products.html (accessed Dec. 8, 2010).

[12]   G. Gibson, B. van Well, J. Hodgkinson, R. Pride, R. Strzoda, S. Murray, S. Bishton, M. Padgett, "Imaging of methane gas using a scanning, open-path laser system", *New Journal of Physics*, Vol. 8, 26, pp. 1-8, 2006.

[13]   T. Kobayashi, N. Sugimoto, H. Kuze, "Laser Remote Sensing Techniques of Leak Gases", *Review of Laser Engineering*, Vol. 33, No. 5, pp. 295-299, 2005 (in Japanese).

[14]   R. Bamha, T. Reichardt, R. Schmitt, R. Sommers, S. Birtola, G. Hubbard, T. Kulp, M. Tamura, K. Kothari, "Development and Testing of a Portable Active Imager for Natural Gas Detection", *Review of Laser Engineering*, Vol. 33, No. 5, pp. 306-310, 2005.

[15]   M. Boies, W. Marinelli, T. Nakamura, B. Green, "Mobile Laser-Absorption Imaging System for Detection of Natural Gas Leaks", *Review of Laser Engineering*, Vol. 33, No.5, pp. 311-315, 2005.

[16]   SRI International, "Remote Gas Leak Sensor for Underground Pipelines", http://www.sri. com/rd/ Remote_Gas_Leak_Sensor.pdf, 2010.

[17]   A. Berezin, S. Malyugin, A. Nadezhdinskii, D. Namestnikov, Ya. Ponurovskii, S. Rudov, D. Stavrovskii, Yu. Shapovalov, I. Vyazov, V. Zaslavskii: "Remote helicopter-borne detector for searching of methane leaks", *Spectrochimica Acta Part A*, Vol. 66, pp. 803-806, 2007.

[18]   T. Berkoff, J. Wormhoudt, R. Miake-Lye: "Measurement of SO$_2$ and SO$_3$ using a tunable diode laser system", *Proc. SPIE*, Vol. 3534, pp. 686-693, 1998.

[19]   F. Tittel, Y. Bakhirkin, A. Kosterev, G. Wysocki: "Recent advances in trace gas detection using quantum and interband cascade lasers", *Review of Laser Engineering*, Vol. 34, No. 4, pp. 275-282, 2006.

[20]   M. Taslakov, V. Simeonov, H. van den Bergh, "Open path atmospheric spectroscopy using room temperature operated pulsed quantum cascade laser", *Spectrochimica Acta Part A*, Vol. 63, pp. 1002-1008, 2006.

[21]   V. Zéninari, L. Joly, B. Grouiez, B. Parvitte, A. Barbe, "Study of SO$_2$ line parameters with a quantum cascade laser spectrometer around 1090 cm^{-1}: Comparison with calculations of the ν_1 and $\nu_1+\nu_2-\nu_2$ bands of ^{32}SO$_2$ and the ν_1 band of ^{34}SO$_2$", *Journal of Quantitative Spectroscopy and Radiative Transfer*, Vol. 105, pp. 312-325, 2007.

[22]   W. Rawlins, J. Hensley, D. Sonnenfroh, D. Oakes, M. Allen, "Quantum cascade laser sensor for SO$_2$ and SO$_3$ for application to combustor exhaust streams", *Applied Optics*, Vol. 44, No. 31, pp. 6635-6643, 2005.

[23]   T. Fukuchi, "Development of infrared laser absorption spectroscopy device for measurement of SOx in flue gas", *The Papers of Technical Meeting on Instrumentation and Measurement, IEE Japan*, IM-10-028, pp. 1-4, 2010.

[24]   O. Bjorøy, I. Linnerud, V. Avetisov, K. Haugholt, "Simultaneous in-situ measurement of O$_2$, HCl, HF, CO and dust in gas from a waste incinerator using diode laser spectroscopy", 5th International Symposium on gas analysis by tunable diode lasers, Freiburg, Germany, February 1998; http://www.neo.no/ news/ neo_freiburg_1998.html (accessed Dec. 8, 2010).

[25]   R. Measures, *Laser Remote Sensing*, Wiley Interscience, New York, 1984.

[26]   S. Svanberg, "Differential Absorption Lidar (DIAL)", in *Air Monitoring by Spectroscopic Technique*s, M. Sigrist, ed., Wiley, New York, 1994.

[27]   G. Gimmestad, "Differential-Absorption Lidar for Ozone and Industrial Emissions", in *Lidar: Range-Resolved Optical Remote Sensing of the Atmosphere,* C. Weitcamp, ed., Springer, New York, 2005.

[28]   M. Sjöholm, P. Weibring, H. Edner, S. Svanberg: "Atomic mercury flux monitoring using an optical parametric oscillator based lidar system", *Optics Express*, Vol. 12, No. 4, pp. 551-556, 2004.

[29]   R. Grölund, H. Edner, S. Svanberg, J. Kotnik, M. Horvat: "Mercury emissions from the Idrija mercury mine measured by differential absorption lidar techniques and a point monitoring absorption spectrometer", *Atmospheric Environment*, Vol. 39, pp. 4067-4074, 2005.

[30] R. Adrain, D. Brassington, S. Sutton, R. Varey, "The measurement of SO_2 in power station plumes with differential absorption lidar", *Optical and Quantum Electronics*, Vol. 11, pp. 253-264, 1979.

[31] C. Weitkamp, "The distribution of hydrogen chloride in the plume of incineration ships: Development of new measurement systems", *3rd International Ocean Disposal Symposium*, Woods Hole, USA, Oct. 12-16, 1981.

[32] A. Zybin, J. Koch, H. Wizemann, J. Franzke, K. Niemax: "Diode laser atomic absorption spectrometry", *Spectrochimica Acta Part B*, Vol. 60, pp. 1-11, 2005.

[33] P. Monkhouse, "On-line diagnostic methods for metal species in industrial process gas", *Progress in Energy and Combustion Science*, Vol. 28, pp. 331-381, 2002.

[34] P. Monkhouse, "On-line spectroscopic and spectrometric methods for the determination of metal species in industrial process", *Progress in Energy and Combustion Science*, in press, 2010.

[35] L. Berzins, T. Anklam, F. Chambers, S. Galanti, C. Haynam, E. Worden: "Diode laser absorption spectroscopy for process control - sensor system design methodology", *Surface and Coatings Technology*, Vol. 76-77, pp. 675-680, 1995.

[36] K. Hagans, J. Galkowski, "The use of laser diode for control of uranium vaporization rates", *Proc. SPIE*, Vol. 2068, pp. 23-27, 1994.

[37] L. Berzins, "Using laser absorption spectroscopy to monitor composition and physical properties of metal vapors", *Proc. SPIE*, Vol. 2068, pp. 28-40, 1994.

[38] H. Park, D. Kwon, Y. Rhee: "Real-time monitoring of Yb vapor density using an extended cavity violet diode laser", *Spectrochimica Acta Part A*, Vol. 60, pp. 3305-3309, 2004.

[39] Y. Lee, *Principles of Terahertz Science and Technology*, Springer, New York, 2009.

[40] F. Theopold, J. Wolf, L. Wöste, "DIAL Revisited: BELINDA and White-Light Femtosecond Lidar", in *Lidar: Range-Resolved Optical Remote Sensing of the Atmosphere,* C. Weitcamp, ed., Springer, New York, 2005.

[41] T. Somekawa, M. Fujita, Y. Izawa, "Direct absorption spectroscopy of CO_2 using a coherent white light continuum", *Applied Physics Express*, Vol. 3, 082401, 2010.

[42] T. Fukuchi, T. Nayuki, H. Mori, N. Goto, T. Fujii, K. Nemoto: "Development of a Differential Optical Absorption Spectroscopy System Using High Luminance LED for Measurement of NO_2", *IEEJ Transactions EIS.*, Vol. 123-C, Vol. 8, pp. 1382-1386, 2003 (in Japanese).

[43] Sensor Electronic Technology, Inc: "UV TOP Technical Data", http://www.s-et.com/ (accessed Dec. 8, 2010).

CHAPTER 4

Gas Sensing Using Raman Scattering

Hideki Ninomiya[*]

Electronics Technology Department, Shikoku Research Institute, 2109-8 Yashima-Nishimachi, Takamatsu-shi, Kagawa 761-0192, Japan

Abstract: The Raman shift is a characteristic property of a molecule, and the intensity of Raman scattering is proportional to the density of the molecule. Therefore, gas sensing using Raman scattering can detect and identify gases. The method is especially useful for detection of hydrogen gas, because the hydrogen molecule does not have absorption bands from the near ultraviolet to near infrared that can be used for optical detection using absorption. On the other hand, the hydrogen molecule exhibits a strong Raman effect, so Raman scattering is a suitable method for hydrogen gas detection. In this chapter, the fundamentals of Raman scattering, detection of hydrogen gas by Raman scattering, development of lidar systems for detection, imaging, and concentration measurement of hydrogen gas, are presented. The concentration of hydrogen gas leaked into the open air can be remotely measured by simultaneous measurement of the Raman scattering signals from the hydrogen gas and atmospheric nitrogen. Hydrogen gas leak detection using Coherent Anti-Stokes Raman Scattering (CARS) is also presented.

Keywords: Raman scatttering, coherent anti-stokes raman scattering, raman lidar, raman spectroscopy, hydrogen gas, imaging, gas concentration, gas flow, shadowgraph, nitrogen.

1. INTRODUCTION

Compared to other noncontact optical detection techniques such as laser-induced fluorescence or infrared (IR) absorption, Raman detection has significant advantages for gas detection. IR absorption is mostly used for environmental gas or exhaust gas analysis, with a highly reflective surface placed at the end of the line-of-sight. Raman detection, on the other hand, relies on light scattered by a materials so Raman detection does not need a reflective surface.

Furthermore, homonuclear diatomic molecules, such as hydrogen (H_2) and nitrogen (N_2) have no absorption bands from the near ultraviolet (UV) to near IR, which makes its optical detection difficult, as conventional methods such as laser-induced fluorescence, absorption spectroscopy, and FT-IR cannot be used. However, these molecules exhibit a strong Raman scattering effect, so the technology readiness level of Raman scattering is considered to be highest among the remote sensing methods applicable to diatomic gas detection.

The Raman scattering process is an inelastic scattering process of light in matter. The energy difference between incoming and scattered light corresponds to a rotational or vibrational energy difference in the molecule.

The efficiency of the Raman scattering process is extremely small, because the cross section is about three orders of magnitude smaller than the corresponding Rayleigh cross section. However, by using a laser system which has high brightness, Raman scattering is a possible method to detect a large variety of gases including diatomic molecules.

In this chapter, experimental works on laser remote sensing of H_2 gas using Raman scattering, which can measure H_2 gas up to a distance of a few to dozens of meters in outdoor conditions, are presented.

H_2 energy is expected to become an important energy source in the future, so laser remote sensing of H_2 gas is expected to become a key technology for H_2 gas leak detection for safety surveillance in H_2 energy facilities, such as H_2 fueling stations.

*Address correspondence to Hideki Ninomiya: Electronics Technology Department, Shikoku Research Institute, 2109-8 Yashima-Nishimachi, Takamatsu-shi, Kagawa 761-0192, Japan; Tel: (+81)87-844-9243; E-mail: ninomiya@ssken.co.jp

Tetsuo Fukuchi and Tatsuo Shiina (Eds)

2. RAMAN SCATTERING

2.1. Origin of Raman Scattering

When light is scattered from an atom, molecule, or particle, most photons are elastically scattered (Rayleigh or Mie scattering), such that the scattered photons have the same energy and wavelength as the incident photons. However, a small fraction of the scattered light is scattered by a change in internal energy of the molecule, with the scattered photons having a frequency different from the frequency of the incident photons. This phenomenon has been discovered by Indian physicist C. V. Raman and is called "Raman scattering" or "Raman effect" [1]. When the electromagnetic wave (*i.e.* incident laser light) interacts with matter, the electron orbits within the constituent molecules are perturbed periodically with the same frequency (ω_0) as the electric field of the incident light. The perturbation or oscillation of the electron cloud results in a periodic separation of charge within the molecules, which is called a dipole moment. The oscillation by the dipole moment acts as a light source. The majority of scattered light has the same frequency (ω_0) as the incident light, and this process is referred to as elastic scattering. On the other hand, a small proportion of the incident light is scattered at different frequencies, and this process is referred to as inelastic scattering. Raman scattering is one such example of inelastic scattering.

The incident light induces a dipole moment during the light and material interaction, whose strength P is given by

$$P = \alpha E , \tag{1}$$

where α is the polarizability and E is the strength of electric field of the incident light. The polarizability is a material property that depends on the molecular structure. The electric field of incident light can be expressed as

$$E = E_0 \cos(2\pi\omega_0 t) . \tag{2}$$

As the ability to perturb the local electron cloud of a molecule depends on the relative location of the individual atoms, the polarizability is a function of the instantaneous position of constituent atoms. For any molecular bond, the individual atoms are confined to specific vibrational or rotational modes, in which the energy levels are quantized in a manner similar to electronic energies.

The physical displacement dQ of the atoms from their equilibrium position due to the particular vibrational mode is expressed as

$$dQ = Q_0 \cos(2\pi\omega_1 t) , \tag{3}$$

where Q_0 is the maximum displacement from the equilibrium position and ω_1 is the frequency of the vibrational or rotational modes. For a typical diatomic molecule (*e.g.* N_2 and H_2), the maximum displacement is about 10% of the bond length. For such small displacements, the polarizability may be approximated by a Taylor series expansion,

$$\alpha = \alpha_0 + (\partial\alpha / \partial Q)dQ , \tag{4}$$

where α_0 is the polarizability of the molecular mode at equilibrium position. From eqs.(3) and (4), the polarizability is

$$\alpha = \alpha_0 + (\partial\alpha / \partial Q)Q_0 \cos(2\pi\omega_1 t) . \tag{5}$$

The dipole moment is given as

$$P = \alpha_0 E_0 \cos(2\pi\omega_0 t) + (\partial\alpha / \partial Q)Q_0 \cdot E_0 \cos(2\pi\omega_0 t)\cos(2\pi\omega_1 t) . \tag{6}$$

Eq. (6) can be rewritten as

$$P = \alpha_0 E_0 \cos(2\pi\omega_0 t) + (\partial\alpha / \partial Q)Q_0 E_0 / 2) \cdot [\cos\{2\pi(\omega_0 - \omega_1)t\} + \cos\{2\pi(\omega_0 + \omega_1)t] \qquad (7)$$

Examination of the above equation reveals that induced dipole moments are created at three distinct frequencies, ω_0, $(\omega_0 - \omega_1)$, and $(\omega_0 + \omega_1)$. The first scattered frequency (ω_0) corresponds to the incident frequency, and the scattered light is referred to as Rayleigh scattering. The latter two frequencies $(\omega_0 - \omega_1)$ and $(\omega_0 + \omega_1)$ are referred to as the Stokes and anti-Stokes components of Raman scattering.

2.2. The Raman Scattering Process in Molecules

In gases, Raman scattering occur with a change in the vibrational, rotational or electronic energy of a molecule. As a result, the Raman scattering signal consists of radiation that has a frequency shift associated with the stationary energy states of the irradiated molecule. The Raman scattering process is shown schematically in Fig. (**1**). The virtual state is not a molecular energy state that actually exists.

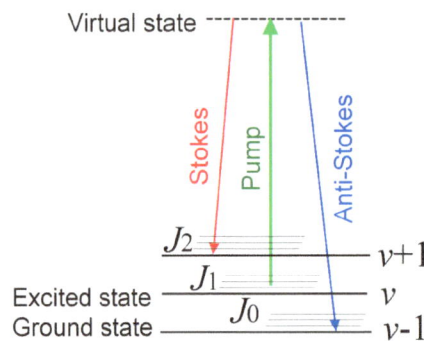

Fig. (1). Schematic diagram of Stokes and anti-Stokes Raman scattering. The Stokes Raman spectrum line is shifted towards longer wavelength from the pumping light. The anti-Stokes line is short than the wavelength of the pumping light.

In the event that the molecule gains energy from the radiation field, the scattered radiation has lower frequency than the incident light, which corresponds to the Stokes component. If the molecule loses energy to the radiation field, the scattered radiation has higher frequency, which corresponds to the anti-Stokes component.

The molecular levels correspond to vibrational and rotational excitation. In the special case of diatomic molecules possessing zero electron angular momentum around the internuclear axis, the selection rules allow vibrational and rotational transitions for which the change in the molecular rotational quantum number J can be only 0 or ±2, and the charge in the vibrational quantum number v can be only 0 or ±1. Therefore, the Raman scattering spectrum consists of the Q-branch of the vibrational band ($\Delta v=1$, $\Delta J=0$), the S-branch ($\Delta v=1$, $\Delta J=+2$) and the O-branch ($\Delta v=1$, $\Delta J=-2$). The Raman band differs in its appearance from the IR band, for which the selection rules are $\Delta v=1$ and $\Delta J=\pm1$ (R- and P-branch).

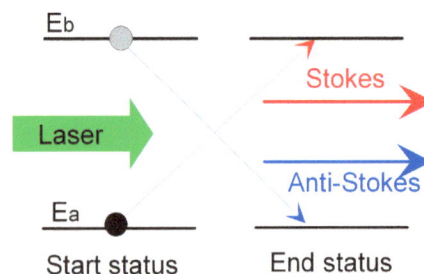

Fig. (2). Two energy levels model of molecule and Raman scattering.

In general, the scattering cross section of the Q-branch is larger than that of the S- or O-branch by two orders of magnitude, so the wavelength of the Q-branch is used for detection of atmospheric gases using Raman scattering.

The relation between the Raman scattering process and the energy status of a laser-irradiated molecule is shown schematically in Fig. **(2)**. The molecule is shown as a model which has two energy levels E_a and E_b ($E_a < E_b$). In the Stokes scattering process, the molecule transfers from the lower energy level E_a to the higher energy level E_b with the result that an incident photon with energy $h\omega_L$ changes to a photon with energy $h(\omega_L - \omega_R)$. Here, h is Planck's constant, and ω_L and ω_R are the frequency of the laser light and Raman shift. The relation $h\omega_R = (E_b - E_a)$ is obtained from the energy conservation law. In the anti-Stokes scattering process, the molecule transfers from the higher energy level E_b to the lower energy level E_a, with the result that an incident photon with energy $h\omega_L$ changes a photon with energy $h(\omega_L + \omega_R)$.

The intensity of the anti-Stokes signal is weaker than the Stokes signal and this tendency becomes more pronounced with increasing Raman shift. The intensity of the Raman signal is proportional to the product of the state transition probability and the number of molecules in the start status. The state transition probabilities at the Stokes and anti-Stokes scattering processes are nearly equal in many cases. For this reason, in the case of the two energy level model as shown in Fig. **(2)**, the ratio of the intensities of the Stokes and anti-Stokes signals is equal to the ratio of the number of molecules in the level E_a and the level E_b. The ratio of the number of molecules in each level is given by a Boltzmann distribution, so the ratio of the anti-Stokes signal (I_{as}) and Stokes signal (I_s) varies with temperature.

$$\frac{I_{as}}{I_s} = \exp\left\{-\frac{E_b - E_a}{kT}\right\} = \exp(-\frac{h\omega_R}{kT}) \tag{8}$$

Here k is Boltzmann constant, and T is the absolute temperature. Therefore, the temperature can be measured from the ratio of intensities of anti-Stokes and Stokes signals, and this measurement method is used to analyze temperatures in flames.

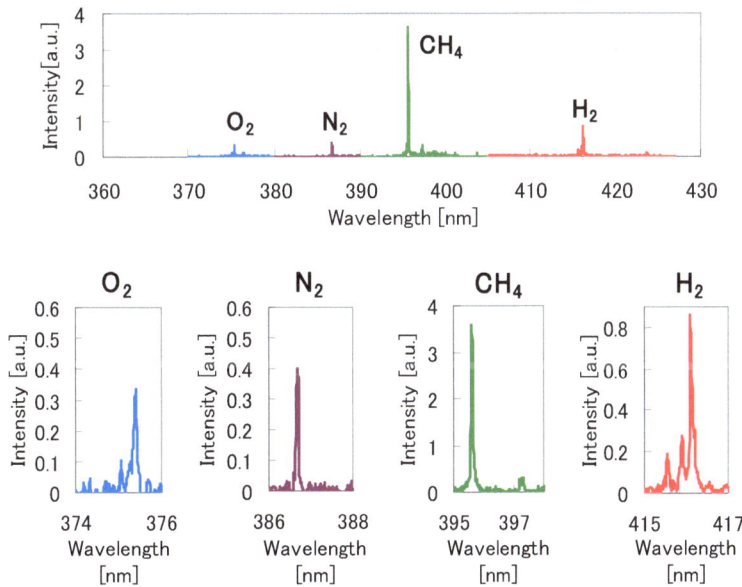

Fig. (3). Examples of Raman Stokes scattering spectra when using a laser wavelength of 355 nm.

When the energy of the incoming laser is adjusted such that it or the scattered light coincides with an electronic transition of the molecule, the intensity of the scattering light is enhanced. This phenomenon is called "resonance Raman scattering".

The Raman scattering signal can be amplified by using a combination of a Raman process with stimulated emission, and this scattering is called stimulated Raman scattering (SRS).

The frequency shifts and the backscattering cross sections of vibrational Raman scattering are listed in Table **1** for molecules of interest for air analysis work involving laser remote sensing [2].

An example of the spectra of the Stokes Raman scattering excited by a Nd:YAG laser operating at 355 nm is shown in Fig. (**3**).

The Raman scattering process is induced by an electric dipole moment which arises from the distortion of a molecule in an electric field. When laser light irradiates an atom or molecule, an electric dipole moment $P=\alpha E$ is induced by the optical electric field E. Therefore, the strength of the electric dipole moment is determined by the amplitude of the irradiated laser light. Raman scattering is strongly radiated in a direction at right angles to the electric field of the irradiated laser light.

For an incident laser wavelength λ, the Raman scattering cross section is proportional to λ^{-4}. Therefore, in order to maximize the Raman scattering efficiency, the laser wavelength should be as short as possible, *i.e.* in the UV range, provided that atmospheric attenuation is at an acceptable level. The most commonly used UV lasers are excimer lasers and higher harmonics of solid-state lasers.

In this scheme, the 1st order Stokes line is used, so the Raman scattering signal is detected on the longer wavelength side of the laser wavelength.

Table 1: Raman shift and backscattering cross sections appropriate for 337.1 nm excitation

Molecule	Raman Shift (cm^{-1})	Cross Section (10^{-30} cm^2 · sr^{-1})
CCl_4	459	26.0
Freon C-318	699	7.8
$NO_2(v_2)$	754	24.0
SF_6	775	12.0
Freon116	807	7.3
Freon114	908	5.3
O_3	1103.3	6.4
SO_2	1151.5	17.0
$CO_2(2v_2)$	1285	3.1
$NO_2(v_1)$	1320	51.0
$CO_2(v_1)$	1388	4.2
O_2(total)	1556	4.6
O_2	1556	3.3(Q)
$C_2H_4(v_2)$	1623	5.4
NO	1877	1.5
CO	2145	3.6
N_2(total)	2330.7	3.5
N_2	2330.7	2.8(Q)
H_2S	2611	19.0
$CH_3OH(v_2)$	2846	14.0
C_3H_8	2886	81.8(C)

Table 1: cont….

C_5H_{12}	2885	124.0(C)
CH_4	2914	32.2(C)
$CH_4(v_1)$	2914	21.0
C_3H_6	2942	63.6
C_2H_5OH	2943	19.0
C_4H_8	3010	89.6(C)
$CH_4(v_3)$	3017	14.0
C_2H_4	3020	28.6
$C_2H_4(v_1)$	3020	16.0(c)
C_6H_6	3072	65.2
NH_3	3334	11.0
C_2H_2	3372	3.36
H_2O	3651.7	7.8(Q)
H_2	4160.2	8.7

*Q indicates the value of the Q-branch Raman backscattering cross section, C indicates a broad multi-peaked structure associated with the C-H stretch mode.

This results in a problem of interference from laser-induced fluorescence which is caused by laser incidence on obstacles behind the gas (such as a wall, equipment or dust). The fluorescence signal is much stronger than the Raman scattering signal, so appropriate measures must be taken to suppress the interference from fluorescence.

In addition, stray laser light and laser light reflected from obstacles may also interfere with the Raman scattering measurement. An effective method to suppress the stray and reflected laser light is to install a notch filter or long pass filter in front of the detector.

2.3. Coherent Anti-Stokes Raman Scattering (CARS)

A possible method to eliminate the interference from fluorescence is the use of the anti-Stokes component of Raman scattering. Since the wavelength of the anti-Stokes line lies on the shorter side of the laser wavelength, whereas fluorescence lies solely on the longer side, the effects of fluorescence can be eliminated.

However, when molecules at room temperature are irradiated, the anti-Stokes component (for vibrational Raman scattering) almost never appears. This is because the Raman shift is much larger than the energy corresponding to room temperature. For example, for N_2, the Raman shift is $\Delta\omega=2331$ cm^{-1}, which corresponds to a temperature of $hc\Delta\omega/k=3357$ K. Here c is the speed of light.

The anti-Stokes line can be efficiently be generated by Coherent Anti-Stokes Raman Scattering (CARS), in which a pump beam and a Stokes beam is directed simultaneously through the target gas. An energy level diagram of CARS in the case of H_2 and a pump wavelength of $\lambda=355$ nm is shown in Fig. **(4)**. The ground and excited states represent the vibrational levels ($v=0,1$) of the ground electronic state ($X^1\Sigma_g^+$). A wave vector diagram for phase matching in CARS is shown in Fig. **(5)**.

The first measurement involving CARS in gases were by Rado [3], who determined the non-resonant susceptibilities of a number of gases using a ruby laser and H_2 Stokes light beam stimulated in a high pressure H_2 gas cell.

The intensity of the CARS signal is proportional to the square of absolute magnitude of the 3rd order non-linear susceptibility $\chi^{(3)}$. In the absence of pressure broadening, the CARS signal intensity I_{CARS} is proportional to $|\chi^{(3)}|^2 I_p^2 I_s$, where I_p is the pump beam intensity and I_s is the Stokes beam intensity. Since $\chi^{(3)}$ is proportional to n, the result $I_{CARS} \propto n^2$ is expected, where n is the density of gas.

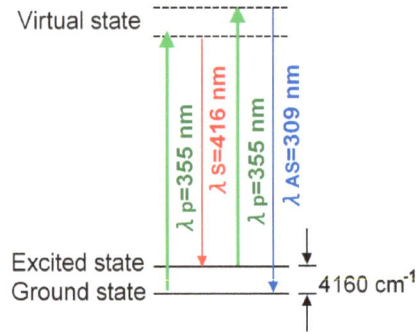

Fig. (4). Energy level diagram of CARS of H_2 when using a laser wavelength of 355 nm.

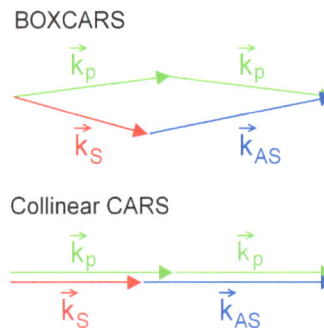

Fig. (5). Wave vector diagram for phase matching in CARS. k_P, k_S, k_{AS} are the wave vectors of the pump, Stokes, and anti-Stokes beams. The phase matching condition is $k_p+k_p=k_S+k_{AS}$.

CARS has been used for thermometry and concentration measurement in discharges and flames [4-7]. In combustion fields, the spatial resolution is important, so a BOXCARS arrangement, in which the pump beam and Stokes beam intersect at a certain position, is usually used. However, this arrangement is only useful if the location of the gas is previously known. In the case of the gas detection, the location of the gas is not known, so it is preferable to use an arrangement in which CARS can occur at an arbitrary location along the beam propagation direction. Therefore, in the case of the gas detection, a collinear arrangement, in which the pump beam and Stokes beam are transmitted along the same axis, is used. The phase matching condition in this case is shown in Fig. **(5)**, where k_p, k_S, k_{AS} are the wave vectors of the pump, Stokes, and anti-Stokes beams, respectively.

3. RAMAN LIDAR SYSTEM

3.1. Optical System

A Raman lidar system is commonly used for detection of gases in outdoor conditions. Raman lidar systems have been extensively used in the area of atmospheric science, mainly for the measurement of water vapor. As the cross section of the Raman scattering process is small, the lidar system uses a laser of high intensity for generating Raman scattering light and a large telescope for receiving the Raman scattering light.

A schematic of the optical head using the coaxial type is shown in Fig. **(6a)**, and that using the biaxial type is shown in Fig. **(7)**, respectively. The coaxial type shown in Fig. **(6b)**, which uses the telescope for both transmitter and receiver corresponds to the inline type described in Chapter 2. The Raman lidar consists of a laser transmitter and a receiver. The receiver consists of a telescope and a detector.

The laser transmitter uses a pulsed laser which has short pulse width, and the Raman scattering signals are measured in synchronization with laser irradiation. Then the location and concentration distribution of gases are obtained by changing the time scale into a distance scale. The telescope can be a reflecting type (such as Newtonian telescope) or refracting type (such as Galileo telescope). The telescope is arranged in a coaxial or biaxial with the transmitted laser beam.

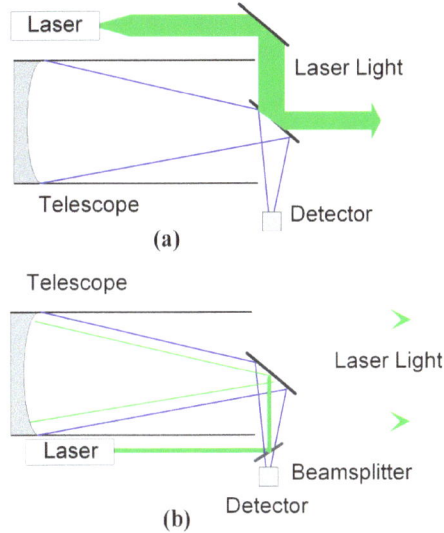

Fig. (6). Two possible coaxial lidar arrangements: (a) separate beam expansion; (b) beam expansion *via* receiving mirror.

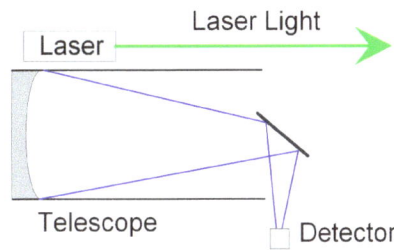

Fig. (7). Schematic diagram of the biaxial lidar system.

As Raman lidar systems commonly use a UV laser in order to enhance the Raman scattering efficiency, the Raman scattering occurs in the UV or visible (VIS) region. Therefore, the primary and secondary mirrors of the telescope are UV coated, with high reflectivity from the UV to VIS region.

Fig. **(8)** shows an example of the Raman lidar using a refracting telescope in a biaxial arrangement.

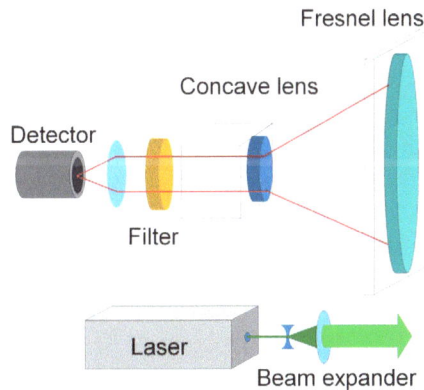

Fig. (8). Schematic diagram of a biaxial lidar system using a refracting telescope.

The backscattering Raman light is collected by the telescope. The collected light passes through a bandpass optical filter to reject sun light and selectively transmit Raman scattering from the target molecule, and directed to a detector such as a photomultiplier tube. In addition, a laser line edge filter (for example, laser light rejection ratio $<10^{-6}$) is installed in front of the detector to reject stray laser light, Rayleigh and Mie scattering light, and Raman scattering from other molecules.

3.2. Intensity of Raman Signal

The Raman return signal depends on the laser power, scattering processes, distance, aperture of the acceptance surface, and efficiency of the light collection optical system, as shown in Fig. **(9)**.

The intensity of the Stokes Raman return signal is given by the lidar equation.

$$S(r) = \gamma P_0 KY(r)S\frac{1}{r^2}\frac{\Delta r}{2}N\sigma T(r)$$

$$T(r) = \exp[-(a_L + a_R)r]$$

(9)

where γ is the quantum efficiency of the detector, P_0 is the laser power, K is the efficiency of the light collection optical system, $Y(r)$ is the efficiency function which depends on the setting of the optical receiver and laser beam, S is the area of the acceptance surface, r is the distance, N is the gas density, σ is the Raman scattering cross section, Δr is the axial range resolution, a_L is the extinction coefficient of laser light, and a_R is the extinction coefficient of Raman scattering.

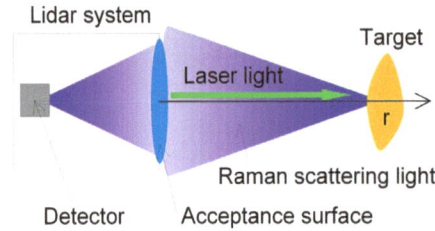

Fig. (9). View showing the measurement method of Raman backscattering light.

In the case of the coaxial lidar system, the axes of the optical receiver and laser beam coincide, so $Y(r)$ is the efficiency function of the optical receiver, such as the focal distance of the receiver and the transmittance of the optical components. However, in the case of the biaxial lidar system, the axes of the optical receiver and laser beam are different, so $Y(r)$ includes not only the efficiency function of the optical receiver but also the overlap function between the receiver-optics field of view and the area of laser irradiation. In the near field, $Y(r)$ depends on the nexus of the laser beam axis with respect to the receiver-optics field, as shown in Fig. **(10a)** when the nexus small and Fig. **(10b)** when the nexus is large. As $Y(r)$ is a system specific factor which depends on the efficiency of the optical receiver and the layout of the laser and optical receiver, we are able to determine $Y(r)$ approximately from the Rayleigh scattering signal. Furthermore, in open air N_2 and O_2 gases are distributed homogeneously, so we can determine $Y(r)$ from the Raman scattering signal from atmospheric N_2 or O_2 molecules.

On the other hand, if the Raman scattering signals from the target gas and atmospheric N_2 are simultaneously measured, the concentration of the target gas can be calculated based on the ratio of these signals.

The Raman scattering signal of atmospheric N_2 is given by

$$S_N(r) = \gamma_N P_0 K_N Y(r)S\frac{1}{r^2}\frac{\Delta r}{2}N_N\sigma_N T_N(r)$$

$$T_N(r) = \exp[-(a_L + a_{NR})r]$$

(10)

where γ_N is the quantum efficiency of the detector at the Raman scattering wavelength of N_2, K_N is the efficiency (transmittance) of the receiver at the Raman scattering wavelength of N_2, N_N is the N_2 density, σ_N is the Raman scattering cross section of the N_2 molecule, and a_{NR} the extinction coefficient of Raman scattering light of the N_2 molecule.

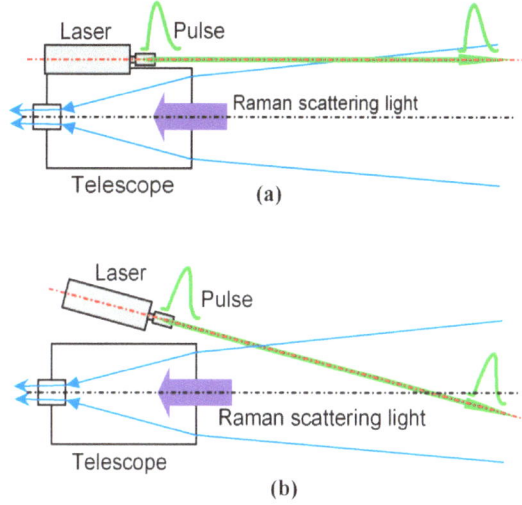

Fig. (10). Schematic diagram of the biaxial lidar system. (a) the nexus of the axes is small; (b) the nexus of the axes is large.

The Raman scattering signal of the target gas is given by

$$S_T(r) = \gamma_T P_0 K_T Y(r) S \frac{1}{r^2} \frac{\Delta r}{2} N_T \sigma_T T_T(r)$$

$$T_T(r) = \exp[-(a_L + a_{TR})r] \tag{11}$$

where γ_T is the quantum efficiency of the detector at the Raman scattering wavelength of the target gas, K_T is the efficiency (transmittance) of the light collection optical system at the Raman scattering wavelength of the target gas, N_T is the target gas density, σ_T is the Raman scattering cross section of the target molecule, and a_{TR} is the extinction coefficient of Raman scattering light of the target molecule.

In the case of near field, $(a_L + a_{NR})r \ll 1$ and $(a_L + a_{TR})r \ll 1$, so the ratio of the signals of the target gas and N_2 gas is given by

$$S_T(r)/S_N(r) = (N_T/N_N) \cdot C$$

$$C = (\gamma_T \cdot K_T \cdot \sigma_T)/(\gamma_N \cdot K_N \cdot \sigma_N) \tag{12}$$

where, C is a constant, because σ_N and σ_T are constant, γ_N and γ_T are the quantum efficiencies of the detector, and K_N and K_T are the efficiencies of the light collection system which can be measured.

When the concentration of the target gas is low compared to the concentration N_N of atmospheric N_2 which is about 80%, the spatial distribution of the concentration N_T of the target gas can be obtained from the ratio $S_T(r)/S_N(r)$.

4. APPLICATION TO GAS SENSING

As Raman lidar is capable of simultaneous measurement of gas concentration and temperature, this system is mostly used for atmospheric observation. The range of observations is from sub km to several km. On the other hand, it has been also reported that the Raman lidar system can be used in near field (up to a distance of a few to tens of meters) in outdoor conditions. Here are some applications of the Raman lidar system.

4.1. Raman Spectroscopy

A Raman spectroscopy system is used for *in situ*, real-time, noncontact detection and identification of unknown gases.

Fig. **(11)** shows a schematic diagram of a system for Raman spectroscopic analysis and the principal specifications are listed in Table **2**. The laser transmitter is a pulsed Nd:YAG laser operating at 266 nm and 6 ns pulse width. The receiver is a UV camera lens that is biaxial with the transmitted laser beam. The spectrometer is a 50 cm system. A gated, intensified CCD captures the Raman spectrum. The synchronizer is used to synchronize the firing of the laser and the CCD gate. In order to suppress stray laser light and Rayleigh and Mie scattering, two laser line edge filters are installed in front of the slit of the spectrometer. The transmission of one edge filter at the laser wavelength is $<10^{-6}$, and the transmission at the Raman scattering wavelengths is $>98\%$.

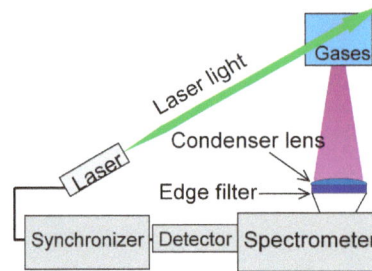

Fig. (11). Schematic diagram of the spectroscopy system for analysis of gases by Raman scattering.

Raman scattering light is collected by the condenser lens and directed into the spectrometer. An example of the Raman spectrum of H_2 gas in air located 3 m from the system is shown in Fig. **(12)**.

Table 2: Principal specifications of the Raman spectroscopy system for analysis of gases

Component	Specification
Laser (transmitter)	
type	Nd:YAG (FHG)
wavelength	266 nm
output energy	30 mJ max
repetition rate	20 Hz
Receiver	
type	UV camera lens
focal length	105 mm
diameter	22 mm
Spectrometer	
focal length	50 cm
grating	1800 grooves/mm
Detector	
type	gated, intensified CCD
image resolution	1024×256 pixels

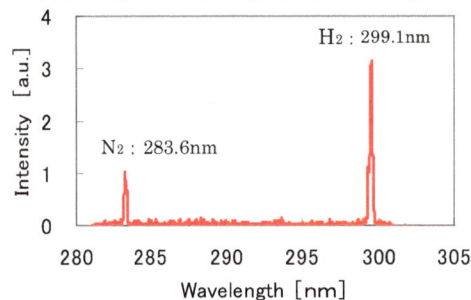

Fig. (12). Raman spectra of H_2 gas and N_2 gas in air located 3 m from the spectroscopy system.

In outdoor conditions, sunlight (background UV radiation) may be a significant source of noise. In order to evaluate the effect of sunlight on Raman spectra measurement, sunlight was deliberately passed through the condenser lens and the edge filter and inserted into the spectrometer. Raman scattering spectra obtained with the gate time of the intensified CCD array set to 10, 50, and 100 ns are shown in Fig. **(13)**. The effect of sunlight appeared for wavelengths above 302 nm, and no effect was observed on the H_2 Raman signal at 299 nm.

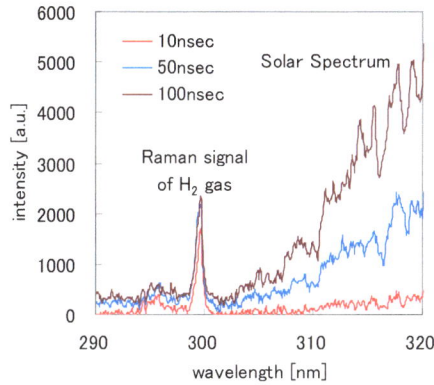

Fig. (13). Raman scattering spectra of H_2 gas and sunlight obtained with the gate time of the intensified CCD array set to 10, 50, and 100 ns.

4.2. Detection of Gas Leakage

A Raman lidar system for detection of H_2 gas has been developed [8, 9]. H_2 gas is considered to be a clean energy source and a possible alternative to fossil fuels for use in transportation and power generation. As H_2 gas is flammable and is fueled at high pressure in the order of several hundred atmospheres, a detection system to detect H_2 gas from a safe distance is necessary for safety precautions. The developed Raman lidar system can detect H_2 gas leaks from a distance of several m to several tens of m. A schematic diagram of the Raman lidar system is shown in Fig. **(14a)** and the interior of the detector assembly is shown in Fig. **(14b)**. The principal specifications of the lidar system are listed in Table **3**.

The beam from the Q-switched, pulsed Nd:YAG laser operating at 355 nm is transmitted along the axis of a Newtonian telescope of aperture 212 mm using two beam steering mirrors. Laser backscatter, including Rayleigh, Mie, and Raman scattering, is collected by the telescope, collimated, and split into two beams by a beamsplitter. Each beam passes through a narrowband interference filter and directed into a photomultiplier tube (PMT). In addition, a laser line edge filter is installed in front of the beamsplitter in order to suppress stray laser light and Rayleigh and Mie scattering. The transmission of the edge filter at 355 nm is about $<10^{-6}$, and the transmission at the Raman scattering wavelengths is $>98\%$. The telescope, laser, and detector are mounted together on a tripod, and can be pointed in an arbitrary azimuthal direction.

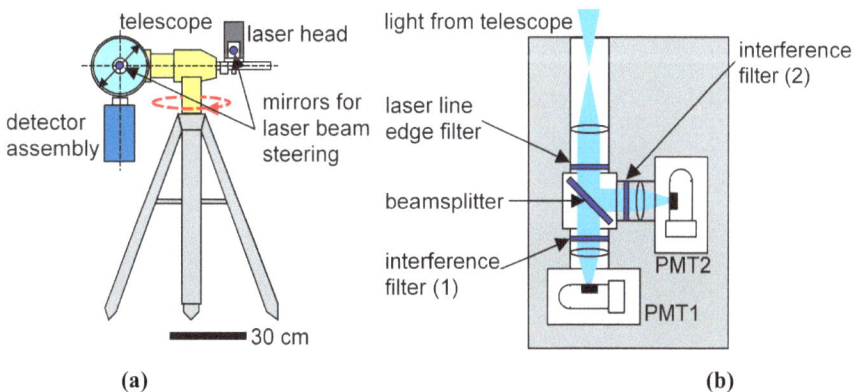

Fig. (14). Schematic diagram of the Raman lidar system for detection H_2 gas leakage: (a) general view of lidar system, (b) layout of the detector assembly. (PMT: photomultiplier tube)

The efficiency function $Y(r)$ (which is the overlap function between the optical receiver and laser beam) of the lidar system as a function of the detection distance r was calculated. The function $Y(r)$ and the predicted signal $Y(r)/r^2$, assuming a $1/r^2$ dependence of the backscatter intensity, are shown in Fig. **(15)**.

Table 3: Principal specifications of the Raman lidar system for detecting gases

Component	Specification
Laser (transmitter)	
type	Nd:YAG (THG)
wavelength	355 nm
output energy	60 mJ max.
repetition rate	20 Hz
Telescope (receiver)	
type	Newtonian
primary mirror diameter	
	212 mm
secondary mirror diameter	
	68 mm
focal length	830 mm
mirror reflectivity	>90% at 362 nm and 416 nm
Detector	
type	photomultiplier tube

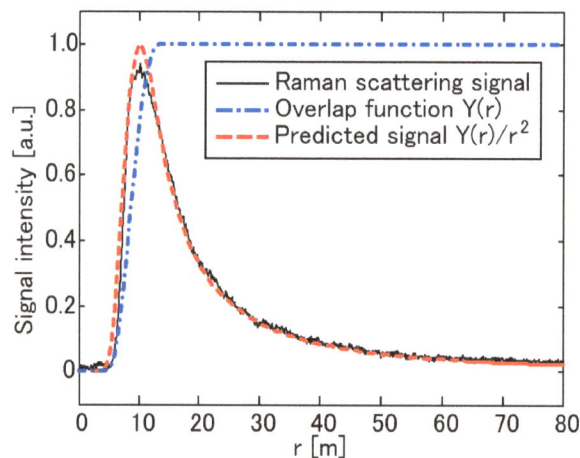

Fig. (15). Calculated overlap function of the lidar system, predicted signal, and measured N_2 Raman backscatter signal.

In order to verify the validity of the above calculation, Raman scattering by atmospheric N_2 at 387 nm wavelength was measured using the lidar system, with the telescope focused at infinity. The result is also shown in Fig. **(15)**, which showed good agreement with the calculated results, except for the deviation near the maximum around r=10 m.

The Raman scattering wavelengths by H_2 (1st order Stokes line) are listed in Table **4**. The present system can measure rotational and vibrational Stokes Raman signals simultaneously when interference filters, whose characteristics are shown in Table **5**, are placed in each channel. The present system can also measure the Raman signal (either the rotational or vibrational Raman signal) and the background at a neighboring wavelength, which allows elimination of any spurious signal resulting from broadband fluorescence when the difference between the signals from the two channels is taken.

Table 4: Raman Scattering Wavelengths (1st order Stokes line for laser wavelength 355 nm)

Vibrational		Rotational	
Raman shift	Raman scattering wavelength	Raman shift	Raman scattering wavelength
4155 cm^{-1}	416.4 nm	587 cm^{-1}	362.5 nm

Table 5: Characteristics of interference filters used for detection of Raman scattering

	Vibrational	Rotational
diameter	25.4 mm	25.4 mm
center wavelength	416.5 nm	362.5 nm
linewidth (FWHM)	0.9 nm	1.6 nm
transmission	29%	21%

Experiments of remote H_2 gas detection were conducted in daytime, outdoor conditions. H_2 gas was released from a gas nozzle located at a distance of 10-50 m from the lidar system, as shown in Fig. **(16)**. A photodiode mounted on the telescope detected the laser light scattered from the steering mirror and provided the trigger signal for detection of the Raman scattering signals.

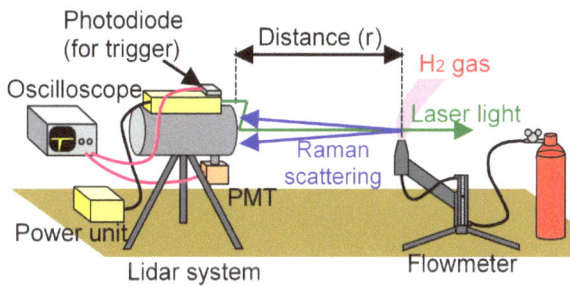

Fig. (16). Experimental setup for H_2 gas detection.

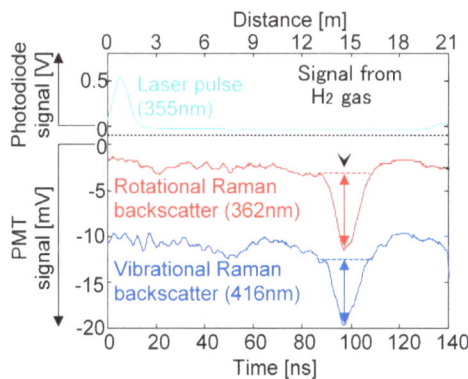

Fig. (17). Example of Raman backscatter signals. Top trace: laser pulse measured by photodiode, middle trace: rotational Raman backscatter signal (362 nm), bottom trace: vibrational Raman backscatter signal (416 nm).

An example of vibrational and rotational Raman scattering signals from H_2 gas is shown in Fig. **(17)**. In this case, the release point of H_2 gas was 14 m from the lidar system and H_2 gas released at a rate of 30 liter/min.

The two signals ware measured simultaneously by the two PMTs. The result shows that the system could detect H_2 by either vibrational or rotational Raman scattering. As the experiment was conducted during daytime, the background level was higher for the longer wavelength (416 nm). Raman backscattering from H_2 gas was obtained at signal-to-noise ratios of over 5 for detection distances of 10-50 m.

In order to investigate the dependence of Raman intensity on the detection distance, the Raman signal was measured with the nozzle placed at r=10-50 m from the lidar system, and H_2 gas released at a rate of 30 liter/min. Fig. **(18)** shows the rotational Raman scattering intensity at different distance. Since the telescope focus was adjusted at 30 m distance, the signal intensity exhibited a $1/r^2$ dependence with an enhancement near at 30 m.

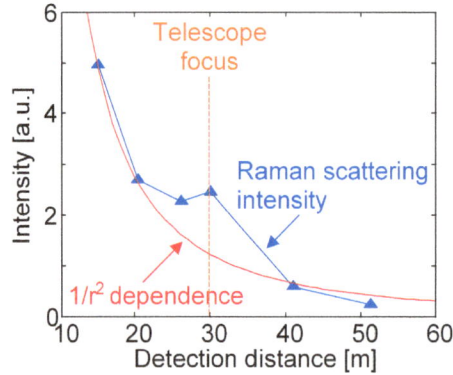

Fig. (18). Raman scattering signal intensity at different distance.

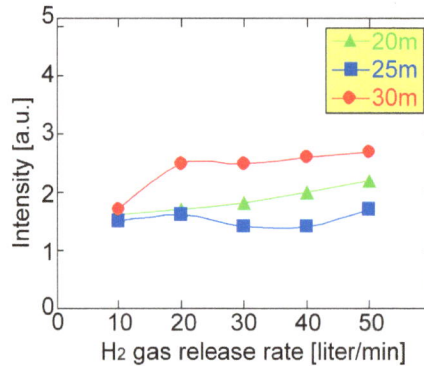

Fig. (19). Raman scattering signal intensity at different release rate of pure H_2 gas.

Next, in order to investigate the dependence of Raman intensity on the H_2 gas release rate, the Raman signal was measured with H_2 gas release rates of 10-50 liter/min, with the nozzle placed at r=20, 25, 30 m. The results are shown in Fig. **(19)**. The Raman signal intensity did not show any marked dependence on the H_2 gas release rate. Provided that mixing of H_2 gas and ambient air is negligible at the point of measurement, the local H_2 gas density is constant with respect to release rate, temperature and pressure being identical.

In order to investigate the sensitivity of the Raman lidar system as a function of distance r, Raman scattering by atmospheric O_2 (about 20% concentration) was measured.

The obtained signal is shown in Fig. **(20a)**. The portion for r=15-60 m is expanded and shown in semi-log scale in Fig. **(20b)**. The red line shows the result of a least squares fit, log $[V(r)/V(15m)]$=-0.039r [m]+0.56, which showed that the signal in this interval could be closely approximated by an exponential decay with an e-folding distance of about 25 m.

This Raman lidar system can measure the temporal variation of the Raman scattering signal from the laser irradiation, and the spatial distribution of the gases can be observed by conversion of the time scale to distance as shown in Fig. **(17)** and Fig. **(20)**.

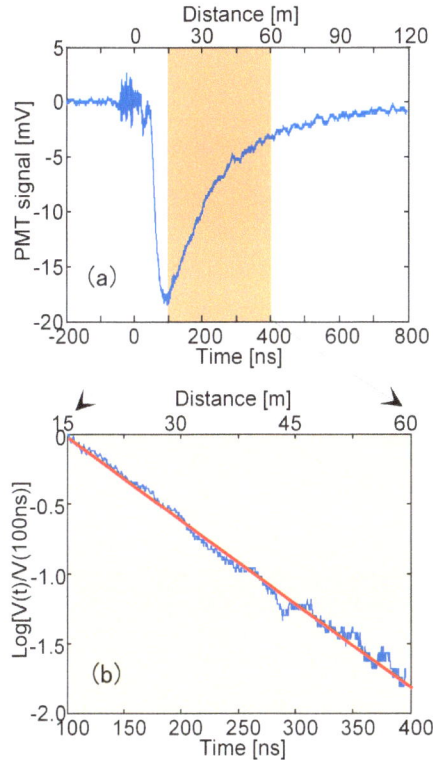

Fig. (20). (a) Raman scattering signal by atmospheric O_2, (b) semi-log plot for distance 15-60 m.

4.3. Imaging of Leakage Gas Using Raman Scattering

Two-dimensional imaging of the spatial distribution of gas using the Raman lidar has been reported [8].

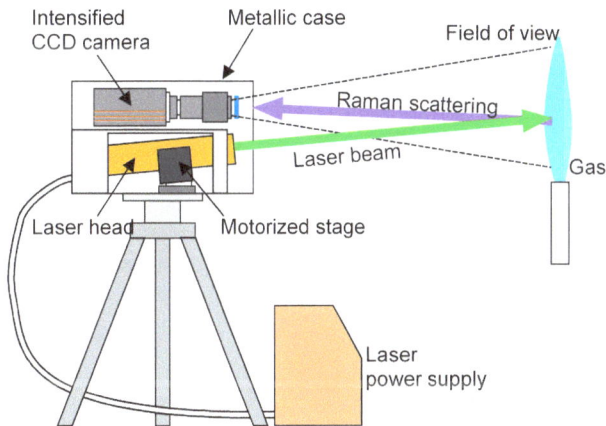

Fig. (21). Schematic diagram of the experimental apparatus for visualization of H_2 gas.

An experimental apparatus for imaging of H_2 gas was constructed using a laser and an intensified CCD. A schematic diagram of the experimental apparatus is shown in Fig. **(21)** and the principal specifications are listed in Table **6**. A smaller Nd:YAG laser operating at 266 nm was used. The laser head and receiver, consisting of an interference filter, UV camera lens, and a gated, intensified CCD camera, were mounted on a common metallic case, which was mounted on a tripod.

The laser head was mounted on a motorized stage inside the metallic case, which enabled fine adjustment of the laser beam direction with respect to the field of view of the camera lens. The interference filter and UV camera lens are the same as those described in the previous section. The metallic case could be pointed in an arbitrary azimuthal angle, and the elevation angle could be adjusted to approximately -4 to +4 degrees with respect to the horizontal. As the laser beam was transmitted off-axis relative to the optical axis of the receiver, the minimum distance at which the laser beam entered the field of view of the receiver was limited to about 1 m. The intensified CCD camera was gated relative to the firing of the laser with a gate width of 50 ns.

Table 6: Principal specifications of the apparatus for visualization of H_2 gas

Component	Specification
Laser (transmitter)	
type	Nd:YAG (FHG)
wavelength	266 nm
output energy	2 mJ max.
repetition rate	10 Hz
Receiver	
type	UV camera lens
focal length	105 mm
diameter	22 mm
Detector	
type	gated, intensified CCD
image resolution	512×512 pixels

H_2 gas visualization experiments were conducted by placing a nozzle of aperture diameter 2.5 mm at a distance of 2-8 m from the above apparatus, and releasing H_2 gas from the nozzle. The gas release rate was controlled by a flow meter installed between the gas cylinder and the nozzle. The direction of the apparatus was adjusted so that the field of view was centered at about 5 cm above an outlet port of the nozzle, and the laser beam direction was adjusted to traverse this point. The camera lens was adjusted so that an object placed at a distance of 5 m was in focus.

Fig. **(22a)** shows an image obtained with no H_2 gas release. From no spot appeared in the image, the rejection of stray laser light, Mie and Rayleigh scattering was confirmed to be sufficient. Fig. **(22b)**, Fig. **(22c)**, Fig. **(22d)**, Fig. **(22e)**, Fig. **(22f)**, Fig. **(22g)**, Fig. **(22h)** show images of Raman scattering from H_2 gas, obtained with the nozzle placed at a distance of 2, 3, 4, 5, 6, 7, 8 m from the apparatus, respectively. The H_2 gas release rate was set to 15 liter/min. Each image represents integration over 500 laser shots.

In order to evaluate the dependence of the sensitivity on distance, the intensity at each distance is needed. Taking the maximum intensity of each image did not yield accurate results, because the spot image had a grainy appearance (especially when it was out of focus) and the intensity profile was not smooth. Taking the number of pixels in each image which exceeded a certain threshold value did not yield accurate results because larger, out of focus spot images resulted in larger values. Therefore, to smooth out the grainy appearance of the spot images by spatial averaging, Gaussian blur was applied to each image. The $1/e^2$ width of the Gaussian function used for smoothing was set to 3 pixels, which was sufficient to eliminate noise and spikes in the intensity profile. The result of applying Gaussian blur to the images in Fig. **(22)** is shown in Fig. **(23)**.

An example of the original intensity profile (vertical cross section of Fig. **(22c)** through the center of the spot image) and the result of applying Gaussian blur (vertical cross section of Fig. **(23c)** at the same position) are shown in Fig. **(24)** as solid and dashed lines, respectively.

The Raman scattering intensity was taken as the peak value of the spot images in Fig. **(23)**. The background intensity was taken as the average value of the intensity of the images over a spatial region excluding the spot image. The Raman scattering intensity and background intensity for each distance are shown in Fig. **(25).**

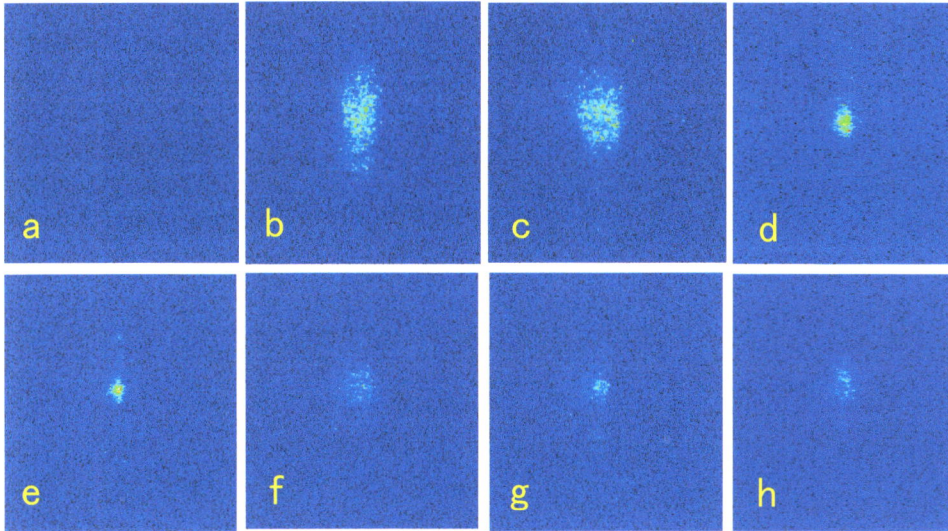

Fig. (22). (a) Image obtained with no H_2 gas release (back ground); images of Raman scattering from H_2 gas released at (b) 2m, (c) 3m, (d) 4m, (e) 5m, (f) 6m, (g) 7m, (h) 8m from the apparatus. The H_2 gas release rate is 15 liter/min.

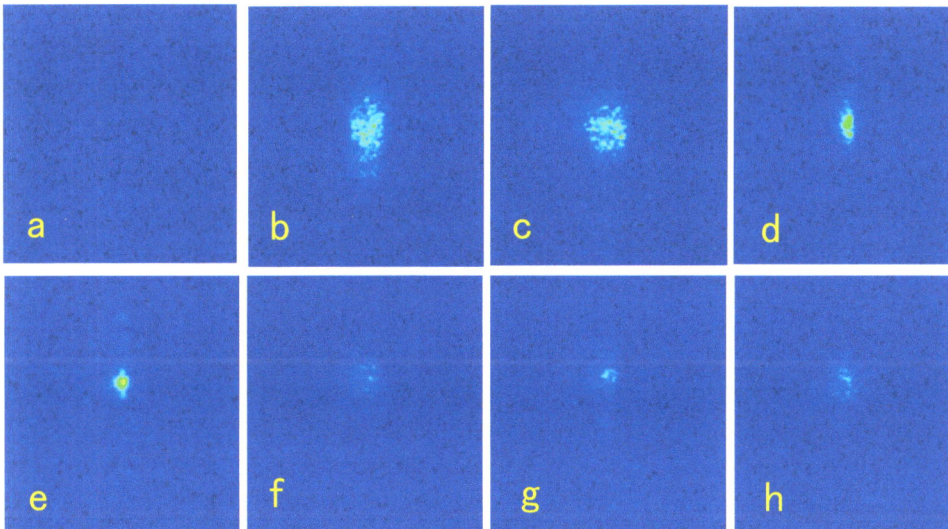

Fig. (23). Result of applying Gaussian blur to Raman scattering images in Fig. **(22)**. The smoothing is set to 3 pixels.

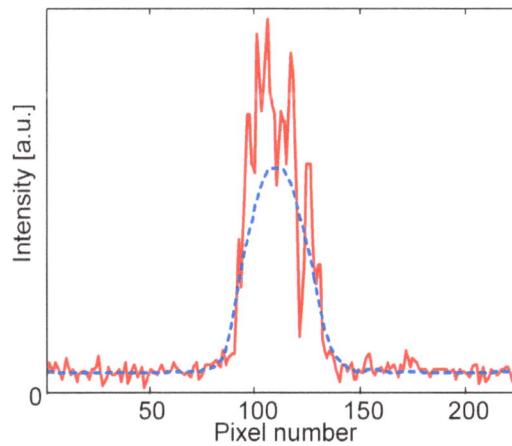

Fig. (24). Example of intensity profile of raw image (solid line) and result of applying Gaussian blur (dashed line).

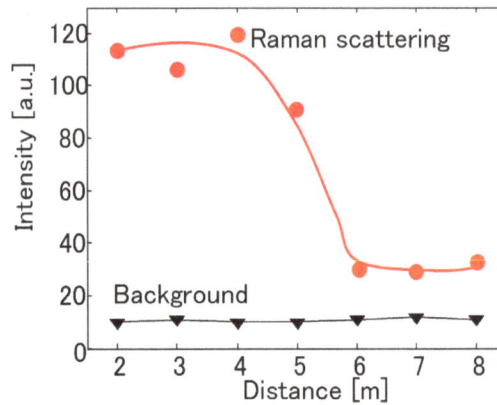

Fig. (25). Dependence of Raman scattering and background intensities on distance.

The results showed that the Raman scattering intensity decreased significantly beyond 5m. The signal-to-noise ratio at distance over 5 m can be improved by suitable image processing for contrast enhancement, by the use of spatial filters other than Gaussian blur, or by increasing the laser pulse energy, receiver aperture, or the quantum efficiency of the imaging device at the Raman scattering wavelength. These improvements are for further investigation.

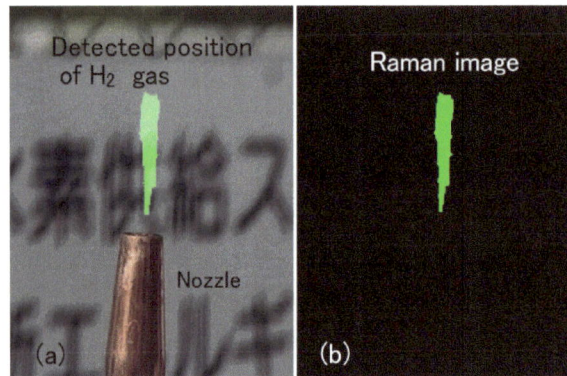

Fig. (26). Example of the visualization image of H_2 gas. (a) Raman image painted on the visual image, (b) Raman image obtained by the system.

An example of the result of two-dimensional imaging of H_2 gas is shown in Fig. **(26)**. Here, H_2 gas was released at a rate of 15 liter/min and at a distance of 5 m from the lidar system.

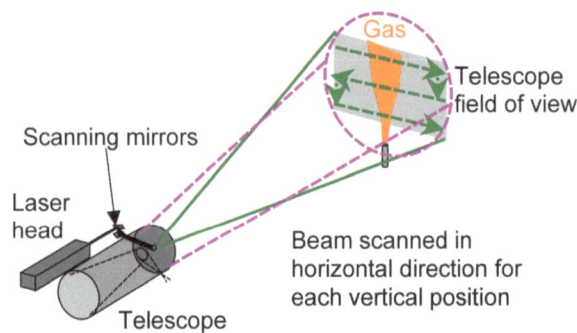

Fig. (27). Schematic diagram of laser beam steering for two-dimensional visualization of gas.

In order to enable the lidar system to perform two-dimensional imaging of gas, a laser beam scanner is installed between the laser head and the beam steering mirror mounted on the telescope axis. The Raman lidar system used in this work is the same system as shown in Fig. **(14)**. The scanner is set so that the beam is scanned and horizontally (10 steps) for each vertical position (10 steps), as shown schematically in Fig. **(27)**. A visual camera, whose field of view included that of the telescope, is mounted near the scanner.

With each of the scanner mirrors at a given angle, the position of the laser spot in the image obtained by the visual camera is recorded. By repeating this procedure, the corresponding position on the visual image for each scanner mirror angle setting is calibrated.

The laser beam is directed at the initial position (top left hand corner), and backscatter signal is obtained. If the backscatter signal exceeds a certain threshold level, the corresponding position in the visual image is marked by color. This is repeated for each position in the horizontal and vertical scan. The resulting image shows in color the region from which Raman backscatter from the gas has been recorded.

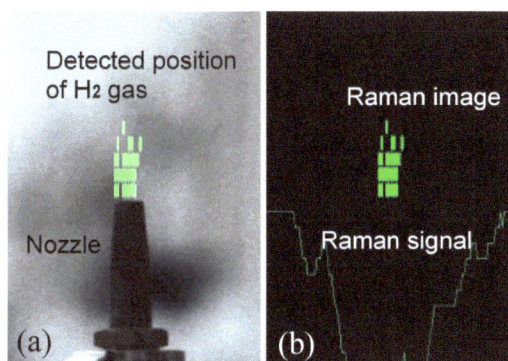

Fig. (28). Example of two-dimensional visualization of H_2 gas. (a) Raman image painted on the visual image, (b) Raman image and signal obtained by the lidar system.

An example of the result of two-dimensional imaging of H_2 gas is shown in Fig. **(28a)**. The Raman signal is shown in Fig. **(28b)**. Here, H_2 gas was released at a rate of 20 liter/min and at a distance of 11 m from the lidar system. The spots indicated in light green indicate the positions (obtained from the scanner mirror setting) from which Raman backscattering light has been obtained.

At the time of the experiment, the transverse wind speed was negligible compared to the initial velocity of the released gas. Therefore, the region of the H_2 gas did not change during the time of scanning, which was about 30 s, as previously mentioned.

4.4. Imaging of Leakage Gas Using CARS

An imaging method of H_2 gas leakage using collinear CARS has been reported [10]. When using a laser wavelength of 355 nm, wavelengths of Stokes and anti-Stokes Raman scattering are 416 nm and 309 nm, respectively, as shown in Fig. **(4)**. The phase matching condition is the collinear CARS, as shown in Fig. **(5)**.

Table 7: Principal specifications of the experimental system for imaging of anti-Stokes Raman scattering from H_2 gas released into air

Component	Specification
Transmitter (laser)	
type	Nd:YAG laser
wavelength	355 nm
output energy	2.2 mJ
repetition rate	10 Hz
Transmitter (Stokes light)	
type	optical parametric oscillator

wavelength	416 nm
output energy	0.6 mJ
condensing lens type focal length f-number	 UV lens 105 mm 4.5
Detector type imaging device imaging resolution	 gated image intensifier CCD camera 640x480 pixels

The experimental configuration is shown in Fig. **(29)**, and the principal specifications are listed in Table **7**.

An experimental study was performed to evaluate the applicability of CARS to imaging of H_2 gas leak into air. A Nd:YAG laser of wavelength 355 nm and repetition rate 10 Hz is used to pump an optical parametric oscillator (OPO), whose output is tuned to the Stokes line of wavelength 416 nm. The pulse energies of the laser beam and the Stokes beam at the OPO output are 2.2 mJ and 0.6 mJ, respectively. The laser beam and Stokes beam are aligned in a collinear configuration, and are focused directly above the center of a nozzle of aperture 1 mm. H_2 gas is released through the nozzle, and the release rate is controlled by a flowmeter.

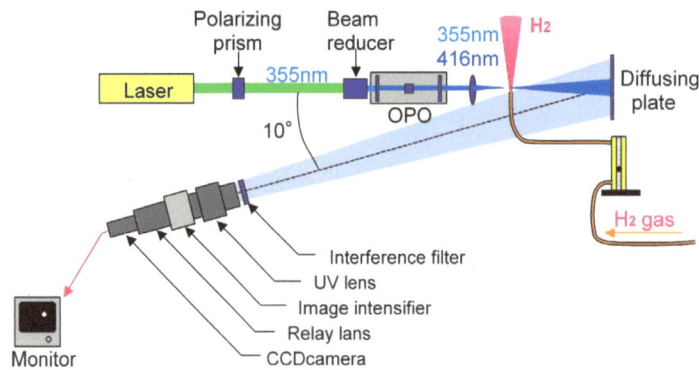

Fig. (29). Experimental configuration for imaging of anti-Stokes Raman scattering from H_2 gas released into air.

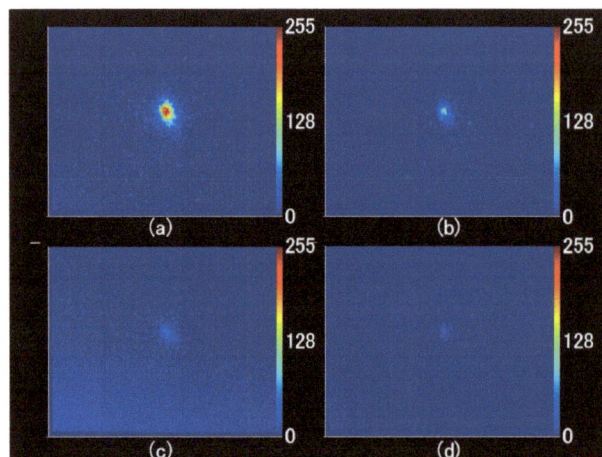

Fig. (30). Images of anti-Stokes light generated by H_2 gas released through a nozzle of diameter 1 mm, at release rates of (a) 8 ml/min, (b) 6 ml/min, (c) 4 ml/min, and (d) 2 ml/min.

The anti-Stokes beam of wavelength 309 nm is generated by CARS. The three beams are incident on an acrylic diffusing plate placed on the downstream side of the nozzle, and the projection of the anti-Stokes beam on the diffusing plate is selectively imaged using an UV camera lens, interference filter (center wavelength 309 nm, bandwidth 1.5 nm), a gated image intensifier, and a CCD camera. The interference filter provides sufficient rejection of light at 355 nm, 416 nm, and at fluorescence wavelengths to provide a clear image at 309 nm.

Examples of the images of the anti-Stokes beam are shown in Fig. **(30a)**, Fig. **(30b)**, Fig. **(30c)**, Fig. **(30d)**, for H_2 gas release rates of 8, 6, 4, 2 ml/min, respectively. The anti-Stokes beam can be imaged down to a H_2 gas release rate of 2 ml/min.

In addition to H_2 gas leak detection, detection of H_2 flames is also important for safety measures of H_2 handling facilities. H_2 flames are difficult to detect with naked eye in daylight, outdoor conditions because of the absence of emission in the visible region. The most prominent emission of H_2 flames is the OH emission in the UV region, whose peak wavelength is 309 nm. Since this coincides with the anti-Stokes Raman scattering wavelength corresponding to a pump wavelength of 355 nm, a system could be constructed to detect both H_2 gas leaks by CARS and H_2 flames by passive optical detection using a single detection wavelength of 309 nm. This is another motivation for using a pump wavelength of 355 nm.

4.5. Measurement of Gas Concentration

A Raman lidar system is developed for measurement of spatial distribution of gas concentration [11]. In this work, a system for measuring Raman scattering light is constructed using a refracting telescope using a Fresnel lens and a pulsed laser. Raman scattering signals from H_2 gas leaked into open air and atmospheric N_2 are simultaneously measured, and the concentration of H_2 gas is calculated based on the ratio of these signals (by eq. (12)).

Table 8: Principal specifications of the Raman lidar system for measuring gas concentration

Component	Specification
Laser (transmitter)	
type	Nd:YAG (THG)
wavelength	355 nm
output energy	6 mJ max
pulse width	4 ns
repetition rate	100 Hz
Telescope (receiver)	
type	Fresnel lens
lens material	ACRYLITE #000
lens diameter	200 mm
focal length	230 mm
total efficiency of the light collection optical system	8% at 387 nm (N_2)
	13% at 416 nm (H_2)
Detector	
type	photomultiplier tube

A schematic diagram of the experimental system for Raman scattering light intensity measurement is shown in Fig. **(31)**, and the principal specifications are listed in Table **8**.

The system uses a Q-switched, pulsed Nd:YAG laser operating at 355 nm. The beam from the Nd:YAG laser is expanded by a beam expander. The diameter of irradiation laser beam is 12 mm and the divergence angle of beam is 1 mrad. The Raman scattering light is collected by a Fresnel lens of aperture 200 mm, collimated, and split into two beams by a beam splitter. Each beam passes through a narrowband interference filter and directed to a photomultiplier tube (PMT). The interference filter of center wavelength 416.3 nm and bandwidth (FWHM) 1.8 nm is used for detecting H_2 Raman scattering light. The interference filter of center wavelength 386.8 nm and bandwidth (FWHM) 2.0 nm is used for detecting N_2 Raman scattering light. In addition, a laser line edge filter (rejection ratio $<10^{-6}$) and an IR cut filter are installed in front of the interference filter to reject stray laser light, Rayleigh scattering, Mie scattering and heat accompanying solar radiation. An optical fiber placed near the laser head collects part of the laser pulse, which provides the trigger signal to the A/D converter.

In order to confirm the ratio of the Raman scattering cross sections of H_2 and N_2, spectral intensities of H_2 and N_2 gas are measured, as shown in Fig. **(32a)**. Taking into account the transmittance characteristics of the interference filters, as shown in Fig. **(32b)**, the above ratio was found to be 3.7:1.

Function tests of the experimental system were conducted in daytime, outdoor conditions. Pure H_2 gas was released from a nozzle located at a distance of 5-30 m from the lidar system, as shown in Fig. **(33)**.

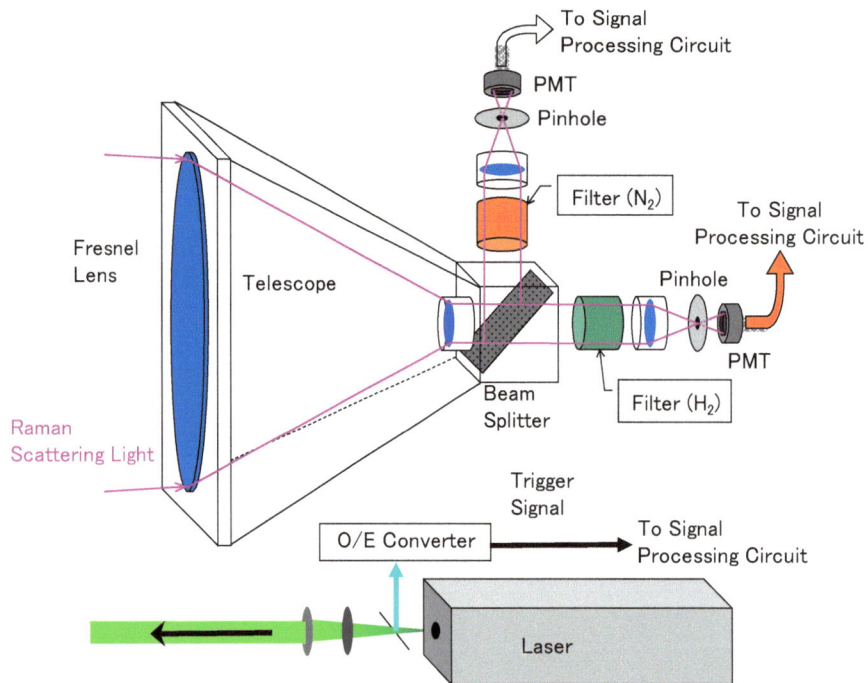

Fig. (31). Schematic diagram of the biaxial Raman lidar system using the refracting telescope.

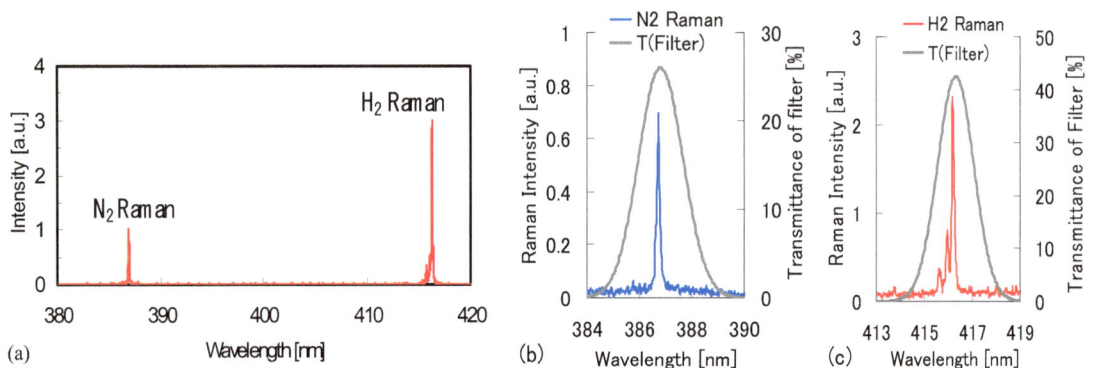

Fig. (32). Raman scattering spectrum and filter transmission: (a) Raman spectra of H_2 gas and N_2 gas at same concentrations, (b) Raman scattering from N_2 gas (filter center wavelength 386.8 nm, FWHM 2.0 nm), (c) Raman scattering from H_2 gas (filter center wavelength 416.3 nm, FWHM 1.8 nm).

Typical Raman scattering signals from N_2 gas and H_2 gas are shown in Fig. **(34a)**, and the spatial distribution of H_2 gas concentration calculated from the ratio of these signals is shown in Fig. **(34b)**.

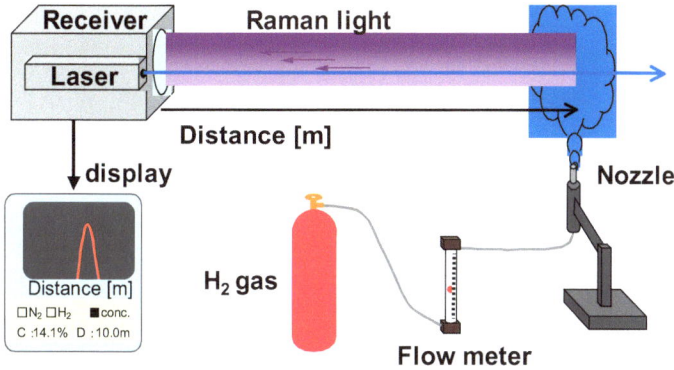

Fig. (33). Experimental setup for measurement of spatial distribution of H_2 gas concentration.

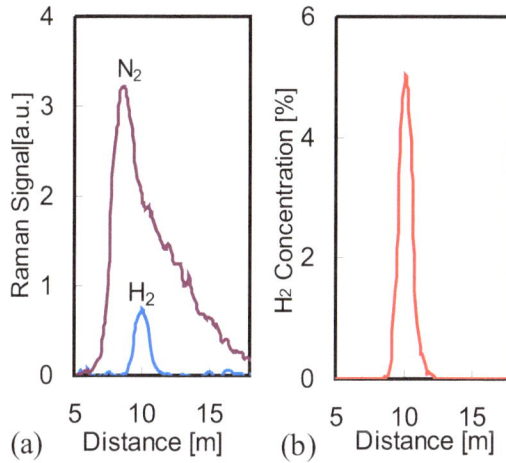

Fig. (34). Example of (a) Raman scattering signal intensity and (b) distribution of H_2 gas concentration, when H_2 gas is released into open air at 50 liter/min.

In this case, H_2 gas was released at a point 10 m from the lidar system. The concentration of H_2 gas at 10 m was about 5 %, and this value coincided with an average value measured by H_2 gas sensor at same point. In this measurement, a minimum detectable concentration of H_2 gas was 0.6 % at 10m distance.

4.5. Imaging of Spatial Distribution of Gas Concentration in Gas Flows

In order to visualize and measure the behavior of the concentration distribution of H_2 gas dispersed into the atmosphere, the visualization of a H_2 gas flow using Raman scattering was performed and the spatial distribution of H_2 concentration in the flow was measured [12]. A schematic diagram of the experimental setup is shown in Fig. **(35a)**, and the principal specifications are listed in Table **9**.

An UV laser light irradiated a H_2 gas flow discharged from a nozzle and Raman scattering was observed using a gated ICCD camera. The image of the H_2 gas flow was observed by shadowgraph in the same condition.

The system uses a Q-switched Nd:YAG laser (wavelength 355nm, pulse width 4 ns) operating with pulse energy of 25 mJ. The laser beam is expanded in a vertical direction by a beam expander which consists of two cylindrical lenses and irradiated a region 5 mm wide and 60 mm high over the nozzle.

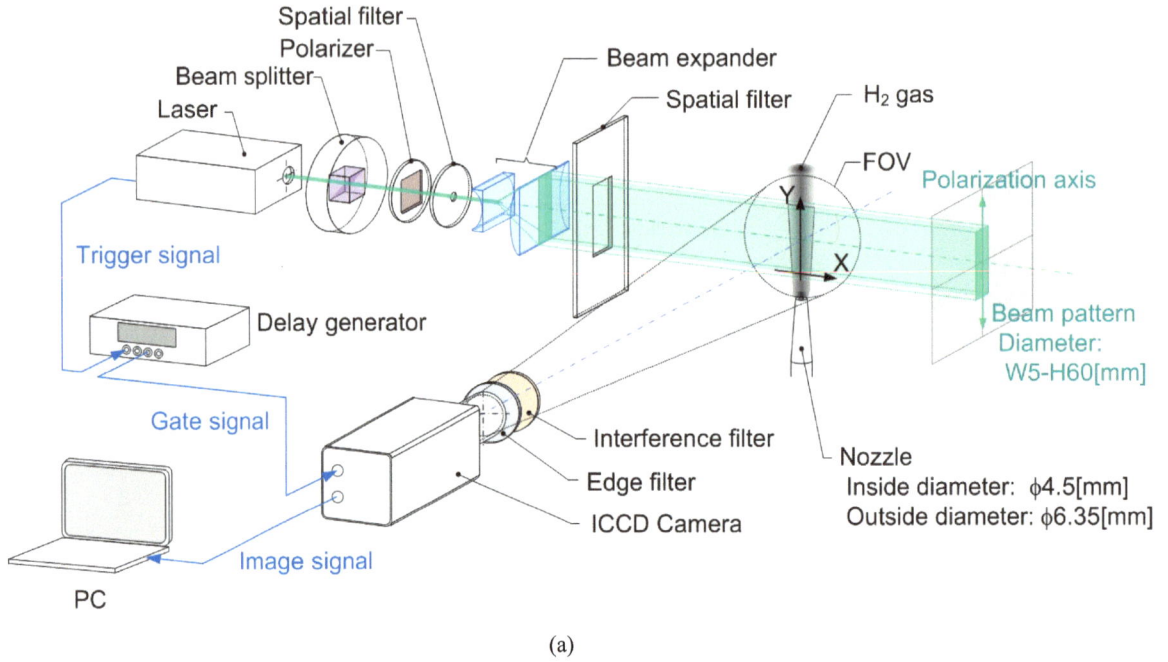

(a)

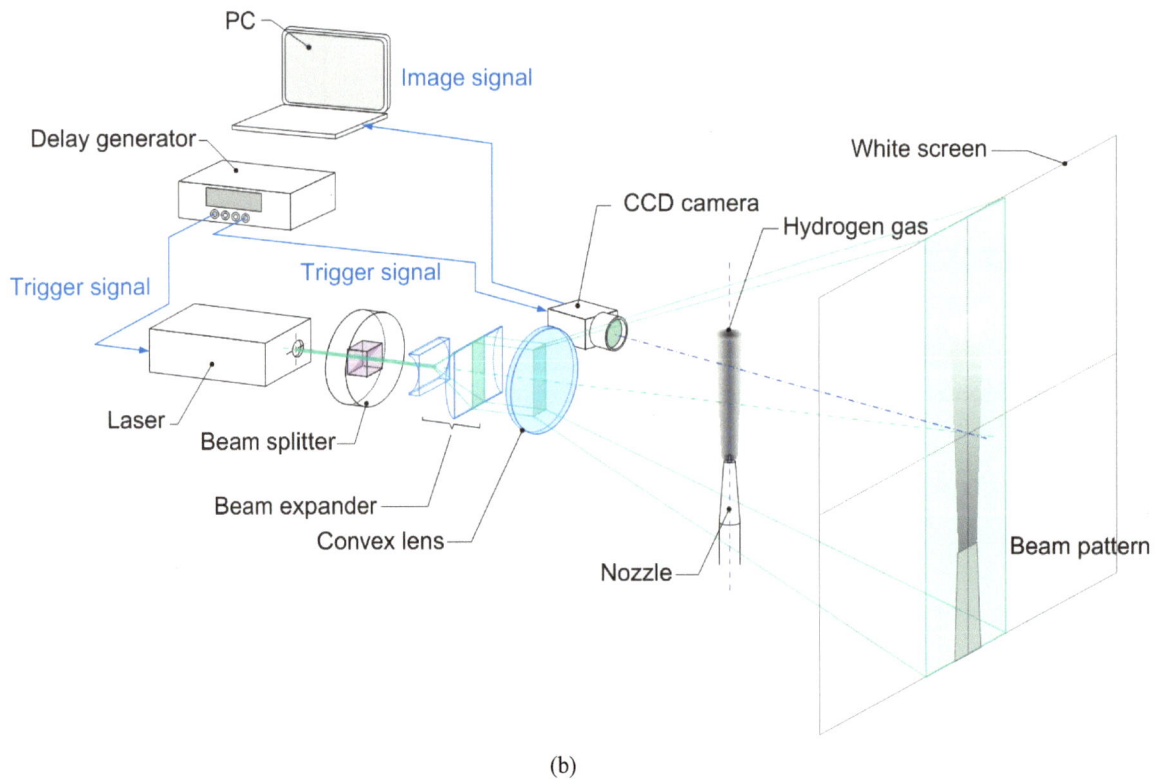

(b)

Fig. (35). (a) Experimental setup for measurement of H_2 concentration distribution in gas flows using Raman scattering, (b) experimental setup for visualization of gas flows by shadowgraph

Raman scattering is observed at a right angle with respect to the laser beam axis. The Raman scattering light is collected by an UV camera lens (105 mm, f/4.5), and it is observed by the gated ICCD camera which is synchronized with the laser irradiation. An interference filter is used to separate the Raman scattering. The interference filter of center wavelength 386.5 nm and bandwidth (FWHM) 1.0 nm is used for detecting N_2 Raman scattering light, and the interference filter of center wavelength 416.5 nm and bandwidth (FWHM) 1.0 nm is used for detecting H_2 Raman scattering light. In addition, a laser line edge filter (laser light rejection ratio $<10^{-6}$) is installed in front of the ICCD camera to reject stray laser light and Rayleigh or Mie scattering light.

Table 9: Principal specifications of the Raman imaging system for measuring gas concentration in gas flows

Component	Specification
Laser (transmitter)	
type	Nd:YAG (THG)
wavelength	355 nm
output energy	70 mJ max
repetition rate	20 Hz
Receiver	
type	UV camera lens
diameter	22 mm
Detector	
type	intensified CCD
image resolution	512×512 pixels

First, the image of Raman scattering from atmospheric N_2 gas was acquired using the experimental setup in Fig. **(35a)** without H_2 gas release and using the filter of center wavelength 386.5 nm, and ambient light suppression was confirmed. Next, the filter was replaced by the filter of center wavelength 416.5 nm, H_2 gas was discharged from the nozzle into the atmosphere, and the image of Raman scattering from H_2 was acquired. Last, the spatial filter was replaced by a convex lens to expand the beam. The expanded beam was projected onto a white screen positioned behind the nozzle, as shown in the experimental setup in Fig. **(35b)**. The shadowgraphs were obtained by imaging the beam pattern on the screen using a CCD camera.

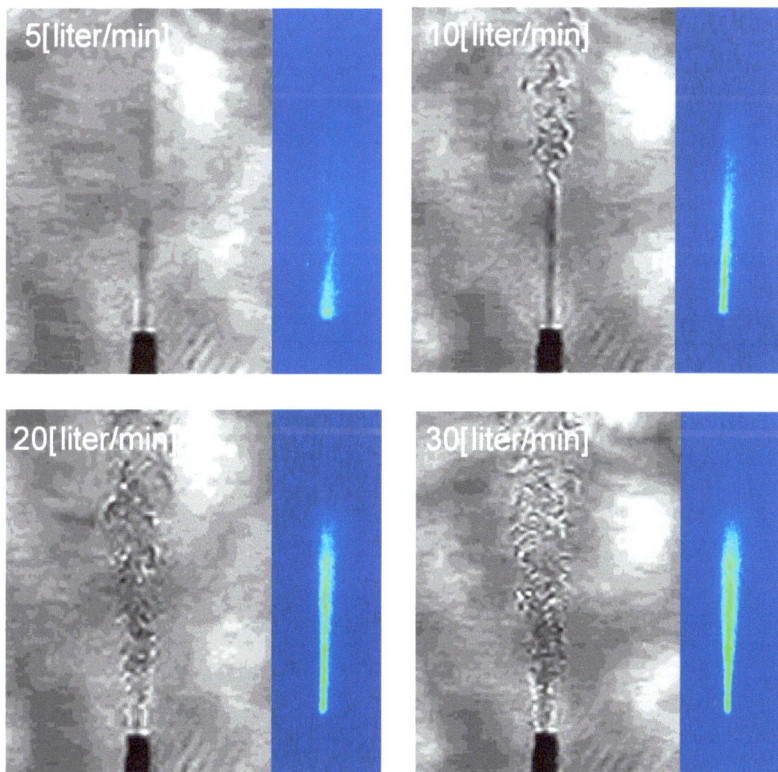

Fig. (36). Images of Raman scattering light (right) and shadowgraphs (left) obtained for different H_2 gas release rates.

Obtained shadowgraphs and Raman images are shown in Fig. (36). From the images of Raman scattering and shadowgraphs, the H_2 gas diffused sideways with increasing release rate. The brightness of each Raman image was subjected to smoothing by a Gaussian filter (3×3 pixels). Then the brightness of N_2 and H_2 Raman images at the same point were compared, and the spatial distribution of the H_2 gas concentration was calculated. The calculated values are shown in Fig. (37), and the images of the concentration distribution are shown in Fig. (38). Here X is the horizontal displacement and Y is the displacement in the direction of gas flow, relative to the center of the nozzle.

The concentration of H_2 at gas release rate 50 liter/min was 99.6% at the outlet of the nozzle, and H_2 gas diffused sideways with increasing distance from the outlet. From the brightness of the images of Raman scattering by N_2 and H_2 at the same point, the spatial distribution of the leakage H_2 gas concentration was measured. Examples of the spatial distribution of the H_2 gas concentration in the gas flow are shown in Fig. (39).

These results indicate that Raman scattering is an effective method to measure the concentration profile or concentration distribution of Raman active molecules in gas flows. The method applied to both laminar and turbulent flows. The obtained information can be used analyze the dispersion of certain gas species in gas flows, and provide experimental verification of simulation results of gas flows.

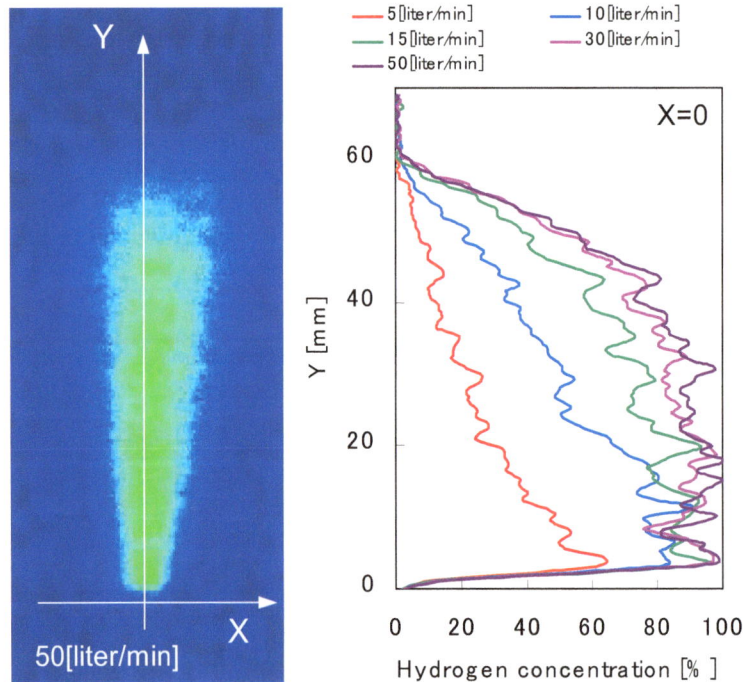

Fig. (37). Raman image and spatial distribution of H_2 concentration along the Y axis (line along the nozzle axis) for different H_2 gas release rates.

5. CONCLUSIONS

In this chapter, the application of Raman scattering to detection, visualization, and concentration measurement of hydrogen gas has been presented. Various lidar systems for hydrogen gas detection were developed.

The first model described in section 4.2, which used a telescope of aperture 212 mm, could detect hydrogen gas up to a distance of 50 m. Furthermore, the region of hydrogen gas could be visualized by scanning the laser within the field of view of the receiver.

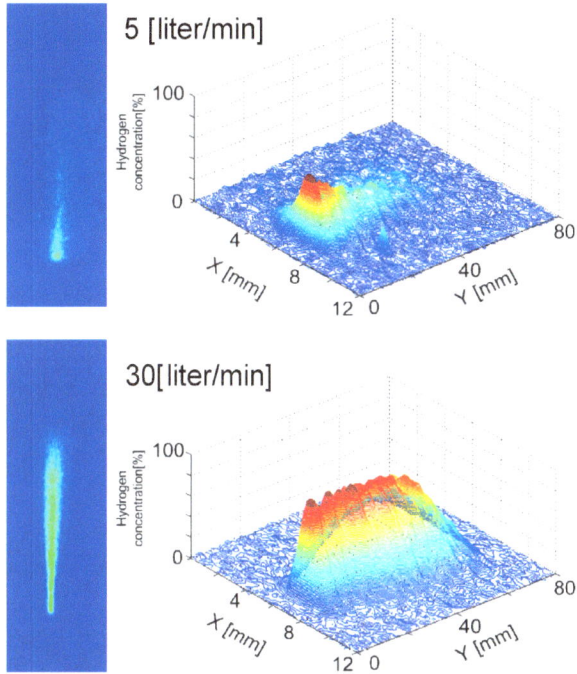

Fig. (38). Concentration distribution of H$_2$ in the gas flow for H$_2$ release rates of 5 liter/min and 30 liter/min.

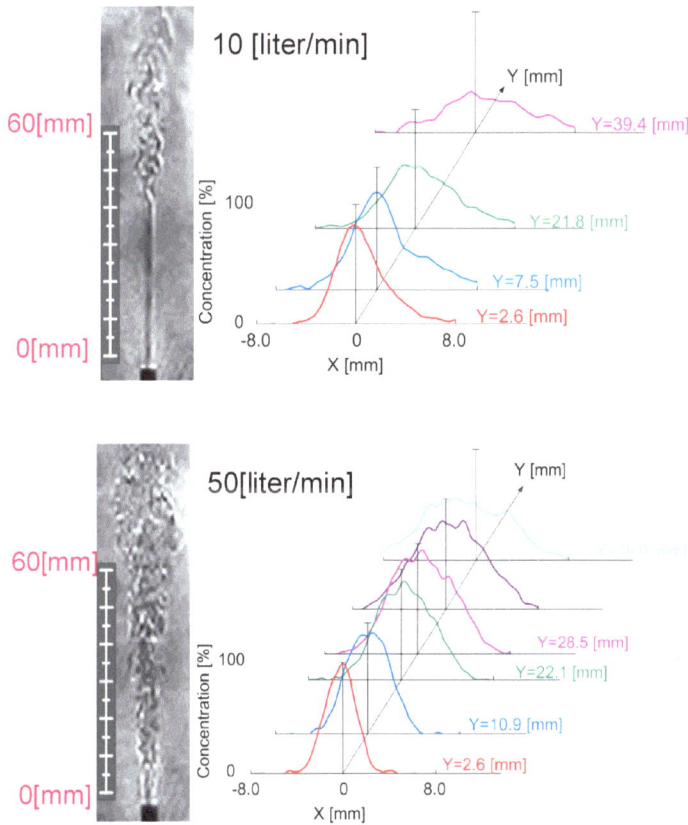

Fig. (39). H$_2$ concentration profiles at different positions of the gas flow, indicating the spread of H$_2$ gas in the direction perpendicular to the flow direction.

The second model described in section 4.3, which used a camera lens for the receiver, could obtain images of Raman scattering up to a distance of about 10 m.

The most recent model, described in section 4.4, used a Fresnel lens of aperture 200 mm to achieve high sensitivity and compactness at the same time. Simultaneous measurement of the Raman scattering signals from hydrogen gas and atmospheric nitrogen enabled hydrogen gas concentration measurement. This was applied to study of hydrogen gas flows.

In addition, hydrogen gas leak detection using Coherent Anti-Stokes Raman Scattering (CARS) was also performed. Using this method, small hydrogen gas leak rates, in the order of several ml/min, could be detected.

With the growing use of hydrogen energy, safety surveillance systems for hydrogen gas leak detection are expected to become more important. Raman scattering offers a method for remote sensing of hydrogen gas, and can be applied to gas leak surveillance in hydrogen handling facilities.

ACKNOWLEDGEMENTS

The author expresses his gratitude to Mrs. Sachiyo Sugimoto, Mr. Ippei Asahi, and Mrs. Saeko Yaeshima for assistance in system development, experiments, and analysis of experimental data.

REFERENCES

[1]   C. V. Raman and K. S. Krishnan," A new type of Secondary Radiation", *Nature*, Vol. 121, p.501, 1928.

[2]   R. Measures, *Laser Remote Sensing*, John Wiley & Sons, New York, 1992.

[3]   W. G. Rado, "The nonlinear third order dielectric susceptibility coefficients of gases and optical third harmonic generation", *Applied Physics Letters*, Vol. 11, pp. 123-125, 1967.

[4]   F. Vestin, M. Afzelius, P. Bengtsson, "Improved temperature precision in rotational coherent anti-Stokes Raman spectroscopy with a modeless dye laser", *Applied Optics*, Vol. 45,pp. 744-747, 2006.

[5]   J. Hussong, W. Stricker, X. Bruet, *et al*, "Hydrogen CARS thermometry in H_2-N_2 mixtures at high pressure and medium temperatures: influence of linewidths models", *Applied Physics B*, Vol. 70, pp. 447-454, 2000.

[6]   R. D. Hancock, K. E. Bertagnolli, R. P. Lucht, "Nitrogen and hydrogen CARS temperature measurements in a hydrogen/air flame using a near-adiabatic flat-flame burner", *Combustion and Flame*, Vol. 109, pp. 323-331, 1997.

[7]   J. W. Nibler, "Coherent anti-Stokes Raman scattering of gases", W. Kiefer and D. A. Long (eds.), *Non-Linear Raman Spectroscopy and Its Chemical Applications*, D. Reidel Publishing Company, pp.261-280, 1982.

[8]   H. Ninomiya, S. Yaeshima, K. Ichikawa, T. Fukuchi, "Raman lidar system for hydrogen gas detection", *Optical Engineering*, Vol. 46, 094301, 2007.

[9]   H. Ninomiya, K. Ichikawa, T. Fukuchi, "Development of a Raman lidar system for hydrogen gas detection", *Proceedings of the 23rd International Laser Radar Conference*, pp. 165-166, 2006.

[10]   T. Fukuchi, H. Ninomiya, "Leak detection of Hydrogen Gas Using Anti-Stokes Raman Scattering", *IEEJ Trans.EIS*, Vol. 128, pp. 1191-1196, 2008 (in Japanese).

[11]   H. Ninomiya, I. Asahi, S. Sugimoto, Y. Shimamoto, "Development of remote sensing technology for hydrogen gas concentration measurement using Raman scattering effect", *IEEJ Trans. EIS*, Vol. 129, 1181-1185, 2009 (in Japanese).

[12]   I. Asahi, H. Ninomiya, "Hydrogen gas concentration distribution measurement by measuring the intensity of Raman scattering light", *IEEJ Trans. EIS*, in press.

Industrial Applications of Laser Remote Sensing, 2012, 89-98

CHAPTER 5

Marine Observation Lidar

Masahiko Sasano[*]

National Maritime Research Institute, Shinkawa 6-38-1, Mitaka, Tokyo181-0004, Japan

Abstract: Remote sensing using pulsed lasers (lidar) has mainly been applied to atmospheric observation, but it can also be applied to marine observation. In this chapter, lidar applications to marine observation, including bathymetry, detection of oil spills on sea surfaces, water quality inspection, monitoring of marine organisms, are presented. The present state and future prospects on these applications are presented.

Keywords: Marine lidar, bathymetry, oceanography, shipborne lidar, helicopter-borne lidar, oil spill, water quality, fluorescence, chlorophyll, marine life.

1. INTRODUCTION

Lidar (Light Detection and Ranging) is an active remote sensing technique which uses a pulsed laser. The distance to the target is measured based on the round-trip time of flight of the light pulse, and the distribution of the target can be measured based on the received light intensity. Therefore, it is commonly used as an observation technique when minute targets which act as light scatterers, such as water droplets and aerosols exist in media with relatively high transmission, such as the atmosphere. Atmospheric lidar observation covers a height range of up to several tens of km, and is used for detailed observation of the atmosphere, as it provides higher range resolution than radar (active remote sensing using radio waves), or sodar (active remote sensing using sonic waves).

The distance to the target in atmospheric lidar observation is given by

$$R = \frac{c(t - t_0)}{2} \tag{1}$$

where R is the distance from the lidar device to the target, t_0 is the time of transmission of the laser pulse, t is the time of reception of the scattered laser pulse. Here c is the speed of light in the atmosphere, and is frequently substituted by the speed of light in vacuum, because the difference is only about 0.03%.

The equation describing the distribution of the target in atmospheric lidar observation is the lidar equation [1],

$$P(R) = \frac{CS(R)A\beta(R)}{R^2} \exp\left(-2\int_0^R \alpha(r)dr\right) \tag{2}$$

Here C is an instrumental constant, $S(R)$ is the overlap between the laser and the receiver field of view, A is the receiver area, α is the attenuation coefficient, and β is the backscatter coefficient. The distribution of the target (such as aerosols) at distance R is proportional to $\beta(R)$. Several methods have been developed for the solution of the lidar equation, depending on the observation method or target [2-5].

In addition, a variety of wavelengths have been used for atmospheric lidar, because the optical transmission of the atmosphere is relatively high from the near ultraviolet to the far infrared, as shown in Fig. **(1)** [6].

*Address correspondence to Masahiko Sasano:** Sensing Technology Research Group, Project Team for Common & Fundamental Technology, National Maritime Research Institute, Shinkawa 6-38-1, Mitaka, Tokyo181-0004, Japan; Tel: +81-422-41-3123; E-mail: sasano@nmri.go.jp

Tetsuo Fukuchi and Tatsuo Shiina (Eds)

The wavelengths are selected depending on the target [7-9].

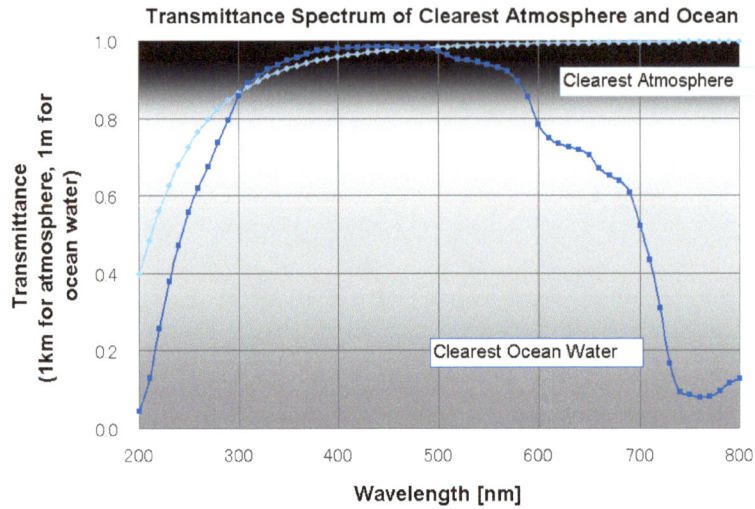

Transmittance Spectrum of Clearest Atmosphere and Ocean

Fig. (1). Transmittance Spectrum of Clearest Atmosphere [/km] and Ocean Water [/m] (Data referred by [6] for atmosphere and [13] for ocean water).

In this manner, atmospheric lidar has been used for observation of clouds, dust, and water vapor in the atmosphere [10-12].

On the other hand, lidar observation of the marine environment (marine lidar) is not as common as atmospheric lidar. One reason is the low optical transmission of seawater, which results in a shortening of the lidar observation range. When the amount of matter is considered, the amount in air from sea level to 10 km altitude is the same as in sewater from the surface to depth 10 m. Also, the water molecule, which is the principal component of seawater, has large optical absorption. Therefore, the transmission through 1 m of seawater is less than 1/1000 of that through 1 m of atmosphere.

The optical transmission of seawater has a wavelength distribution as shown in Fig. (1). Only visible light centered in the blue is transmitted, and ultraviolet light, and red to infrared light are not transmitted [13].

The merit of atmospheric observation using lidar is the ability to continuously observe the atmosphere from the ground. Direct observation is difficult, as it requires *in situ* sampling using aircraft or sonde. On the other hand, the observation distance of marine lidar only extends to few tens of m, and is limited to platforms such as aircraft and vessels, so long-term, continuous observation is difficult. Therefore, there are no significant merits over direct observation such as water samplers and CTD sensors. Also, the transmission of sonic waves through seawater is high [14], so sonar (active remote sensing using sonic waves in water) has been put into practical use. Therefore, comparatively speaking, there are few merits for marine lidar observation.

In spite of the technical difficulties accompanying marine lidar observation, there has recently been an increase in the use of marine lidar. This is because observation over a wide region using aircraft or vessels have drawn attention because of the growing importance of marine observation in view of the climate changes on a global scale [15]. Also, pulsed lasers resistant to vibration or shocks have become available [16, 17], and this is another reason for advances in the development of marine lidar.

The ranging of targets in marine lidar observation is described by the equation

$$R = \frac{c(t-t_0)}{2n} \qquad (3)$$

Here the lidar instrument is assumed to be above the sea surface. R is the distance from the sea surface to the target, t_0 is the time of reception of the laser pulse scattered from the sea surface, t is the time of reception of the laser pulse scattered from the target, c is the speed of light in vacuum (3×10^8 m/s), and n is the index of refraction of seawater (about 1.34). Note that the propagation speed of light in seawater is about 25% slower than that in air.

In marine lidar, the distribution of the target inside the sea is described by the equation (hereafter referred to as the marine lidar equation) [18]

$$P(R) = \frac{CS(R)A\beta(R)}{(R+nH)^2} T^2 \exp\left(-2\int_0^R \alpha(r)dr\right) \qquad (4)$$

Here H is the distance from the sea surface to the lidar instrument, T is the sea surface transmission of the laser light and scattered light from the water across the sea surface. Since waves always exist on the sea surface, the angle of incidence of laser light with respect to the sea surface is constantly changing, so the sea surface transmission also changes.

When compared to atmospheric lidar, the effect of multiple scattering in marine lidar is very large. Therefore, solving the marine lidar equation to obtain the distribution of targets is difficult compared to atmopsheric lidar. In actual use of marine lidar, as shown in the following examples, the distribution of targets in the sea are not measured, and alternative methods of use are common.

2. BATHYMETRY

At present, marine lidar is commonly used for bathymetry (water depth measurement) in shallow waters. The second harmonic of the Nd:YAG laser (wavelength 532 nm) is most commonly used as the laser source, and the water depth is obtained from the time difference between the observation times of the scattering from the sea surface and from the sea bottom using eq.(3). Since this observation only obtains the position, it suffices to find the peaks in the water surface signal and the sea bottom signal, and it is not necessary to solve the marine lidar equation (4). This method is effective for depth measurement of shallow waters in which navigation of measurement vessels is difficult. In order to obtain many observation points per scan, the lidar observation direction is scanned by the use of scanning mirrors.

Commercial aircraft-mounted bathymetry devices are available from several manufacturers, such as Optech (Canada) [19], Airborne Hydrography (Sweden) [20], and Fugro LADS (Australia) [21]. Water depth data have been obtained using this method. In Japan, the Agency of Maritime Safety (Japan Coast Guard) possesses a SHOALS-1000 made by Optech [22]. The schematic of the system and an example of observation using this system are shown in Fig. **(2)** and Fig. **(3)**.

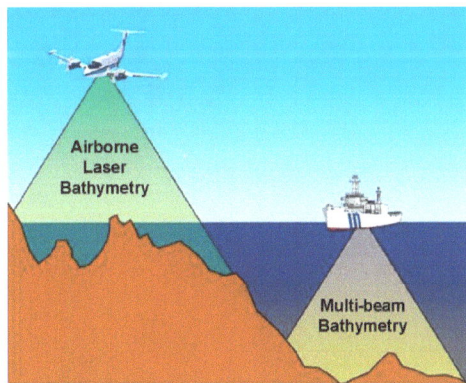

Fig. (2). Schematic of the Air-borne Laser Bathymetry for Shallow sea area and Ship-borne Multi-beam Bathymetry for Deep sea area [22].

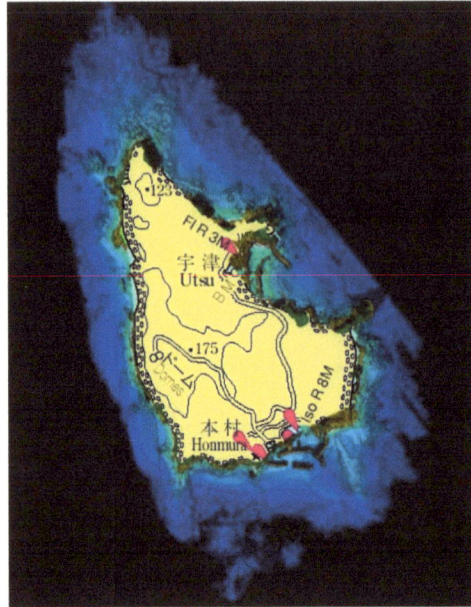

Fig. (3). Survey Results at Mishima, Yamaguchi-Pref., Japan by Japan Coast Guard Airborne Laser Bathymetry [22].

The authors also conducted water depth measurement experiments in an experimental basin using the third harmonic of the Nd:YAG laser (wavelength 355 nm) [23]. The basin was located indoors, filled with pure water, and the surface was serene. Experimental results showed that the water depth measurement error (compared to the actual depth of the basin) was about 1 cm in the range 5.5 m to 30 m. An overview of the experiment is shown in Fig. (4): the layout of the basin (upper left), placement of the bathymetry apparatus (upper right), objects (coral) placed at the bottom of the basin (lower left), and the top view of the bathymetry apparatus (lower right). The relationship between the lidar observation time and the water depth is shown in Fig. (5).

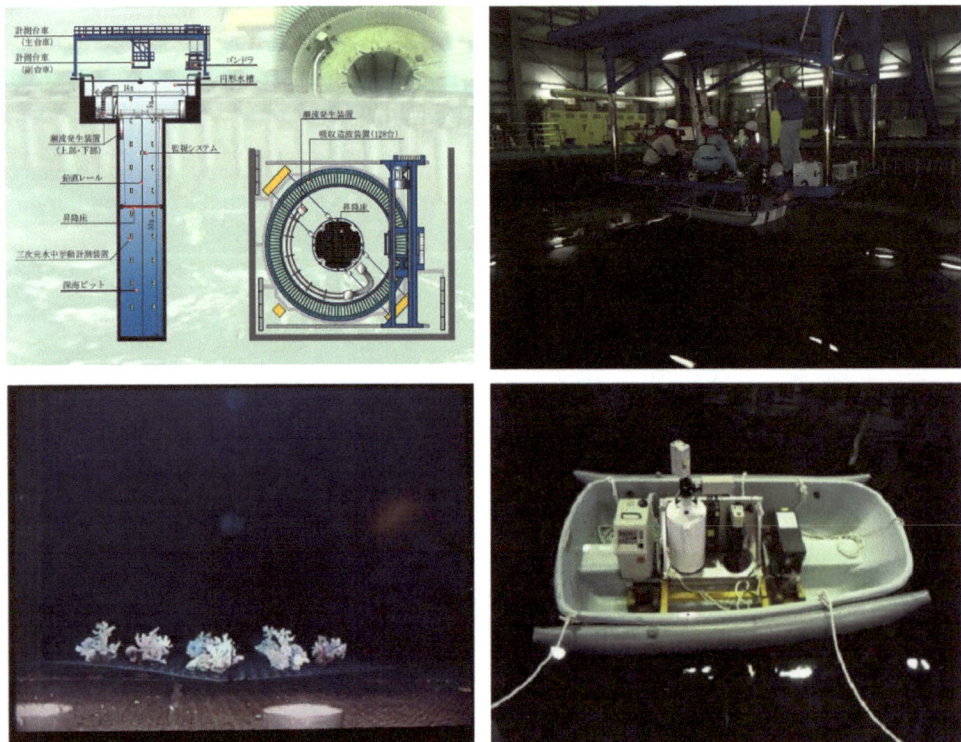

Fig. (4). Overview of the lidar sounding experiment at Deep Sea Basin, Japan (movable floor with 35 m of max. depth).

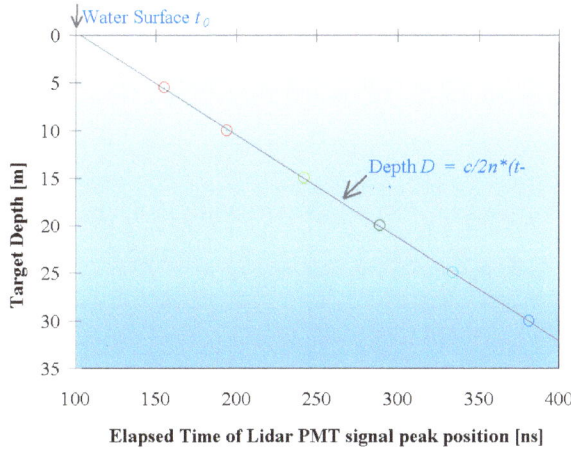

Fig. (5). Linear Relationship between Target Depth and Lidar Peak Signal.

Lidar bathymetry uses eq. (3), so there is a dependence on the index of refraction n. The index of refraction of seawater n varies according to the observation wavelength, water temperature of the observation region, and salt concentration [24, 25]. However, the effect of these variations on bathymetry is below 0.5%, and it is possible to improve the accuracy of the depth measurement by using a more accurate value for the index of refraction (rather than using a representative value for seawater).

3. DETECTION OF OIL SPILLS AND WATER QUALITY MEASUREMENT

Marine lidar can be mounted on aircraft and observe the water surface. Especially, when fuel oil or crude oil spills onto the sea surface due to an accident, it is necessary to monitor the location and area of the spill. This monitoring must be done over a relatively large area during day and night. For this purpose, as monitoring methods from aircraft, hyperspectral cameras, infrared cameras, radar have been used, in addition to visual observation [26]. The methods which are applicable to nighttime are infrared cameras and radar, but the detection sensitivity of oil spills is not very high. Lidar is expected to provide accurate monitoring during nighttime, and is under development by several firms, such as Optimare (Germany) [27] and LDI3 (Estonia) [28]. A helicopter-mounted fluorescence lidar is also under development at the National Maritime Research Institute [29].

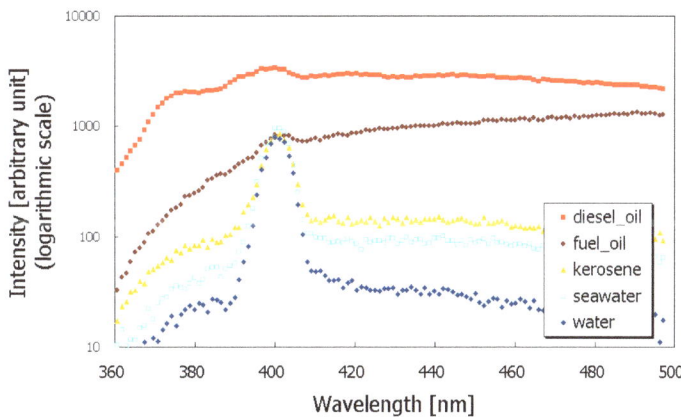

Fig. (6). Emission spectrum of water, seawater, and oil film upon irradiation by a ultraviolet laser at wavelength 355 nm.

Sea surface oil spill detection lidar is an instrument which optically judges whether the matter on the sea surface is water or oil. By irradiating the sea surface with ultraviolet light, Raman scattering from water surface is observed when the surface is water, and strong blue fluorescence is observed when the surface is oil. Therefore, by spectral analysis of the emission upon excitation by a near ultraviolet laser, oil spills can be detected. In this case, there is no need to solve eq.(3) or eq.(4). The emission spectra of water and different types of oil when irradiated by near ultraviolet light of wavelength 355 nm is shown in Fig. **(6)**.

This oil spill detection method is an active remote sensing method, so it can observe during day or nighttime. The distinction between water and oil is clear, so the probability of misjudgement is low. Ultraviolet pulsed lasers of wavelength 308 nm, 337 nm, and 355 nm have been used. Fluorescence from light oil constituents (kerosene, gasoline) increases for shorter excitation laser wavelength, and the detection sensitivity tends to be higher [30].

In addition, in order to cover a large area with a single sweep, the lidar observation direction is scanned by the use of scanning mirrors.

In the observation result of the oil spill detection using lidar, the ratio of the water Raman signal and the fluorescence signal is superimposed on a map, and is used for monitoring of the position and extent of the oil spill. As an example of the lidar observation of oil spills on the sea surface, the observation result using a FLS LiDAR system (manufactured by LDI3) of an oil spill which occurred in 2004 in French territorial waters is shown in Fig. **(7)** [31].

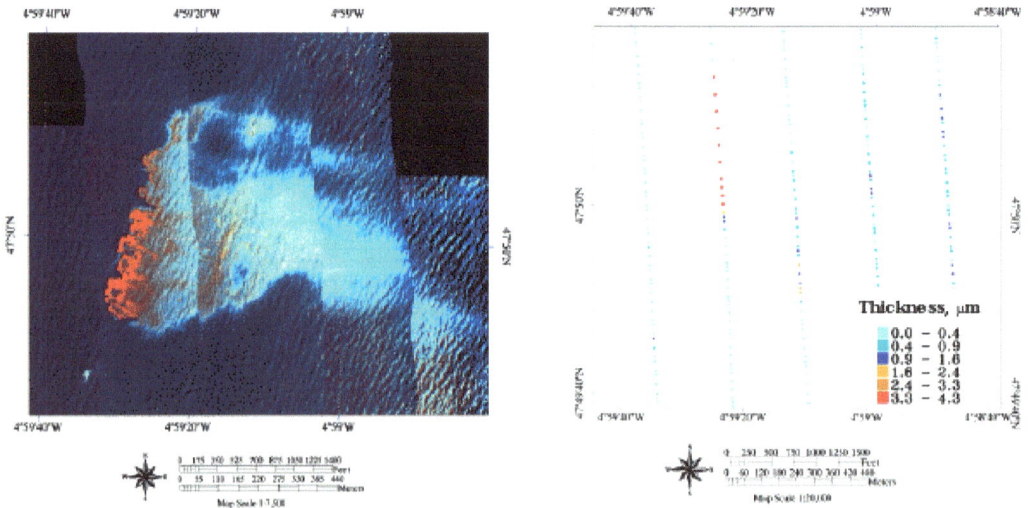

Fig. (7). Example of the Oil Slick detection by (left) hyperspectral camera and (right) fluorescence lidar system [31].

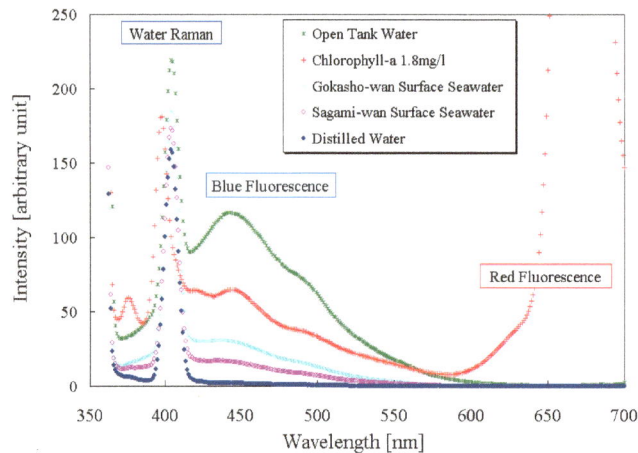

Fig. (8). Emission Spectra of water, seawater, chlorophyll solution, and green tea, observed upon ultraviolet excitation at 355 nm.

Sea surface observation by aircraft-borne or ship-borne lidar also enables water quality investigations. By the same method as the oil spill detection described above, blue fluorescence is observed when ultraviolet light is irradiated onto seawater containing phytoplakton or DOM (dissolved organic matter), although the fluorescence is not as intense as oil, as shown in Fig. **(8)**. Therefore, by measuring the fluorescence intensity by lidar, it is possible to observe the relative concentration of substances such as DOM on the sea surface. This research is conducted by agencies such as NASA [32] and ENEA (Italy) [33].

The authors have also performed water quality observations in Tokyo Bay and Sagami Bay (Japan) using the ship-borne and helicopter-borne fluorescence lidar [34]. The method is the same as in oil spill observation, in which the ratio of the water Raman lidar signal and the fluorescence lidar signal is superimposed on a map. Observation results showed that, the fluorescence from seawater (relative concentration of DOM and other substances) could be observed during day or night time, and the tidal position at the mouth of Tokyo Bay could was verified. Fig. **(9)** shows an example of observation.

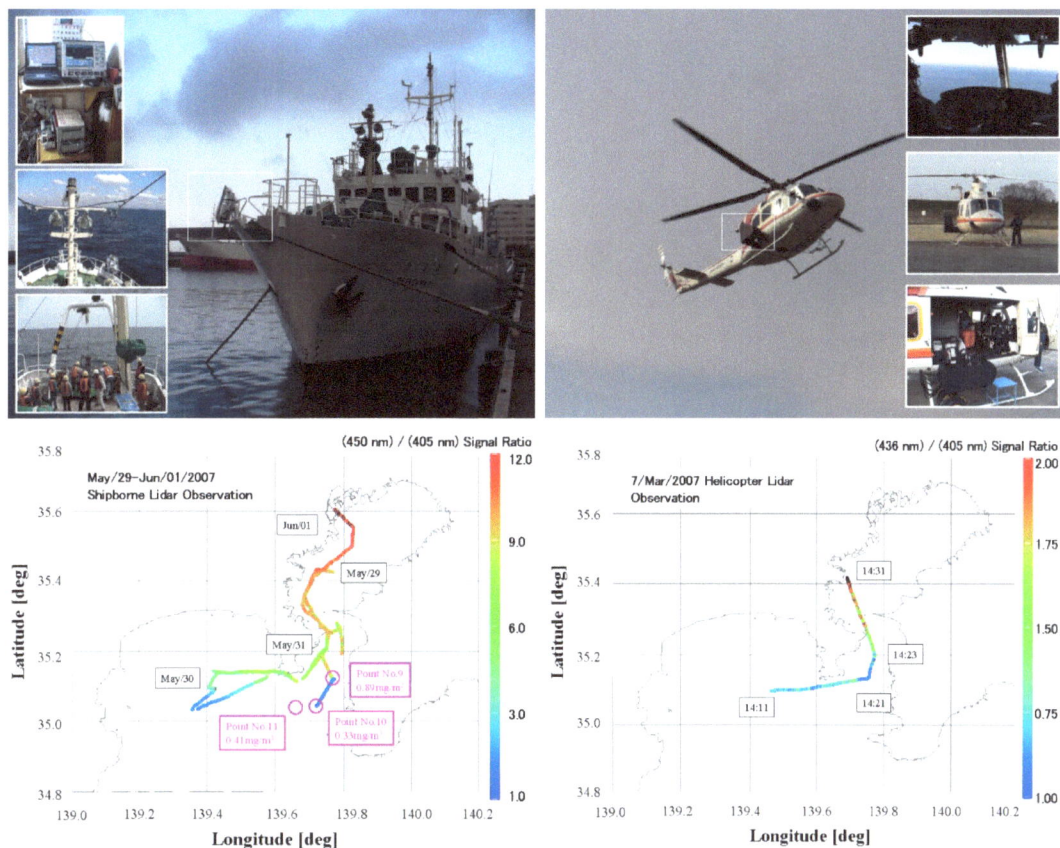

Fig. (9). Examples of water fluorescence detection by ship-borne fluorescence lidar and helicopter-borne fluorescence lidar in Tokyo Bay and Sagami Bay.

4. MARINE LIFE MEASUREMENT

Aircraft-borne or ship-borne marine lidar can also be used for observation of marine life. In this case, the method commonly used is imaging lidar. Since lidar is an optical observation, it is possible to image the marine organism in question by using a two-dimensional optical sensor for the receiving sensor. Gated ICCD cameras are commonly used for the two-dimensional optical sensor. Gating allows very short exposure times (in the order of 100 ns) in synchronization with the laser. In this manner, the image at a target depth beneath the surface can be obtained. The effects due to strong laser scattering and reflection of sunlight at the surface can be largely suppressed. The short exposure time also has the advantage of avoiding image blur due to motion by the organisms in question or movement of the platform.

NOAA (USA) is developing an imaging lidar for imaging of fish schools, using a pulsed laser of wavelength 532 nm [35]. An example of a group of salmon observed by this lidar is shown in Fig. (10).

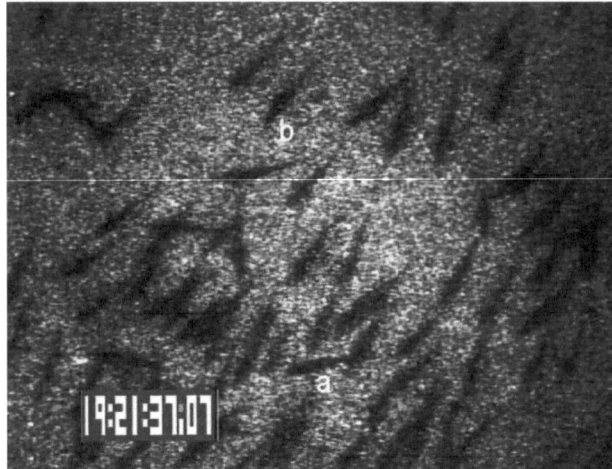

Fig. (10). Example of the lidar image of a group of salmon [35].

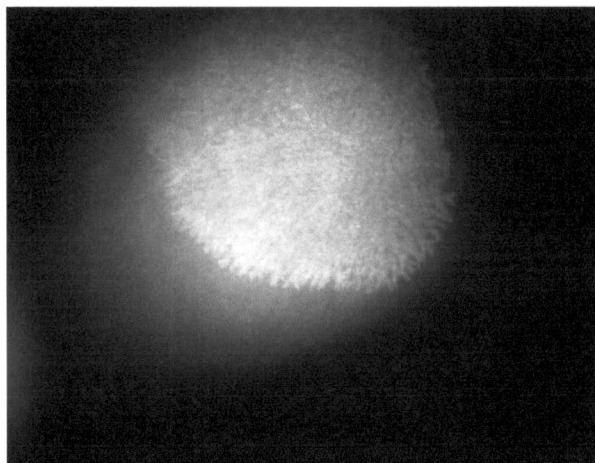

Fig. (11). Example of fluorescence lidar image of table coral.

The authors are currently developing a fluorescence imaging lidar for imaging of living coral [23]. This lidar takes advantage of the fact that a large portion of living corals possess fluorescent proteins, and emit fluorescence from the blue to green when excited by near ultraviolet light. An image of table-like coral obtained using this system is shown in Fig. (11).

5. CONCLUSION

Aircraft-borne or ship-borne marine lidar is a remote sensing instrument which can cover a large marine area, and its use is expanding. Applications such as bathymetry (water depth measurement), oil spill detection, water quality investigation, marine life observation have been presented.

The observation range of the marine lidar is from the sea surface to a maximum depth of several tens of m. At times, only the surface is observed. This is very small compared to the average ocean depth of 3,700 m, and is far from marine observation for understanding the ocean as a whole. This is the major difference when compared to atmospheric lidar. In spite of this, marine lidar is a tool which can provide new observation data for oceanography, in which observation has been difficult. The development of marine lidar is expected to advance in many disciplines related to the marine environment.

Although this chapter only provided examples for oceanic observation, marine lidar can also be applied to other bodies of water, such as lakes, rivers, reservoirs, and marshes. The main advantage of this technique is the ability for remote sensing, which greatly reduces manpower and cost compared to *in situ* sampling methods.

REFERENCES

[1]   R. Measures, *Laser Remote Sensing: Fundamentals and Applications*, Wiley-Interscience, New York (1984).

[2]   G. Kunz, G. de Leeuw, "Inversion of lidar signals with the slope method", *Applied Optics*, Vol. 32, pp. 3249-3256 (1993).

[3]   M. Sandford, "Laser scatter measurements in the mesosphere and above", *Journal of Atmospheric and Terrestrial Physics*, Vol. 29, pp. 1657-1662 (1967).

[4]   J. Klett, "Stable analytical inversion solution for processing lidar returns", *Applied Optics*, Vol. 20, pp. 211-220 (1981).

[5]   F. Fernald, "Analysis of atmospheric lidar observations: some comments", *Applied Optics* Vol. 23, pp. 652-653 (1984).

[6]   C. Tomasi, V. Vitale, B. Petkov, A. Lupi, A. Cacciari, "Improved algorithm for calculations of Rayleigh-scattering optical depth in standard atmosphere", *Applied Optics*, Vol. 44, pp. 3320-3341 (2005).

[7]   G. Ancellet, F. Ravetta, "Compact airborne lidar for tropospheric ozone: description and field measurements", *Applied Optics*, Vol. 37, pp. 5509-5521 (1998).

[8]   V. Matthais *et al.,* "Aerosol lidar intercomparison in the framework of the EARLINET project. 1. Instruments", *Applied Optics*, Vol. 43, pp. 961-976 (2004).

[9]   L. Fiorani, F. Colao, A. Palucci, "Measurement of Mount Etna plume by CO_2-laser-based lidar", *Optics Letters*, Vol. 34, pp. 800-802 (2009).

[10]   Z. Wang, P. Wechsler, W. Kuestner, J. French, A. Rodi, B. Glover, M. Burkhart, D. Lukens, Wyoming Cloud Lidar: instrument description and applications, *Optics Express*, Vol. 17, pp. 13576-13587 (2009).

[11]   C. Xie, J. Zhou, N. Sugimoto, Z. Wang, "Aerosol Observation with Raman LIDAR in Beijing, China", *Journal of the Optical Society of Korea*, Vol. 14, pp. 215-220 (2010).

[12]   J. Machol *et al.,* "Preliminary measurements with an automated compact differential absorption lidar for the profiling of water vapor", *Applied Optics*, Vol. 43, pp. 3110-3121 (2004).

[13]   R. Smith, K. Baker, "Optical properties of the clearest natural waters (200-800 nm)", *Applied Optics*, Vol. 20, pp. 177-184 (1981).

[14]   W. Thorp, "Analytic Description of the Low-Frequency Attenuation Coefficient", *Journal of the Acoustical Society of America*, Vol. 42, pp. 270-271 (1967).

[15]   IPCC AR4 WG1.

[16]   Quantel Corporation, http://www.quantel-laser.com/industrial-scientific-lasers/media/produit/fichier/47_Ultra CFRVA0209.pdf, accessed January 31, 2011.

[17]   Continuum, http://www.continuumlasers.com/ products/pdfs/Inlite_III.pdf, accessed January 31, 2011.

[18]   A. Bunkin, K. Voliak, *Laser Remote Sensing of the Ocean Methods and Applications*, Wiley-Interscience, New York (2001).

[19]   Optech, http://www.optech.ca/prodshoals.htm, accessed January 31, 2011.

[20]   Airborne Hydrography, http://www.airborne hydrography.com/hawkeyeii, accessed January 31, 2011.

[21]   Fugro LADS, http://www.fugrolads.com/ ladsmkII.htm, accessed January 31, 2011.

[22]   Japan Coast Guard, *Annual Report 2009*, http://www.kaiho.mlit.go.jp/info/books/report2009/Degital/honpen/ p086.html, accessed January 31, 2011 (in Japanese)

[23]   M. Sasano, A. Matsumoto, N. Kiriya, H. Yamanouchi, K. Hitomi, K. Tamura, "A New Method for Coral Monitoring using Boat-based Fluorescence Imaging Lidar", *Proceedings of the Techno-Ocean 2010*, Kobe, Japan, SS6-1 (2010).

[24]   M. Daimon, A. Masumura, "Measurement of the refractive index of distilled water from the near-infrared region to the ultraviolet region", *Applied Optics*, Vol. 46, pp. 3811-3820 (2007).

[25]   X. Quan, E. Fry, "Empirical equation for the index of refraction of seawater", *Applied Optics*, Vol. 34, pp. 3477-3480 (1995).

[26]   M. Jha, J. Levy, Y. Gao, "Advances in Remote Sensing for Oil Spill Disaster Management: State-of-the-Art Sensors Technology for Oil Spill Surveillance", *Sensors*, Vol. 8, pp. 236-255 (2008).

[27] Optimare, http://www.optimare.de/cms/en/ divisions/fek/fek-products/imaging-airborne-laser-fluorosensor.html, accessed January 31, 2011.

[28] LDI3 http://www.ldi3.com/index.php?main=3, accessed January 31, 2011.

[29] M. Sasano, K. Hitomi, H. Yamanouchi, "The Fluorescence Lidar for Monitoring of Oil Spill", *Journal of the Visualization Society of Japan*, Vol. 28, pp. 9-14 (2008). (in Japanese)

[30] University of Oldenburg: Catalogue of Optical Spectra of Oils. http://las.physik.uni-oldenburg.de/data/spectra/spectra.htm, accessed January 31, 2011.

[31] M. Lennon, S. Babichenko, N. Thomas, V. Mariette, G. Mercier, A. Lisin, "Detection and Mapping of Oil Slicks in the Sea by Combined use of Hyperspectral Imagery and Laser Induced Fluorescence", *EARSeL eProceedings*, Vol. 5, pp. 120-128 (2006).

[32] F. Hoge, A. Vodacek, R. Swift, J. Yungel, N. Blough, "Inherent optical properties of the ocean: retrieval of the absorption coefficient of chromophoric dissolved organic matter from airborne laser spectral fluorescence measurements", *Applied Optics*, Vol. 34, pp. 7032-7038 (1995).

[33] R. Barbini, F. Colao, R. Fantoni, A. Palucci, S. Ribezzo, "Shipborne laser remote sensing of the Venice lagoon", *International Journal of Remote Sensing*, Vol. 20, pp. 2405-2421 (1999).

[34] M. Sasano, K. Hitomi, T. Morinaga, H. Yamanouchi, "Development of Shipborne and Helicopter-based Oceanographic Fluorescence Lidar", *Proceedings of the 24th Internaional Laser Radar Conference*, Boulder, USA, pp. 983-986 (2008).

[35] J. Churnside, J. Wilson, "Airborne lidar imaging of salmon", *Applied Optics*, Vol. 43, pp. 1416-1424 (2004).

CHAPTER 6

Plant and Vegetation Monitoring Using Laser-Induced Fluorescence Spectroscopy

Kazunori Saito[*]

Faculty of Engineering, Shinshu University, 4-17-1 Wakasato, Nagano City, Nagano 380-8553, Japan

Abstract: This chapter describes the application of laser-induced fluorescence (LIF) spectroscopy to monitoring of living plants. Several applications in agriculture, horticulture, and forestry are described along with specific LIF monitoring systems. The following topics are covered. 1) The basics of LIF spectroscopy for plant monitoring. If an ultraviolet laser is used as an excitation source, the LIF of the plant receiving the light consists of blue-green fluorescence and red-far red fluorescence. Because LIF is a physiologically based optical phenomenon exchanging absorbed optical energy among living molecules, it can include information about living status. 2) A lettuce leaf and a sasanqua (*Camellia sasanqua*) leaf were monitored using long–term LIF. The growth status of the lettuce was shown by variation in the LIF intensity at 460 nm and 530 nm. Symptoms of water stress in the sasanqua leaf also appeared at the same wavelengths. Such variation can be used for quality control of their products and understanding the process of the development of stress in plants. 3) A laser-induced fluorescence spectrum (LIFS) light detection and ranging (lidar) was developed for monitoring a large tree. The entire LIF spectrum of a zelkova (*Zelkova serrata Makino*) tree growing outside was monitored at different growth stages. The formation of chlorophyll is discussed. 4) A LIF imaging system and LIFS imaging lidar were developed for laboratory use and for remote monitoring, respectively. LIF images of a spinach leaf indicated plant activity and productivity. The LIFS imaging lidar successfully created a chlorophyll distribution map of a whole poplar tree. The effectiveness of LIF imaging is described. 5) Based on these results, "a Green-Cross; general hospital for plants" and "Optical Farming" are proposed for future development.

Keywords: Laser-induced fluorescence, chlorophyll, photosynthesis, fluorescence spectroscopy, fluorescence lidar, plants, leaf, imaging lidar, agriculture.

1. INTRODUCTION

All living things need energy to live and almost of all this energy comes from the sun. The energy in sunlight can be converted to more useful sources of energy such as food and fuel through photosynthesis. With few exceptions, photosynthesis is carried out only by plants. Plants are also an important part of the global ecosystem due to their generation of oxygen, assimilation of carbon dioxide, sugar production, and production and extinction of atmospheric components by emission of volatile organic compounds such as isoprene and terpene to name some roles. Green landscape and scenery, colored leaves, and many different flowers and foliage plants enrich our lives.

On the other hand, these functions that only plants have are influenced from contemporary environmental destruction. For example, ozone depletion in the stratosphere increases harmful ultraviolet (UV-B) radiation reaching the earth's surface, the greenhouse effect is exaggerated by increased level of carbon dioxide and methane, acid rain and pollution change the quality of soil and water, and long-range transportation of particulate aerosols caused by desertification affects the distribution of vegetation. These and other environmental problems are great concern, and preserving our ecosystem depends in part on monitoring vegetation status.

Furthermore, the rapid increase in the global population, especially in the developing and underdeveloped countries, is likely to cause a global food shortage in the future. In order to increase the food productivity,

Address correspondence to Kazunori Saito: Shinshu University, 4-17-1 Wakasato, Nagano City, Nagano 380-8553, Japan ; Tel:+81-26-269-5457; Email:saitoh@cs.shinshu-u.ac.jp

Tetsuo Fukuchi and Tatsuo Shiina (Eds)

new innovations in agriculture will be required. A possibility for efficient, mass production of vegetables is an industrial manufacturing facility, using artificial lighting (such as light-emitting diodes) and growing vegetables in a controlled environment. In this case, measurement and control instrumentation will be necessary for optimal production, so methods to monitor plant activity will become important. In this view, although vegetation monitoring may not be directly related to industrial applications at the present, there is a high probability that it may become part of it in the future.

In this chapter, laser-induced fluorescence (LIF) spectroscopy and its application to plant monitoring in the field is described. LIF is a novel technique that is able to quickly detect negative effects on plant growth and diagnose the plant's physiological conditions.

2. BASIC CONCEPT OF LIF SPECTROSCOPY USED FOR PLANT MONITORING

2.1. Idea of Using Plant LIF for Plant Monitoring

When we attempt to obtain information on plants, we should consider that plants are living systems. They contain a lot of different organisms and molecules that work cooperatively and exclusively, and that support their life.

Both chemical and physical methods are available for the plant monitoring. The former can offer sophisticated data on every molecule, but it is not useful for describing a living system because it requires sampling plant material and treatments such as crushing, separation, and processing in a chemical liquid. In contrast, optical methods are candidates for this purpose, especially when we consider the features of plants. The most important feature of plants that discriminates them from other living organisms is photosynthesis. This phenomenon starts as absorption of a photon of light by plant pigments and then a certain part of the absorbed light energy can be reemitted. If the process of absorption and emission can be monitored using an optical method, a non destructive method that does not require sampling, living plant information will be obtained.

Three different processes of light-depended reactions in plants are available for such an optical method; reflection, transmission and emission. Reflection is widely used for plant monitoring, from monitoring at the leaf level using a handheld spectrometer to monitoring a global area with a satellite-borne sensor. Reflection is based on the colors of plants. Because the appearance of a change in colors of the plant surface is due to a final stage of physiological change, faster and more direct information that reflects ongoing plant activity is desired. Transmission is not suitable for application in the field, because a monitoring device or system requires two separate light configurations, one for a light source and another for a detector. Some of the systems pinch leaves and light cannot penetrate through thick or densely colored leaves. On the other hand, emission offers a favorable condition for field applications, because the light source and detector can be placed on the same side. Since plants do not emit light naturally, an external light source is required to cause the emission, which occurs in the form of fluorescence or phosphorescence. A laser is suitable for the light source, since it can provide monochromatic light, which enables fluorescence detection at specific wavelengths.

In the next section, we consider the use of fluorescence emission for plant monitoring, which can meet troublesome requirements.

2.2. Mechanism of Fluorescence

Some plant pigments emit fluorescence in response to light irradiation as a mechanism of dissipating excess energy that is not used for plant activities. The process of fluorescence of a single molecule is shown in Fig. **(1)** with a schematic energy diagram. A detailed explanation of the process is not described here, but can be found, for example, in some books [1-4].

The most important aspect of plant fluorescence that other organisms do not possess is that fluorescence is directly connected to photosynthesis, so that the spectral shape and lifetime of fluorescence can reveal information about photosynthetic activity and related physiological status. Photosynthesis has two steps, a light-dependent reaction and a light-independent reaction. Plant fluorescence is related to the former reaction involving Photosystems I and II. The photosystems do not work well under some growth stress, such as lack of water or nutrition, exposure to high or low temperature, attack by harmful insects and other deleterious conditions, so the absorbed photon energy cannot be transferred among molecules and is not used for photosynthesis. The excess energy is subsequently reemitted as fluorescence. This is the basic concept of fluorescence monitoring for living plant information.

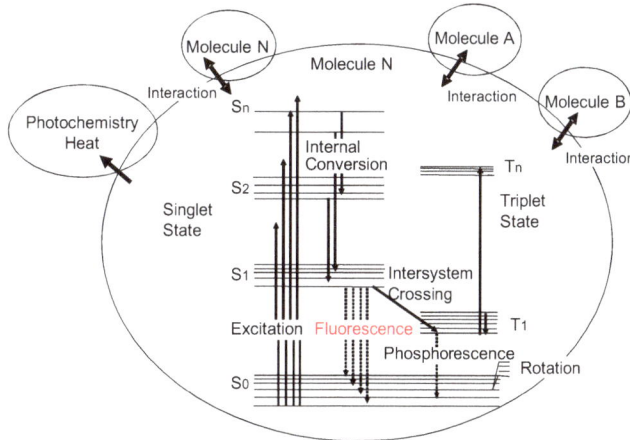

Fig. (1). Florescence of a single molecule in an energy level diagram (Jablonski diagram) with interaction between other molecules: S, singlet state; T, triplet state; S_0, ground state; suffix 1, 2 ...n, excited state.

Fig. **(2)** shows a diagram of energy transfer of absorbed light energy among major photosynthetic pigments and also shows representative plant fluorescence. Molecules of carotenoids, chlorophyll a and b, and several secondary metabolites absorb UV and visible light. Energy transfer among them through an excited state shown in Fig. **(1)** occurs. Use of a UV laser has made it possible to induce fluorescence across a wide spectrum from the visible to infrared regions. The fluorescence spectrum shown in Fig. **(2)** is of a morning glory (*Pharbitis hederacea Choisy*) leaf excited by a 355-nm laser pulse [5]. The fluorescence spectrum is mainly separable into blue-green (B-G) fluorescence and Red-Far Red (R-FR) fluorescence. B-G fluorescence appears in 400- to 650-nm range with a peak at around 460 nm and with a small shoulder at approximately 530nm. R-FR fluorescence appears in 650- to 800-nm range with two peaks at approximately 685 nm and 740 nm. The main fluorophores of B-G LIF are ferulic acid derivatives, other phenylpropanoids and NAD(P)H while that of R-FR LIF is only

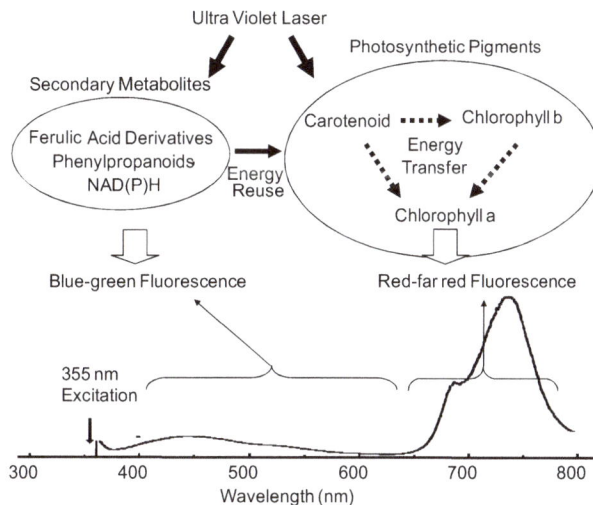

Fig. (2). Energy flow of UV laser absorption and representative laser-induced fluorescence spectrum of a morning glory leaf.

chlorophyll a [6, 7]. it is important to note that the intensity ratio of the two forms of fluorescence can offer information on chlorophyll content. These two fluorescent signals are associated with two photochemical complexes of Photosystem II (685 nm) and I (740 nm) which drives the light-dependent reaction for photosynthesis. The absorption band of chlorophyll for Photosystem I partially overlaps the emission band of Photosystem II, so that Photosystem I reabsorbs a part of the R fluorescence at longer wavelengths and reemits it in the FR region [8]. The amount of absorption and emission depends on the chlorophyll concentration that is obtained by the intensity ratio [9, 10].

To develop and improve plant and vegetation monitoring by LIF spectroscopy, it is indispensable to investigate and understand the details of the spectra of intact living plants. Collecting and compiling a database of the basic LIF spectral data depending on the plant's health status in response to various environmental conditions is important. A series of studies by Chappelle *et al.* were advanced for this reason. They reported the identification of plant types and the effects of growth conditions by LIF spectroscopy [11-13].

3. AGRICULTURAL APPLICATIONS OF LIF SPECTROSCOPY

In agriculture, previous experience of individuals diagnoses the growth status of agricultural products and determines the appropriate harvest time of them. More objective criteria are required for cultivation management. Such a technique would also be helpful to grow a standard product with certain quality parameters. Application of LIF spectroscopy to the agricultural industry is discussed here.

3.1. Growth Monitoring of Agricultural Products with LIF Spectroscopy

Long-term monitoring of lettuce in a farm with LIF spectroscopy was attempted [14]. The observation term was approximately fifty days from seeding to harvest time. The compact mobile fluorescence monitoring system consisted of a third-harmonic Q-switched pulse Nd-YAG laser as an excitation source with a wavelength of 355 nm, a pulse energy of 0.2 mJ, a pulse duration of 7 ns, and a repetition rate of 10 Hz, a spectrometer having an intensified charge coupled device (CCD) array sensor with a wavelength range of 300 to 800 nm and a spectral resolution of 5 nm, and a personal computer. The system is shown in Fig. (3a). The laser was directed to lettuce with a transmitting optical fiber, and the fluorescence from the leaves was delivered to the spectrometer by a receiving optical fiber. The output shape of the laser from the transmitting fiber was a circle, and the input shape of the fluorescence to the receiving fiber was also a circle, but the output from the fiber was a rectangular shape to fit the slit of the spectrometer. The use of optical fibers improved the manipulation of the light. Data was transmitted *via* cell phone from the lettuce field to a laboratory in Shinshu University. A 10-W generator provided all electric power. The whole system was carried by a van into a lettuce farm. A photograph of the system in the field is shown in Fig. (3b) [15].

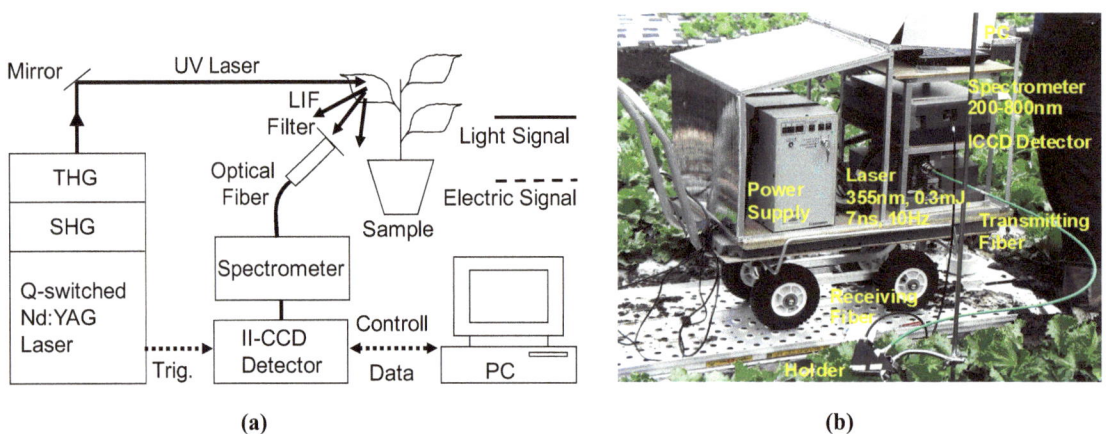

Fig. (3). (a) Schematic diagram of a mobile LIF spectrum monitoring system. SHG and THG are the second and the third harmonic generators. Solid line shows the laser or light path and the dashed line showed the electric signal. (b) A photograph of the system in the lettuce field.

The fluorescence spectrum of lettuce (*lactuca sativa* L var.; Shinano summer) was similar to that of the morning glory leaf. Fig. **(4a)** [15, 16] is the daily variation of relative fluorescence intensity, which is the ratio of the fluorescence intensity at 460 nm (F460) to that at 685 nm (F685), as Cerovic *et al.* reported that the ratio of blue- LIF intensity to red-LIF intensity appeared to be the most sensitive parameter [7]. The variation patterns were quite similar for four observations in different years, which can be explained as follows. 1) During the first four weeks after planting, there was a constant decrease in the ratio. This suggested that the content of chlorophyll, which is the origin of 685-nm fluorescence, increased in correspondence to lettuce growth. 2) The basic growth process probably ended and lettuce development went into the next stage. The accumulation of secondary metabolic products in leaves increased B-G fluorescence.

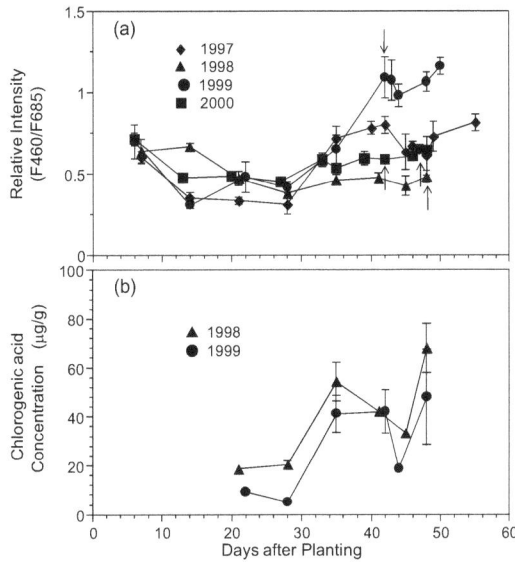

Fig. (4). **(a)** Variation in relative intensity of outer lettuce leaves at 460 nm to 685 nm. The arrow indicates the harvest time as judged by an expert, and **(b)** that of the chlorogenic acid concentration of the lettuce.

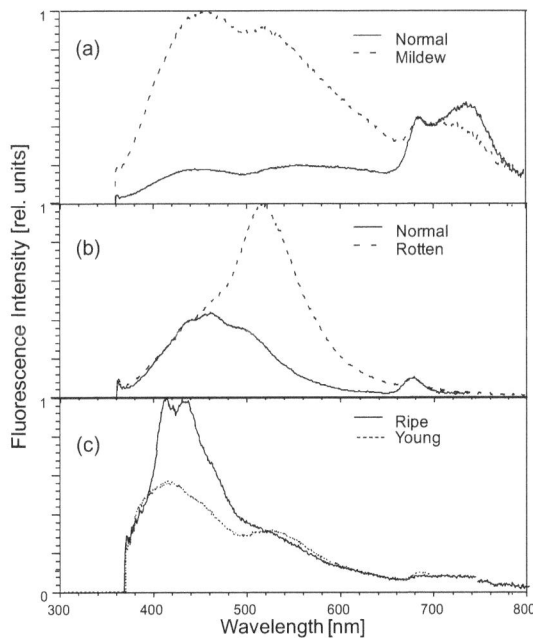

Fig. (5). Examples of LIF spectrum of agricultural products of different status: **(a)** a healthy green oak tree leaf and another with whitish surface mildew, **(b)** napa cabbage with a normal or rotten cores, and **(c)** a ripe pear surface and a young one.

3) The relative fluorescence intensity peaked at approximately 41 days after planting and had a minimum at approximately 45 days. The reason for the decrease is unclear, but this may be related to a transition in the life cycle of the lettuce, for example, from the vegetative period to the reproductive period.

4) A subsequent increase appeared, which is assumed to be due to withering. Four years of observation showed the same patterns in variation, indicating that fluorescence information can be useful for cultivation. Fig. **(4b)** shows the variation in chlorogenic acid concentration of the lettuce measured by high-performance liquid chromatography. A comparison of the area intensity of the fluorescence spectrum at 460 nm analyzed by spectral separation software and the concentration of chlorogenic acid showed a correlation coefficient of 0.8. Other examples of LIF spectra of agricultural products are given in Fig. **(5)** [17]. Fig. **(5a)** was obtained from a healthy green oak tree leaf and another with whitish surface mildew, Fig. **(5b)** was obtained from napa cabbage with a normal or rotten cores, and Fig. **(5c)** was obtained from a ripe pear surface and a young one. As can be seen, the LIF spectrum changed its shape depending on the status.

These results show that LIF spectroscopy has a high potential application to quality control of agricultural products during the period of crop growth and development, potentially leading to the ability to replace empirical judgment with objective LIF data.

3.2. Detection of Tree Sapling Water Stress by LIF Spectroscopy

If information on the health status of plants can be obtained before the symptoms leading to withering or death are apparent, appropriate treatment can be administered at earlier stages when the cause of disease stays within the plant. LIF spectroscopy has the ability that induces the cause by laser as fluorescence and pulls the information to outside. There are several reports that describe fluorescence as an index of plant stress [3, 4].

Water stress detection by LIF spectroscopy has been attempted [14]. The sample were potted sasanqua (*Camellia sasanqua*) tree leaves grown in a pot in a green house. They were grown with different irrigation conditions: 100 ml every day defined as no water stress, 100 ml per half week (mild water stress) and 100 ml per half month (high water stress). The monitoring system was the same as that shown in Fig. **(3)**. As spectral changes appeared particularly at 460 nm and 530 nm, the difference in LIF spectra depending on degree of water stress was investigated at those two wavelengths. Fig. **(6)** shows the daily variation in relative LIF intensity at 530 nm to that at 685 nm after stress treatment.

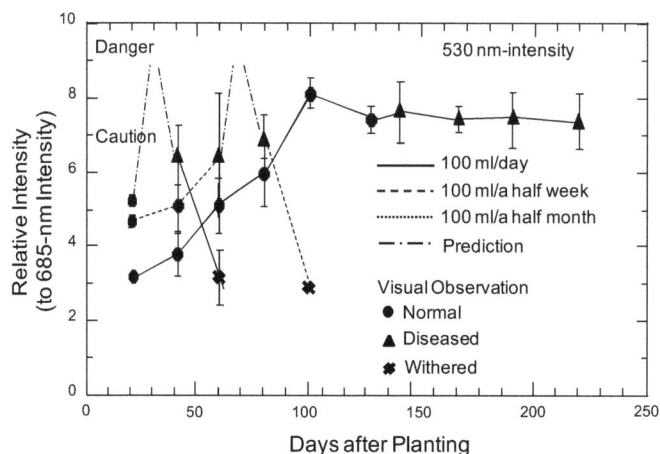

Fig. (6). The variation in relative LIF intensity of water stressed sasanqua leaves at 530 nm to that at 685 nm after water stress treatment was started.

Monitoring started on September 1st (21 days after stress treatment was initiated). LIF intensities at the wavelength with high water stress exhibited an initial increase but then suddenly decreased. Leaves with mild water stress showed a similar tendency to this with high water stress, but the increase was slower than in this with high water stress. However, LIF intensity with no water stress increased for approximately 80 days. The sudden drop shown in leaves with high and mild water stress did not appear in the leaves with no water stress, they kept a constant value of approximately 7. The variation under no water stress included seasonal   variations. Predictive lines were added because the monitoring interval of 3 weeks did not follow such rapid LIF spectrum changes. In other words, LIF spectrum monitoring could detect rapid changes occurring inside a plant that could not be followed by the naked eye because the surface showed no color change. The lines suggest that the health of the tree is poor if the relative intensity exceeds 6, and that the tree is in a critical condition if the value reaches 9.

These results describe B-G fluorescence as is an index for a quick onset and on-going disease and R-FR fluorescence is an index for terminal symptoms. In addition, the combined information from the rate of increase of the 460-nm and/or 530-nm intensity, and the 685-nm emission is meaningful.

4. APPLICATION TO REMOTE MONITORING OF TREES

In this section, we consider the application of LIF spectroscopy to remote sensing. If LIF from trees, forests, and vegetated areas can be detected remotely, it would offer valuable data on vegetation distribution, such as plant biomass distribution, tree height and global forest canopy profile with the trees' health status. A combination of LIF spectroscopy with lidar, LIFS (laser-induced fluorescence spectrum) lidar is one of the most feasible techniques.

To the best of the author's knowledge, plant monitoring using LIFS lidar was first attempted in 1973 [18], A full-dress investigation of LIFS lidar for plant monitoring has awaited development of a practical multi-channel detector for a spectrometer that can measure the entire shape of a spectrum without scanning a wavelength selection device. Research groups such as NASA Goddard [19], IROE [20], Lund Institute of Technology [21], University of Karlsuruhe [22], and Shinshu University [23] performed earlier studies in this area.

The essential configuration of a LIFS lidar for plant monitoring is shown in Fig. **(7a)**, and a photograph of a LIFS lidar developed by Shinshu University is shown in Fig **(7b)** [15]. A compact Q-switched Nd-YAG laser (355 nm, 6 ns, 5 mJ, 10 Hz) situated on a telescope was directed to the tree target away from the LIFS lidar. The laser beam was 3 mm in diameter, which was less than the width of the target tree leaves. A large-diameter telescope (Schmidt-Cassegrainian type, 25.4-cm diameter, 2.0-m focus length) was prepared to collect the leaf fluorescence, because of the concern that the received intensity was much smaller than that of the agricultural product described in the section 3 due to the longer observation range and the fluorescence emission expanding hemispherically from the leaf surface.

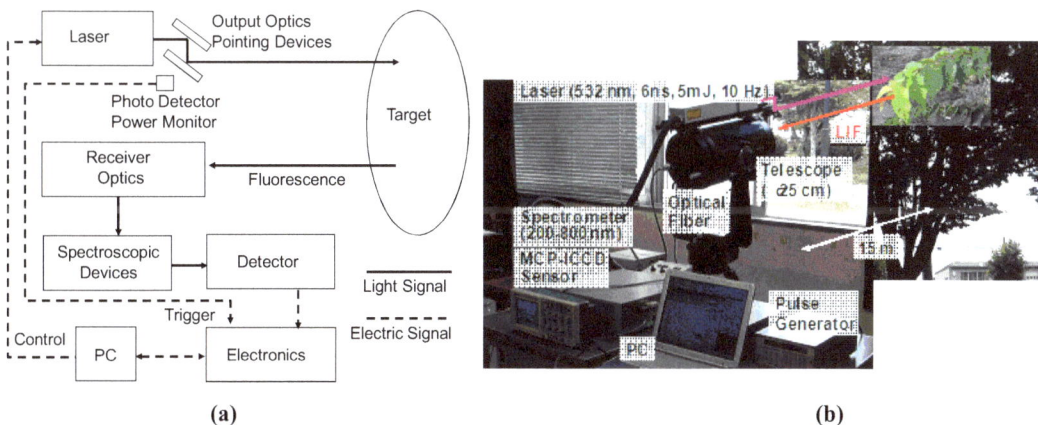

(a)   (b)

Fig. (7). (a) An essential configuration of the laser-induced fluorescence spectroscopy (LIFS) lidar. **(b)** A photograph of the LIFS lidar.

A bundled optical fiber was set at the focal point of the telescope and the fluorescence was delivered to the detection system which consisted of a spectrometer (3-nm wavelength resolution) and an intensified CCD array detector (1024 channels). Connecting the telescope to the spectrometer by an optical fiber simplified the optical handling. The Multi-Channel Plate (MCP) of the ICCD detector was electronically gated for a period on the order of 10-100 ns and the gate opening time was delayed to coincide with the fluorescence arrival time to the detector,l as shown in Fig. **(8)**. This short synchronized detection using a short gate time sufficiently reduced the noise due to the solar background that accumulated during the gate opening time, and made detection of the weak fluorescence of plants naturally growing outdoor possible.

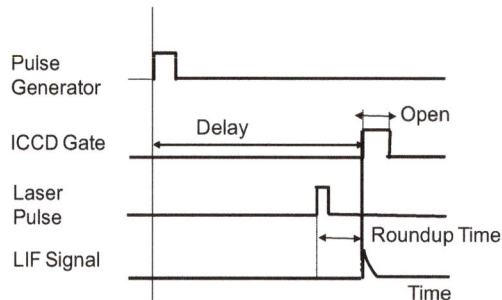

Fig. (8). Operation of synchronized detection using a gated CCD detector. The vertical bar at the left shows the time corresponding to the system trigger.

The target was leaves of zelkova tree (*Zelkova serrata Makino*) located 15 away from the system. The LIF spectra for different growth stages were obtained as shown in Fig. **(9)** [24]. The lower spectrum is for a leaf bud with a red-yellowish green coloration located at the top of the tree branch. The middle spectra are for a younger yellowish-green leaf and for a young green leaf, and the top spectrum is for a mature leaf having the deepest green color and the biggest size among the four.

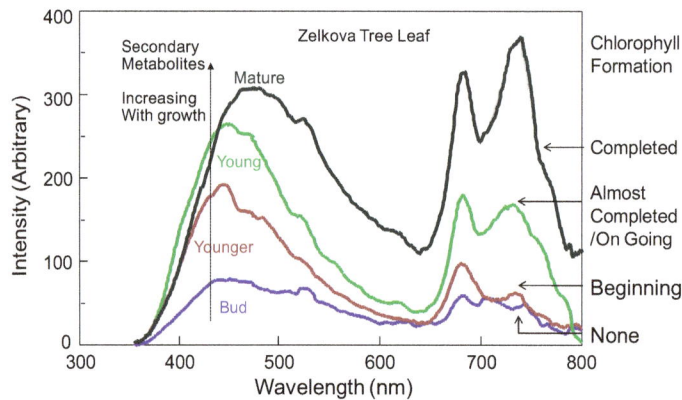

Fig. (9). Remote monitoring of LIF spectrum of zelkova tree leaves corresponding to their growth stages.

The spectrum of each leaf has its own shape depending on the leaf growth or age. The intensity of B-G fluorescence of the leaves constantly increased with time, but the spectral shape did not change. Secondary metabolites that are source pigments of B-G fluorescence were produced and accumulated smoothly during the growth period. The intensity of R-FR fluorescence increased gradually with growth but the spectral shape varied and showed some complexity. The leaf bud did not have the shape of R-FR fluorescence. A small peak of R fluorescence appeared in the younger leaf and became larger with growth. The young leaf showed a normal shape of R-FR fluorescence that had a dip between the R fluorescence and the FR one. The intensity of FR fluorescence exceeded that of R fluorescence. The idea that the ratio of fluorescence intensity at 740 nm to 685 nm is proportional to chlorophyll concentration offers information on chlorophyll formation [9, 10]. That is, chlorophyll formation did not start in the leaf bud and it just started in the next older leaf. This leaf continued the chlorophyll formation with growth and accumulated chlorophyll. Finally, the mature leaf completed the chlorophyll formation with a higher intensity of FR fluorescence than of R fluorescence, so that the ratio was the highest of the four. A high enough chlorophyll content ensured good photosynthetic activity. Production of secondary metabolites was also activated and B-G fluorescence of the mature leaf increased its intensity as well.

The spectrum pattern leading to withering from the early autumn to the late autumn showed the opposite trend. During senescence, specific peaks appeared at 460 nm and 530 nm that were frequently observed in red and yellow leaves [25, 26], and reflected secondary metabolites such as carotenoids, anthocyanins, and other derivatives. The FR fluorescence intensity decreases during withering, showing destruction of chlorophyll. The destruction produced another molecule such as anthocyanin and the lower chlorophyll content unveiled the existence of carotenoids masked by the abundant chlorophyll in summer. These molecules accounted for the specific LIF fluorescence spectrum in B-G region.

These results confirm that LIFS lidar has great potential for monitoring living plants in remote, non-destructive, real-time, and non-chemical ways. The principal advantage of the LIFS lidar is the ability to scan a target area and obtain information on individual trees or plants, which greatly reduces the required manpower in comparison to sampling or *in situ* detection, in which the operator needs to approach every tree or plant to obtain the same data.

5. LIFS IMAGING AND LIFS IMAGING LIDAR

A combination of LIFS spectroscopy with an imaging device offers another possibility for plant monitoring. Most fluorescence monitoring systems have been used for small-scale monitoring such as cells or small organisms, in which certain fluorophores (fluorescent indicators) are injected. Here we describe LIFS imaging which can be applied to larger scales such as a whole leaf or a whole tree. As it is difficult to inject fluorophores uniformly at such a large scale, auto-fluorescence described in the previous sections are used that can give raw information on living tissue without treatment.

5.1. LIFS Imaging in a Laboratory Experiment

A compact but practical LIFS imaging system was developed for laboratory use [27]. A photograph of the system is shown in Fig. **(10)**. The excitation source was a continuous wave UV laser diode (398 nm with 9-nm width, 30 mW max). The operating temperature was controlled by a Peltier device. The laser diode including the Peltier device was 35 mm in diameter and 103 mm long, and the power supply was 166 mm (L) × 80 mm (H) × 180 mm (W). Introduction of the laser could successfully downsize the volume of the system. The beam was magnified by a beam expander to approximately 17 cm in diameter, which could cover the whole area of a sample leaf placed inside a box. The distance from the laser output to the leaf was approximately 930 mm. The beam was tilted with respect to the leaf surface to prevent direct reflection into the detection device. A Liquid Crystal Tunable Filter (LCTF) was used to select the wavelength of leaf fluorescence. The tuning range was from 430- to 750-nm.

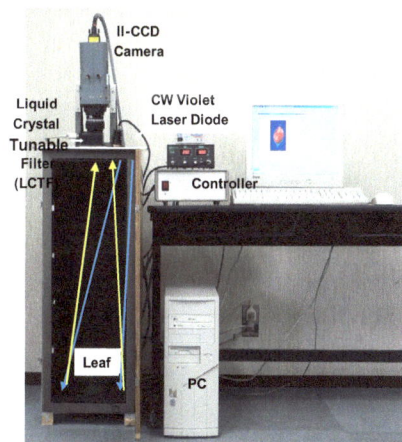

Fig. (10). A photograph of the LIFS imaging system.

The tentative values of the spectral width and the transmittance were approximately 10 nm and 10 % at 550 nm, which became wider and higher with longer wavelength. The large aperture of 35-mm in diameter was suitable for imaging. Tuning of the transmission wavelength was electrically by changing the applied voltage to the filter. This electrical wavelength tuning method was supposed to be ideal for image analysis because it was a mechanical vibration-free technique, so that precise positioning in the detection device was achievable. Positioning directly influences the results of image analysis, which processes the data among pixels. Most plant fluorescence imaging systems reported previously had a mechanical wheel on which several filters were set to select distinct fluorescence wavelengths, and the wavelength selection was made by rotating the wheel. The fluorescence wavelength selected by the LCTF was collected by a 50 mm diameter lens and was detected by a highly sensitive image intensified CCD matrix with 1024×1024 pixels, in which the dark current was reduced to 3 electrons/pixel by cooling with a Peltier device to -40 °C.

The practicality of the LIF imaging system was investigated using a spinach (*Spinacia oleracea* L.) leaf, which was purchased at a super market. LIF images are shown in Fig. **(11)** [14]. Examples of images in the wavelength region from 660 nm to 750 nm, which corresponds to chlorophyll fluorescence, are shown. Each image is described as the LIF intensity relative to that at 630 nm, which was close to a minimum in the spectral region.

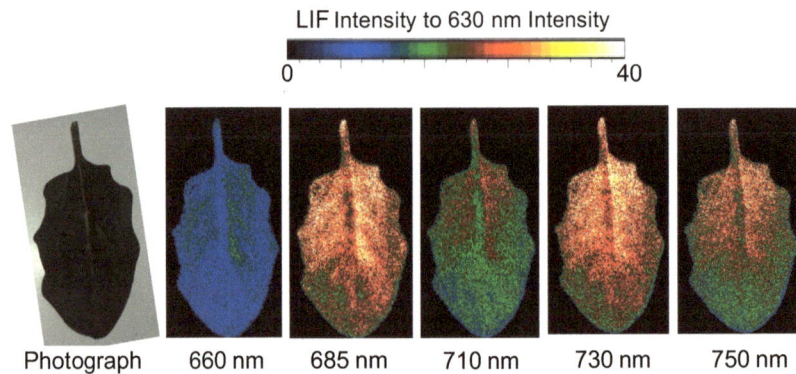

Fig. **(11).** LIFS images of a spinach leaf at different wavelengths in the R-FR region.

The images generally reflected the standard shape of the chlorophyll LIF spectrum as shown in Fig. **(2)**: a weak fluorescence at 660 nm, a first peak at 685 nm, a small dip at 710 nm, a second peak at 730 nm, and then a decrease at the longer wavelength of 750 nm. The intensity distribution of the fluorescence clearly appeared over the leaf, which could not be observed by the naked eye. The fluorescence intensity at 740 nm was almost the same as that at 685 nm over the whole area of the leaf, so it can be inferred that the leaf had a good status with efficient photosynthesis. In every image, the intensity at the basal part of the leaf is large and decreases toward to the top. A coffee tree leaf showed the same pattern [27]. If fluorescence intensity increases with the amount of pigments that are produced and accumulated during the growth process, then the basal area contains a larger amount of pigments.

Experiments confirmed that the imaging system had an advantage in visualization, which could show the spatial variability of plant physiological activities.

5.2. LIF Imaging Lidar for Field Use

Monitoring large plants, like trees, naturally growing outside was attempted using the LIFS imaging lidar shown in Fig. **(12)** [15]. The purpose of the experiment was to estimate chlorophyll concentration remotely and to make a distribution map of the concentration.

A pulse laser (532 nm, 10 mJ, 6 ns, 10 Hz) was directed outside and irradiated a ginkgo tree (*Ginkgo biloba* L.). The beam was magnified at an angle of 150 mrad by a negative lens to cover the whole area of the tree of 5-m height × 6-m width. The receiver was composed of a 42-mm diameter camera lens and a gated intensified CCD camera. The CCD camera had 510 (H) × 492 (V) chips cooled to -30 °C by a Peltier device. The size of the chip was equivalent to 2.24 cm (H) × 1.74 cm (V) at the location on the tree at 62 m, which gave adequate spatial resolution in relation to leaf size. Interference filters centered at 685 nm and 740 nm, which had the same spectral width and transmittance of 20 nm and of 80 % respectively, were inserted alternately to obtain chlorophyll fluorescence. The camera was operated in gated mode with a gate time width of 40 ns, which corresponded to a range resolution of 6 m in the beam pass direction. The gate was delayed and opened just before the fluorescence form the very front of the tree arrived at the camera. This synchronized delay setting with a suitable gate time allowed the LIFS imaging lidar to obtain LIF signals from the tree. It is also possible to capture a range-resolved fluorescence image by sweeping the delay time [28].

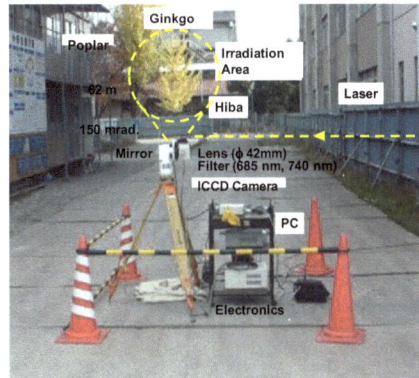

Fig. (12). A photograph of the LIFS imaging lidar.

Chlorophyll content was calculated from the 740-nm to 685-nm intensity ratio, which was compared to the concentration in sampled leaves measured by a high performance liquid chromatography. The correlation coefficient was 0.94, as shown in Fig. **(13)** [10]. The distribution of chlorophyll concentration throughout the whole ginkgo tree, calculated by applying the relationship of the ratio and the concentration, appears in Fig. **(14)** [15] and the average concentration of the whole tree is shown in Fig. **(15)** [10]. Three tree images are shown in each picture; from left to right, a portion of poplar (*Poplus nigra var. italica*),a gingko and a hiba (*Thyjopsis dolabranta var. Hondae Makino*).

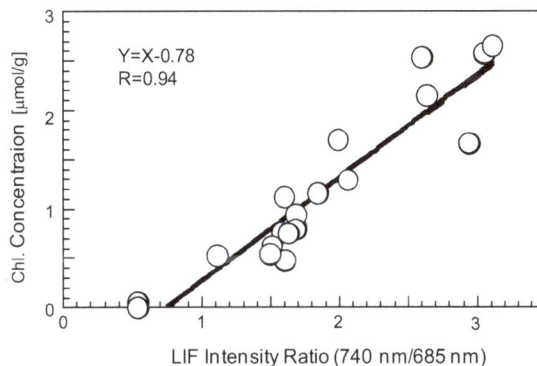

Fig. (13). Relationship between the ratio of the LIF intensity at 740 nm to 685 nm and the chlorophyll concentration measured by high performance liquid chromatography.

The tree in May was small and the concentration of chlorophyll was not high. From images in July, August and September, the tree was crowded with leaves and the concentration was higher than in other months. In October images, a low concentration area appeared at the bottom of the tree (Oct. 10) and at the tip of the branches (Oct. 31). When the senescence process was well underway, only a small portion of the tree had detectable chlorophyll content (Nov. 06), which was absent (Nov. 12). Finally, no chlorophyll fluorescence could be detected on Nov. 21 when almost all leaves had fallen and the remainder had turned completely yellow and looked dry or dead. The images clearly capture the natural variation by season.

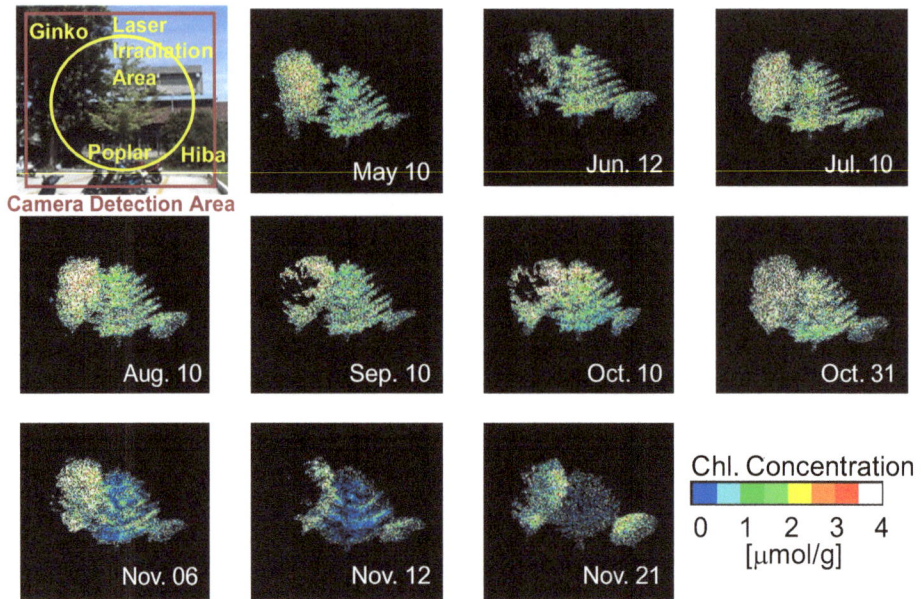

Fig. (14). Image showing distribution of chlorophyll concentration throughout a gingko tree, (from left to right, a portion of a popular, gingko, and a Japanese hiba tree).

Visualization of the chlorophyll concentration of the whole tree as a distribution image is a much more effective way to understand the status of the tree. This result is sufficient to show that the LIFS imaging lidar is a powerful tool for analysis of plant physiology and of forest ecology and management.

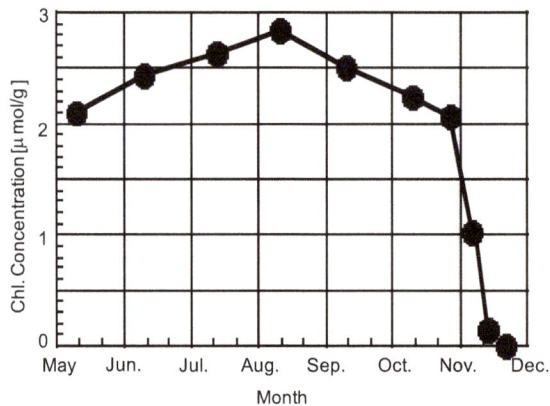

Fig. (15). Monthly variation of the average chlorophyll concentration calculated from the images.

6. CONCLUSION AND FUTURE PROSPECTS

We have described the basics of LIF spectroscopy and its application to plant monitoring. Plant pigments emit visible fluorescence following laser-irradiation. This plant LIF can offer much information on plants' living status.

Several kinds of LIFS monitoring systems were developed to realize the concept. Experimental results obtained in the laboratory and in the field confirmed the effectiveness of the LIF spectral monitoring technique and the practicality of the developed systems.

A combination of the non-destructive optical monitoring systems, LIFS monitoring and imaging developed here and LIF lifetime (LIFL) monitoring and optical absorption spectroscopy, not described in this chapter, will even more powerful for plant and vegetation monitoring and will form the basis of a "Green-Cross; plant general hospital" proposed for future development. Fig. **(16)** is a conceptual image of the "Green-Cross". The UV LIFS imaging lidar determines the general health status of a whole plant, tree, agricultural field, grass land, forest or mountainside. The LIFS lidar using multiple wavelengths with excitation-emission-matrix data investigates details at the smaller or pinpoint level if zones of abnormal symptoms are detected in the image. The excitation-emission-matrix data are used as a database of disease. Environmental monitoring of biological aerosols and water quality in and around agricultural fields by a LIFS lidar is included [29].

Plant *in vivo* information monitored by them is collected using an advanced information communication technology (ICT) system such as a Field Server/AgriServer system [30] with environmental parameters such as solar intensity, temperature and humidity that the server monitors. All data can be delivered through the internet. The LIF technique is an optophysical measurement, which does not require chemical liquid treatment, is compatible with electronics, so it can be easily incorporated into an ICT system network, allowing real-time data delivery. A network of "optical traceability" based on optical check and evaluation can be formed on ground base.

For global vegetation monitoring, satellite monitoring that uses reflection of natural sunlight is becoming popular. However, for managing plant cultivation, the data must contain physiological information. Reflection only offers information about the physical properties of plants. Space and airborne LIFS lidars are the only candidate for this purpose.

Samples undergo a thorough health examination in the "Green-Cross" if more detailed diagnosis is required, where sophisticated spectroscopic devices are prepared. More dynamic diagnosis would be possible with LIF lifetime monitoring [5, 15]. Laser-induced breakdown spectroscopy offers another function for diagnosis by focusing a laser beam into a diameter of a micrometer and operation in an ultrashort time that is achievable by a femtosecond laser.

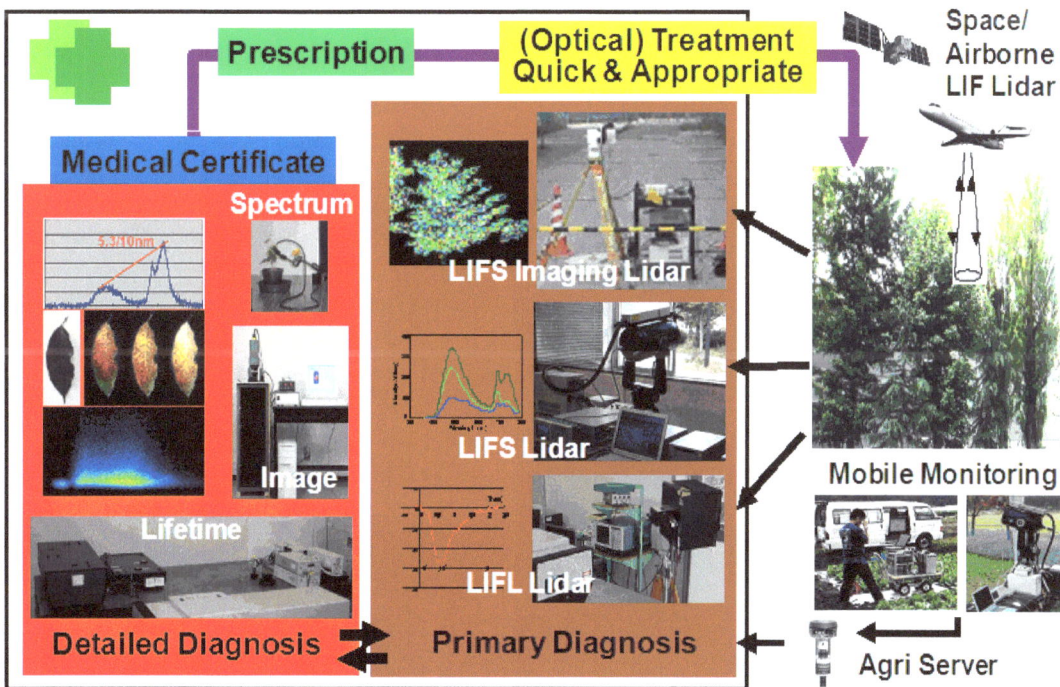

Fig. (16). A conceptual image of the "Green-Cross; plant general hospital".

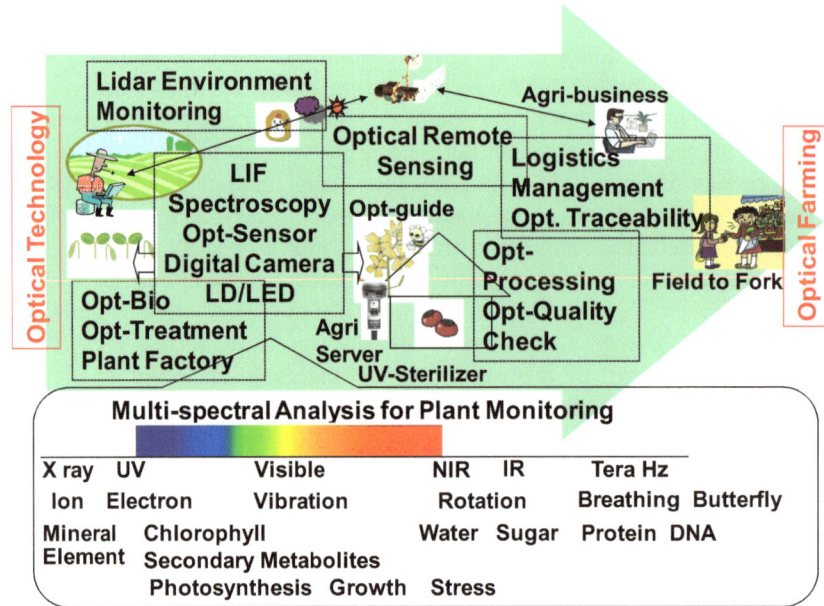

Fig. (17). Explanation of "Optical Farming". Many optical technologies beside these are also available.

The concept of "Optical Farming" is also proposed in Fig. **(17)**. This approach involves diagnosing the growth status of agricultural and horticultural products. Some treatments might be performed by the light itself. A quality check by LIF and other optical methods after planting and during growth, at harvest time, at inspection and shipment locations, and during transportation to the market from the field is an interesting and realistic application. The non-destructive and non-chemical approach offered by the optical method is a priority for food processing, especially for online automated processing. And network of monitoring and traceability based on optical check and evaluation of agricultural products can connect the field to customers, namely "Field to Folk".

Progress in optical technology surely promises realization of these future plans.

ACKNOWLEDGEMENTS

The author wishes to thank Dr. Fumitoshi Kobayashi of the Faculty of Engineering, Shinshu University for his continuous support in developing the systems, Assistant Professor Kazuki Kobayashi of the Division of Graduate School for his extensive ability to work with software, Associate Professor Hiroaki Ishizawa of the Faculty of Textile Science and Technology and Professor Naoto Inoue of the Faculty of Agriculture for offering their knowledge on plants. Part of the research was supported by Shokubutsu Kenkyu Josei (2009-2010) from The New Technology Development Foundation, and by a Grant-in-Aid for Scientific Research (B), 17360191 from the Ministry of Education, Science, Sports, Culture and Technology, Japan. The open-door data access system in field use was partially appropriated from AgriServer data handling system developed by the Strategic Information and Communications R&D Promotion Programme (SCOPE) project, 102304002, supported by the Ministry of Internal Affairs and Communications, Japan.

REFERENCES

[1]   M. Zude, Ed., Optical monitoring of fresh and processed agricultural crops, CRC Press, Bota Raton, 2009.

[2]   N. Wada, M. Mimuro, Eds., Recent progress of bio/chemiluminescence and fluorescence analysis in photosynthesis, Research Signpost, Kerala, India, 2005.

[3]   H. Lichtenthaler, Ed., Application of chlorophyll fluorescence, Kluwer Academic Publishers, Dordrecht, Netherlands, 1988.

[4]   G. Papageorgiou and Govindjee, Eds., Chlorophyll fluorescence (a signature of photosynthesis), Springer, 2004.

[5]   A. Takeuchi, Y. Saito, T. Kawahara, A. Nomura, T. Suzuki, "Possibility of disease monitoring of plants by laser-induced fluorescence method -development and evaluation of LIF measurement system-", Proceedings of SPIE, Vol. 4153, pp. 22-29, 2001.

[6]   C. Bushman, H. Lichtenthaler, "Principles and characteristics of multi-colour fluorescence imaging of plants", Journal of Plant Physiology, Vol. 152, pp. 297-314, 1988.

[7]   Z. Cerovic, G. Samson, F. Marales, N. Tremblay, I. Moya, "Ultraviolet-induced fluorescence for plant monitoring: present state and prospects", Agronomie, Vol. 19, pp. 543-578, 1999.

[8]   G. Krause, E. Weis, "The photosynthetic apparatus and chlorophyll fluorescence" in ref. [3], pp. 3-11.

[9]   H. Lichtenthaler, "Chlorophyll fluorescence signatures of leaves during the autumnal chlorophyll breakdown", Journal of Plant Physiology, Vol. 131, pp. 101-110, 1987.

[10] Y. Saito, K. Kurihara, H. Takahashi, F. Kobayashi, T. Kawahara, A. Nomura, S. Takeda, "Remote estimation of the chlorophyll concentration of living trees using laser-induced fluorescence imaging lidar", Optical Review, Vol. 9, pp. 37-39, 2002.

[11] E. Chappelle, F. Wood Jr., J. McMurtrey III, W. Newcomb, "Laser-induced fluorescence of green plants. 1: a technique for the remote detection of plant stress and species differentiation", Applied Optics, Vol. 23, pp. 34-138, 1984.

[12] E. Chappelle, J. McMurtrey III, F. Wood Jr., W. Newcomb, "Laser-induced fluorescence of green plants. 2: LIF caused by nutrient deficiencies in corn", Applied Optics, Vol. 23, pp. 74-80, 1984.

[13] E. Chappelle, F. Wood Jr., W. Newcomb, J. McMurtrey III, "Laser-induced fluorescence of green plants. 3: LIF spectral signatures of five major plant types", Applied Optics, Vol. 24, pp. 74-80, 1985.

[14] Y. Saito, "Laser-induced fluorescence as an index for monitoring plant activity related to photosynthesis" in ref. [2], pp. 235-251.

[15] Y. Saito, "Laser-induced fluorescence spectroscopy/technique as a tool for field monitoring of physiological status of living plants", Proceedings of SPIE, Vol. 6604, pp. 66041W-1 - 66041W-12, 2007.

[16] H. Ishizawa, Y. Saito, T. Amemiya, K. Komatsu, "Non-destructive monitoring of agriculture products (lettuce) based on laser-induced fluorescence", Nougyou Kikai Gakkaishi, Vol. 64, pp. 89-94, 2002 (in Japanese).

[17] Y. Saito, "Monitoring raw material by laser-induced fluorescence spectroscopy", in ref.[1], pp. 319-336.

[18] R. M. Measures, W. Houston, M. Bristow, "Development and field tests of a laser fluorosensor for environmental monitoring", Canadian Aeronautics and Space Journal, Vol. 19, pp. 501-506, 1973.

[19] F. Hoge, R. Swift, J. Yungel, "Feasibility of airborne detection of laser-induced fluorescence emissions from green terrestrial plants", Applied Optics, Vol. 22, pp. 2991-3000, 1983.

[20] G. Cecchi, P. Mazzinghi, L. Pantani, R. Valentini, D. Tirelli, P. Angelis, "Remote sensing of chlorophyll a fluorescence of vegetation canopies. 1: Near and far field measurement technique", Remote Sensing of Environment, Vol. 47, pp. 18-28, 1994.

[21] H. Edner, J. Johansson, S. Svanberg, E. Wallinder, "Fluorescence lidar multicolour imaging of vegetation", Applied Optics, Vol. 33, pp. 2471-2479, 1994.

[22] M. Sowinska, F. Heisel, J. Miehe, M. Lang, H. Lichtenthaler, F. Tomasini, "Remote sensing of plants by streak cameralifetime measurements of the chlorophyll a emission", Journal of Plant Physiology, Vol. 148, pp. 638-644, 1996.

[23] Y. Saito, K. Takahashi, E. Nomura, K. Mineuchi, T. Kawahara, A. Nomura, S. Kobayashi, H. Ishii, "Visualization of laser-induced fluorescence of plants influenced by environmental stress with a microfluorescence imaging system and a fluorescence imaging lidar system", Proceedings of SPIE, Vol. 3059, pp. 190-198, 1997.

[24] Y. Saito, M. Hara, K. Morishita, T. Tenpaku, K. Ichihara, F. Kobayashi, T. Kawahara, "Fluorescence lidar for remote monitoring of plant information", Proceedings of the International Conference Silvilaser (Matsuyama, Japan), pp. 55-59, 2006.

[25] Y. Saito, M. Kanoh, K. Hatake, T. Kawahara, A. Nomura, "Investigation of laser-induced fluorescence of several natural leaves for application to lidar vegetation monitoring". Applied Optics, Vol. 37, pp. 431-437, 1998.

[26] K. Takahashi, K. Mineuchi, T. Nakamura, N. Sakurai, A. Komatsu, M. Koizumi, H. Kano, "Laser induced fluorescence of tree leaves: spectral changes with plant species and seasons", Proceedings of International Geoscience and Remote Sensing Symposium (Tokyo, Japan), pp. 1985-1987. 1993.

[27] Y. Saito, T. Matsubara, T. Koga, F. Kobayashi, T. Kawahara, A. Nomura, "Laser-induced fluorescence imaging of plants using a liquid crystal tunable filter and charge coupled device imaging camera", Review of Scientific Instruments, Vol. 76, 106103-1 - 106103-3, 2005.

[28] Y. Saito, K. Hatake, E. Nomura, T. Kawahara, A. Nomura, N. Sugimoto, T. Itabe, "Range-resolved image detection of laser-induced fluorescence of natural trees for vegetation distribution monitoring", Japan Journal of Applied Physics, Vol. 36, pp. 7024-7027, 1997.

[29] Y. Saito, F. Kobayshi, F. Kobayashi, "Laser-induced fluorescence spectrum (LIFS) LIDAR for remote detection of biological substances surrounding the "Livingsphere"", Proceedings of the 8th International Symposium on Advanced Environmental Monitoring (Sapporo, Japan), pp. 54-57, 2010.

[30] Y. Saito, S. Takanobu, K. Kobayashi, K. Sato, M. Hirafuji, T. Fukatsu, R. Ichimura, R. Yashiro, S. Takeuchi, K. Yuasa, S. Watanabe, F. Kobayashi, T. Kawahara, T. Kameoka, "Field server monitoring system for construction of IT farming and agri-tourism –Trial report from Obuse-town, Nagano, Japan-". Proceedings of SICE-ICASE International Joint Conference (Busan, Korea), pp. 4848-4851, 2006.

CHAPTER 7

All-Fiber Coherent Doppler Lidar System for Wind Sensing

Shumpei Kameyama[*], Toshiyuki Ando, Kimio Asaka and Yoshihito Hirano

Information Technology R&D Center, Mitsubishi Electric Corporation, 5-1-1 Ofuna, Kamakura, Kanagawa, 247-8501, Japan

Abstract: Coherent Doppler Lidar (CDL) systems are suitable for wind sensing of localized regions under clear weather conditions. Especially, a 1.5 μm all-fiber wind sensing CDL system has many advantages since telecomm fibers and fiber-based components, which have reached technical maturity, can be used. We have developed all-fiber CDL systems during this decade, and in this chapter we introduce our development in addition to the basic theory for system design. After the demonstration of wind sensing performance using the prototype model, we have developed the product model by refining the prototype model. The product model consists of a compact fiber-based optical transceiver unit, an optical antenna unit that includes a wedge scanner, and a Field Programmable Gated Array (FPGA)-based real-time signal processing unit. All-fiber CDL systems can be used for various industrial applications, such as meteorological monitoring, wind survey for wind power generation, and aviation safety.

Keywords: Doppler lidar, coherent laser radar, turbulence, fiber laser, fiber amplifier, acousto-optic modulator, polarization control, wind sensing, wind velocity, wind direction.

1. INTRODUCTION

Coherent Doppler Lidar (CDL) [1, 2] is a kind of laser radar which detects a Doppler frequency shift of backscattered laser light using coherent (homodyne or heterodyne) detection. This sensor can be utilized for wind sensing, as shown in Fig. (1). A transmitted laser light is backscattered by aerosols which are drifting in the air. Since aerosols move with the local wind velocity, a Doppler frequency shift appears in the backscattered light. A Line-Of-Sight (LOS) wind velocity can be detected by detection of the shift frequency. Ranging is possible by using a pulsed laser light as a transmitted light.

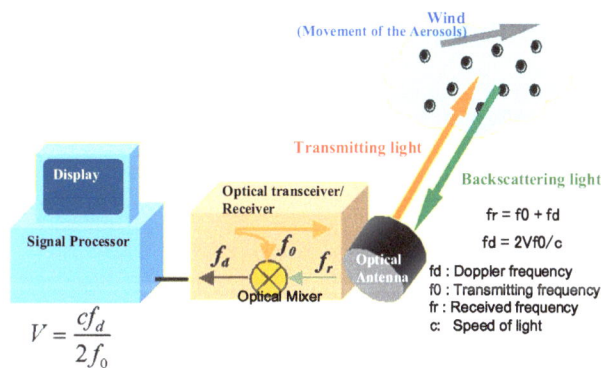

Fig. (1). Principle of wind sensing CDL system.

Since the wavelengths used in CDL systems are much shorter than those used in electromagnetic wave radars, CDL systems are suitable for wind sensing of localized regions because of the very narrow directivity of the transmitted beam. They are also suitable for wind sensing under clear weather conditions because of the high backscatter coefficient even for invisible aerosols. Owing to the above mentioned advantages, CDL systems have been developed for a long time by many researchers.

*Address correspondence to Shumpei Kameyama: Information Technology R&D Center, Mitsubishi Electric Corporation, 5-1-1 Ofuna, Kamakura, Kanagawa, 247-8501, Japan; E-mail: Kameyama.Shumpei@dn.MitsubishiElectric.co.jp

The first wind-sensing CDL system was reported by Huffaker *et al.* [3] for wake vortex detection in 1970, and a 10.6 μm CW CO_2 laser was used in the system. The 10.6 μm pulsed CO_2 laser systems were developed for airborne Clear Air Turbulence (CAT) sensors [4, 5] and for meteorological wind sensing [6, 7]. These CDL systems with technologically mature CO_2 lasers have made fruitful results in many applications for a long time. Since the late 1980s, CDL systems with newly developed solid-state lasers attracted high attention because of advantages of size, weight, reliability, and lifetime. In addition, shorter wavelengths of the solid state laser systems provide the higher backscattering coefficients and higher velocity resolutions for the CDL systems. Kavaya *et al.* [8] completed the 1.06 μm pulsed CDL system using a Nd:YAG master oscillator power amplifier (MOPA) system[9], which realized the measurable range of a few tens of kilometers, and this was used for launch-site wind sensing [10]. Henderson *et al.* reported eye-safe 2 μm pulsed CDL systems with flashlamp-pumped, Ho, Tm:YAG lasers and with diode-pumped Tm:YAG lasers [11, 12]. These have been tested for many applications including airborne wind sensing [13, 14]. Asaka *et al.* [15] proposed the 1.5 μm pulsed CDL system with an Er, Yb:phosphate Glass laser [16], which was superior to 2 μm systems in terms of its eye-safety.

Recently, a 1.5 μm CW all-fiber wind sensing CDL system, which utilized optical fiber components used in communication systems, was reported by Karlsson *et al.* [17], and a pulsed all-fiber system was reported by Pearson *et al.* [18]. The all-fiber CDL has many advantages especially in short range wind sensing applications, although this is not suitable for long range sensing because of the limited transmitted peak power caused by the nonlinear effect in an optical fiber. The advantages are realized by using technologically mature telecomm fibers and fiber-based components. It is stable under vibration, shock, and temperature variation. It provides long lifetime without maintenance. In addition, the system arrangement is made flexible. Moreover, it is suitable for production because the required skill level for manufacturing are not so high. Thus, all-fiber system is expected to become a practical CDL system for many applications including meteorological monitoring, wind survey for wind power generation, and aviation safety. During this decade, we have developed a complete and compact system of all-fiber pulsed CDL for wind sensing [19, 20].

In the development of CDL systems, a design procedure is important. In the system design, the signal-to-noise ratio (SNR) of a heterodyne-detected signal is predicted by using the CLR equation; the line of sight (LOS) velocity estimation precision is then predicted from the relation between the SNR and the performance of velocity estimators. We have derived a simple theory of system design [21, 22] in addition to the hardware development.

In this chapter, we describe the development of all-fiber CDL systems after the explanation of the CLR equation and the signal processing performance which we have derived.

2. COHERENT LASER RADAR EQUATION

2.1. Historical Background

The first CLR equation for wind sensing was reported by Sonnenschein and Horrigan in 1971; they presented an analytical expression for cases of both no turbulence and no beam truncation [23]. Frehlich and Kavaya denoted an equation for a system including the turbulence effect [24]. Their equation was analytically derived by assuming that the beam truncation appearing at the telescope aperture has a transmittance with a Gaussian spatial distribution. Although this is a good approximation to derive the analytical expression, the truncation is known to appear at circular apertures in actual cases. Therefore, numerical calculations for beam propagation, such as numerical Fresnel integration (NFI), are required to consider the effect of beam truncation. For the analytical expression of the truncation effect, nearest Gaussian approximation (NGA) is an interesting concept for the CLR equation. System efficiencies of CLR obtained using NGA have been compared with NFI solution for several beam truncation ratios [25], and the differences were shown to be less than 1 dB for the far-field. We have combined the concepts of NFI and NGA, and derived a semianalytic pulsed CLR equation for coaxial and apertured systems [21]. We introduce this CLR equation in the following section.

2.2. Semianalytic Pulsed CLR Equation

This equation is based on the conventional analytical equation for pulsed systems [23], [24] and is expressed for a soft target as

$$SNR(L) = \frac{\eta_D(L)\lambda E\beta K^{2L/1000}\pi D^2}{8hBL^2},$$

(1)

where h is Planck's constant (Js), E is the transmitting pulse energy (J), D is the diameter of the circular aperture (m), B is the receiver bandwidth (Hz), L is the range (m), λ is the laser wavelength (m), β is the atmospheric backscatter coefficient (m^{-1} sr^{-1}), and K is the atmospheric transmittance (km^{-1}). For hard targets, $2R/(v\tau)$ is substituted for $\lambda\beta\pi$ in Eq. (1), where R is the target reflectivity, v is the laser frequency (Hz), and τ is the pulse width (s). $\eta_D(L)$ is the system efficiency given by

$$\eta_D(L) = \frac{\eta_F}{\left\{1+\left(1-\dfrac{L}{L_F}\right)^2\left[\dfrac{\pi(A_C D)^2}{4\lambda L}\right]^2+\left(\dfrac{A_C D}{2S_0(L)}\right)^2\right\}},$$

(2)

where A_C is the correction factor obtained by NGA and $A_C D$ is the beam diameter of the nearest Gaussian beam. L_F is the phase curvature (focal range) of the transmitting beam, $S_0(L)$ is the transverse coherent length (m), and if C_n^2 (m$^{-2/3}$) is constant along the beam path, $S_0(L)$ is approximated by $\left(1.1k^2 L C_n^2\right)^{-3/5}$ where C_n^2 is the atmospheric refractive index structure constant (m$^{-2/3}$), k is the wave number ($= 2\pi/\lambda$), and η_F is the system efficiency at focal range given by

$$\eta_F = 128F^2\rho^{-4}\times\int_0^\infty\left|\int_0^1\exp\left[-\rho^{-2}x^2\right]J_0(2Fxy)xdx\right|^4 ydy,$$

(3)

where $x = r/D$, $y = r'/D$, r is the distance from the center at the aperture plane (m), r' is the distance from the beam center at the target plane (m), as shown in Fig. (2). $\rho = D_b/D$ is the beam truncation ratio, where D_b is the $1/e^2$ intensity diameter of the Gaussian beam (m), and $F = \pi D^2/(4\lambda L)$. $J_0(Z)$ is the Bessel function of the zeroth order and argument Z.

The calculated result for the system efficiency at focal range η_F and correction factor A_C corresponding to each truncation ratio is shown in Fig. (3).

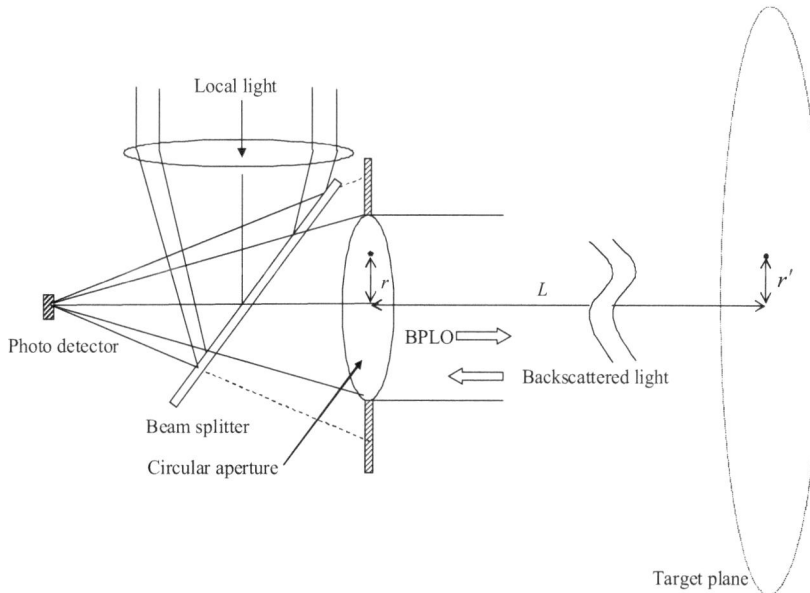

Fig. (2). Schematic of receiving geometry. The transmitted beam is not shown but it is identical with the BPLO.

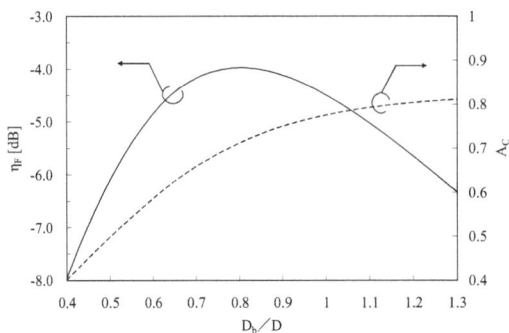

Fig. (3). System efficiency at given values of focal range and correction factor versus truncation ratio. The system efficiency is obtained using NFI, and the correction factor is obtained using NGA.

As denoted in [26], the optimum condition for the truncation ratio is 0.8. Fig. **(3)** shows that the correction factor is 0.71 when the truncation ratio is this optimum condition.

2.3. Optimum Beam Truncation Ratio Considering Atmospheric Turbulence

Although the optimum beam truncation ratio when C_n^2 is negligible is denoted, the optimum ratio is considered to change depending on the C_n^2 condition since there is a limit to the improvement in the SNR when the beam diameter is increased. Fig. **(4)** shows the system efficiency at focal range as obtained by Eq. (1) when $L = L_F$ versus the beam truncation ratio for some values of $S_0(L_F)/D$, which are determined using the value of C_n^2.

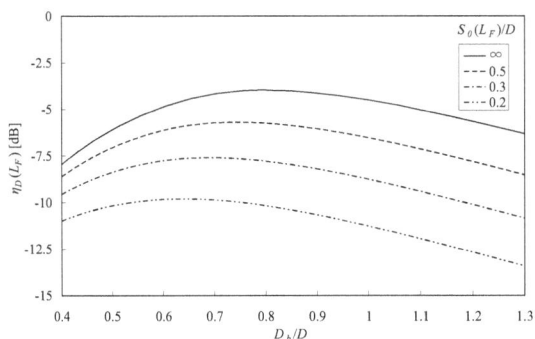

Fig. (4). System efficiency for focal range obtained by Eq. **(6)** when $L = L_F$, as a function of the beam truncation ratio D_b/D for some values of $S_0(L_F)/D$ as determined by C_n^2.

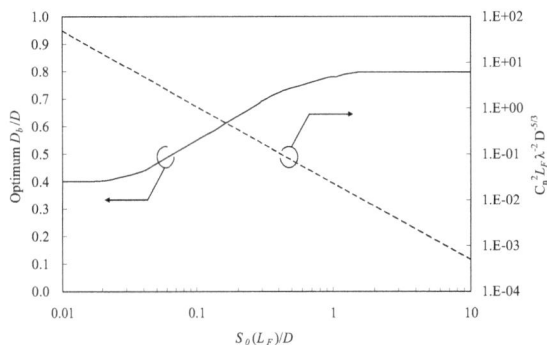

Fig. (5). Optimum truncation ratio considering the influence of C_n^2 versus transverse coherent length normalized by aperture diameter (*i.e.*, $S_0(L_F)./D$). The second vertical axis shows C_n^2 normalized by the range, wavelength, and aperture diameter, which corresponds to the value of the horizontal axis.

The optimum ratio as a function of $S_0(L_F)/D$ is shown in Fig. **(5)**. The corresponding C_n^2 normalized by range, wavelength, and the aperture diameter is also shown in the figure; the optimum ratio decreases as C_n^2 increases.

3. SIMULATION STUDIES

3.1. Performance of Signal Processing for Velocity Estimation

For designing wind sensing CDL systems, the Line Of Sight (LOS) velocity estimation precision is predicted from the SNR and the velocity estimator (*i.e.*, digital signal processing algorithm). Therefore, a simulation tool to estimate the performances (relationships between the SNR, estimation precision, and signal detection probability) of the estimators is required to realize end-to-end simulations for the design. Because it is difficult to express the performance analytically, it has been studied using computer simulations.

The first study conducted on the performances of velocity estimators for a CDL system was reported in [27] and the performance of the Discrete Fourier Transform (DFT)-based velocity estimator was studied and compared with that of the Pulse Pair (PP) estimator. The difference between the estimation precision of the DFT-based estimator and the approximated Cramer-Rao Lower Bound (CRLB), which is the theoretical limit on estimation, was discussed by Rye and Hardesty [28]; they compared the performance of the DFT-based estimator with that of other estimators, including the Maximum Likelihood (ML), PP, and Signal Matching (SM) estimators in [29]. The exact CRLB was derived by Frehlich [30], and the heterodyne-detected signal model used in the simulation was refined by Frehlich and Yadlowsky [31]. Further, in [31], the performances of the PP, ML, and time series estimators of the Auto Regressive (AR) and Minimum Variance (MV) time series models were investigated and compared with the exact CRLB. The performances of these estimators for low SNR values were investigated in [32].

All the studies described above were conducted on the basis of the approximation that the spectrum or the covariance of the heterodyne-detected signal was Gaussian; the realistic wind field was not assumed. A well-known, realistic wind field regime is the Kolmogorov turbulence. The performances of the ML and MV (or Capon) estimators in the Kolmogorov turbulence regime and the effects of turbulence were investigated in [33]. The performance of the MV estimator for general wind regimes and wavelengths was presented in [34] and shown to be valid for a weak to moderate signal regime.

The studies described above offer meaningful information on the performances of velocity estimators. However, from a system designer's point of view, the results of these studies are not convenient or sufficient to be applied directly to system design because of the following reasons. First, in studies other than [34], the concrete wavelength regions of 2 µm and 10 µm were considered but, the 1.5 µm wavelength region, which was not commonly used for CDL systems when the studies were conducted, was not considered. Second, the performances of simple and real-time DFT-based estimators were not studied in the realistic wind field regime (*i.e.*, Kolmogorov turbulence), although these estimators were the major estimators used in many CDL systems [11-14, 17, 19, 20].

We have summarized the performances of DFT-based velocity estimators for wind sensing CDL systems in the Kolmogorov turbulence regime by using the Monte-Carlo simulations [22]. The following sections describe the simulations.

3.2. Simulation Procedure

In the Monte-Carlo simulations, the conditions for laser wavelength, sampling frequency, range resolution, velocity search range, mean velocity, statistical property of wind field, pulse width, wideband SNR, accumulation number, and iteration number are set first. The wind field is generated by a computer and gated heterodyne-detected signals are generated also by a computer. Then, signal processing and velocity estimation are performed according to the algorithm corresponding to the selected estimator. Following the iterations of this simulation procedure, the histogram of the estimated velocity is obtained. The histogram is fitted to the modeled PDF and the estimation precision is defined as the SD of the good estimates. The signal detection probability is defined as 1 minus the fraction of random bad estimates.

The modeled histogram is the single Gaussian model which is defined in [21] and expressed as

$$PDF(v_e) = \frac{b}{v_{search}} + \frac{(1-b)}{(2\pi)^{/2} g} \exp\left[-\frac{(v_e - v_m)}{2g}\right],$$

(4)

where b is the fraction of bad estimates and g is the SD of the good estimates (m), v_e is the estimated velocity (m/s), v_m is the mean velocity (m/s) which is assumed to be zero in this simulation, and v_{search} is the width of the velocity search range (m/s). The parameters b and g of the model PDF are determined by minimizing the means-square difference between the histogram of the estimates from the simulated result and the predicted histogram based on the model PDF.

3.3. DFT-Based Velocity Estimator

Here, we introduce a basic DFT-based estimator in the following manner. This estimator uses only DFT, peak detection, and moment operation around the peak. Here, the signal processing algorithms of this estimator are briefly explained.

In the basic DFT estimator, first of all, periodograms are calculated by using DFT, or in some cases, Fast Fourier Transform (FFT); subsequently, the periodograms are accumulated incoherently. The accumulated periodogram is generally expressed as

$$P(k,i) = \sum_{n=0}^{N-1} \left| \frac{T_S}{M} \sum_{m=0}^{M-1} S(m,n,i) \exp\left(-j\frac{2\pi km}{M}\right) \right|^2,$$

(5)

where N denotes the total accumulation number (*i.e.*, the number of shots), and k denotes the velocity bin number. The noise level is cancelled from the accumulated periodogram, and the peak bin number is detected. The first moment is obtained by a moment operation around the peak where the intensities after the noise cancellation are positive. The mean velocity is obtained as

$$v_e = \frac{\lambda k_m}{2T_S}.$$

(6)

where k_m denotes the first moment described above.

3.4. Simulation Conditions

The simulation conditions used in this study are summarized in Table **1**. The parameters shown in the table are typical values used for wind sensing CDL systems. Three conditions for ε listed in the table correspond to Light, Moderate, and Severe turbulence conditions defined in the International Civil Aviation Organization (ICAO) ANNEX3 Standards. The accumulation number is set to have a constant value of 10.

As denoted in [32], the estimation precision is approximately proportional to $N^{1/2}$ when there are few random bad estimates and N is sufficiently large (≥ 10). For the given high detection probability of ~90%, the threshold value for $(SNR)N^{1/2}$ is approximately constant in the case of large N.

3.5. Simulation Results

We simulated the relationships between the SNR, estimation precision, and detection probability according to the procedure, estimation algorithms, and simulation conditions.

Fig. (**6**) shows one of the simulation results; the simulation parameters are denoted in the figure caption. It is shown that estimation performance is improved with the increase in the SNR but the improvement of the estimation precision is not so large in the region of the high detection probability.

Table 1: Simulation conditions

Parameter	Value
Laser wavelength (μm)	1.55, 2.02, 10.61
Velocity search range ($\sim \pm$ m/s)	38.5
Mean velocity in the range gate (m/s)	0
Range resolution (m)	25, 50, 75, 100, 150
Pulse FWHM/ Range resolution	1
$\varepsilon\,(m^2/s^3)$	0.014 (Light turbulence) 0.032 (Moderate turbulence) 0.322 (Severe turbulence)
Accumulation number	10
Sampling frequency (MHz)	100
Sampling points for DFT	128 (with zero padding)
Wideband SNR (dB)	$-30 \sim 0$, 0.5 step
Iteration number	10,000

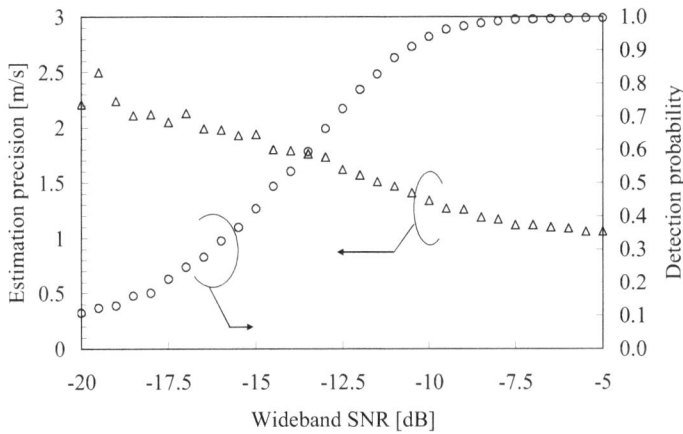

Fig. (6). Relationship between wideband SNR and LOS velocity estimation precision and signal detection probability. The laser wavelength is 1.55 μm, the range resolution is 100 m, and the turbulence condition is severe.

Table 2: Wideband SNR required for the basic DFT estimator to obtain a detection probability of > 90 % for wavelength of 1.55 μm (unit: decibels)

Range resolution [m]	Turbulence conditions		
	Light	**Moderate**	**Severe**
25	−8.0	−8.0	−8.0
50	−10.5	−10.0	−9.5
75	−11.5	−11.0	−10.0
100	−12.0	−12.0	−10.5
150	−13.0	−12.5	−10.5

The SNR required for the basic DFT estimator to realize a detection probability of >90% is shown in Table **2**. The estimation precision at the SNR listed in Table **2**, and the estimation precision at high SNR (=0dB) are listed in Table **3**.

Table 3: LOS velocity estimation precision for the basic DFT estimator at the snr listed in table 2 (left side value) and the snr of 0 db (right side value) for wavelength of 1.55 μm (unit: m/s)

Range resolution [m]	Turbulence conditions		
	Light	Moderate	Severe
25	1.3, 0.7	1.3, 0.8	1.6, 1.0
50	0.8, 0.5	0.9, 0.6	1.4, 0.9
75	0.7, 0.5	0.8, 0.6	1.4, 1.0
100	0.7, 0.5	0.8, 0.6	1.4, 1.0
150	0.6, 0.5	0.8, 0.6	1.5, 1.1

4. PROCEDURE OF SYSTEM DESIGN

Using the derived semianalytic equation [21] and the summary on performance of velocity estimator [22], the design for the coaxial CLR for wind sensing is optimized by the following steps.

i.   Determine the requirements for the system (*i.e.*, wavelength, required measurable range, resolution, velocity search range, detection probability, and velocity estimation precision).

ii. Determine the receiving bandwidth, pulse width, and gate width in signal processing. The receiving bandwidth can be determined by the velocity search range. It is preferred that the pulse width and gate width be the same as the range resolution by default. Determine the focal range such that it falls in the required measurable range.

iii. Determine the atmospheric condition (C_n^2, β, K)

iv. Determine the aperture diameter by considering the size limitation. In addition, determine the corresponding optimum beam diameter and correction factor by referring to Figs. **(3)** and **(5)**.

v.   Obtain the required wideband SNR to satisfy the required detection probability and velocity estimation precision for the required range resolution, velocity search range, wavelength, gate width, and pulse width by referring to the simulation results for the signal processing performance. The simulation results denoted in Table **2** and **3** can be directly applied to determining the required SNR for almost all cases in wind-sensing applications.

vi. Determine the required pulse energy at the required measurable range using Eq. (1) and confirm that the SNR is larger than the value required from the nearest range to the required measurable range. If the SNR is lower than the value required in the near-field, increase the pulse energy to obtain the required SNR in the near-field.

5. DEMONSTRATION OF ALL-FIBER CDL USING PROTOTYPE MODEL

5.1. System Configuration

The configuration of the system is shown in Fig. **(7)**. The system consists of a transceiver, an optical antenna, and a signal processor. In this system, optical components are connected with single mode optical fibers, and this makes the system arrangement flexible. The transceiver consists of a laser source, optical couplers, an Acousto-Optic modulator (AOM), a fiber amplifier, a polarization controller, a balanced receiver, a diode switch, and a polarization control circuit. The optical antenna, which consists of a polarization insensitive optical circulator, and a telescope, is separated from the transceiver, and connected by an optical fiber. A personal computer (PC) with A/D converter, and digital signal processing software is used as a signal processor.

The appearance of the proposed system is shown in Fig. **(8)**. The transceiver has width of 450 mm, height of 416 mm, and depth 400 mm. The optical antenna has width 120 mm , height 120 mm, and depth 300 mm. The optical fibers connecting the transceiver and the optical antenna are 10 m long.

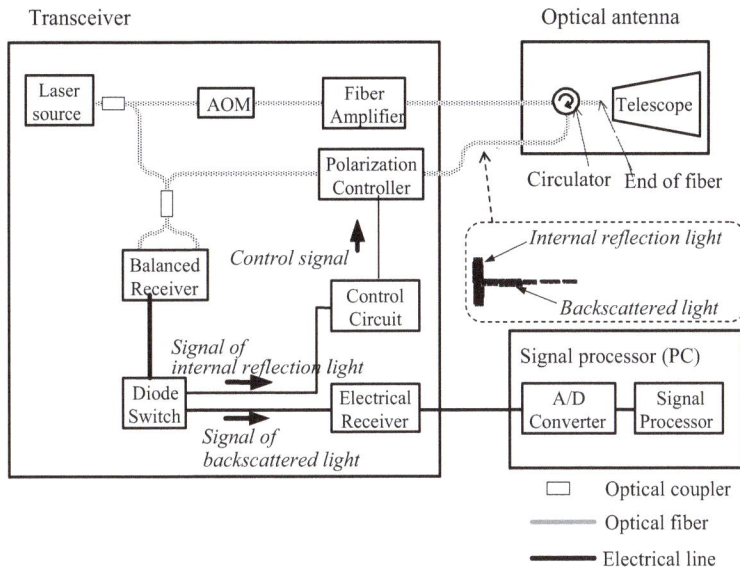

Fig. (7). System configuration.

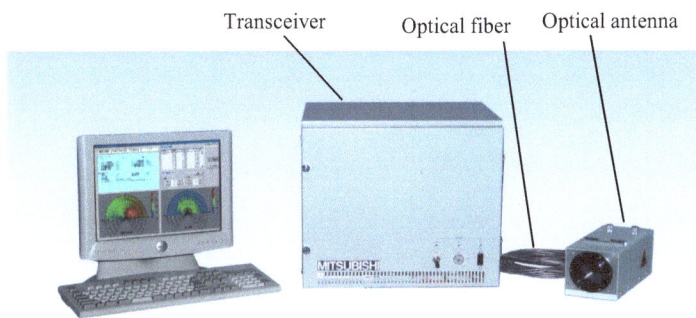

Fig. (8). Appearance of system.

CW light from laser source of wavelength 1.5 μm is divided by the optical coupler. One part is sent to the balanced receiver and is used as the reference light for heterodyne detection. The other is pulsed, frequency-shifted by the AOM, and used as a seed pulse of transmitted light. It is amplified by the fiber amplifier and transmitted to the atmosphere through the optical antenna. The backscattered light from aerosols in the air is received by the same optical antenna and sent to the balanced receiver through the polarization insensitive circulator. The received light is polarization-controlled by the polarization controller, is combined by the reference light using an optical coupler, and is heterodyne-detected by the balanced receiver. In this system, light reflected from the end of the fiber in the optical antenna is used to compensate polarization variations throughout the CDL system. The light reflected from the fiber end is detected by the balanced receiver, and the backscattered light is detected continuously. These two detected signals are temporally divided by the diode-switch. The first signal, including the signal of the internal reflection light, is sent to the polarization control circuit. Since variations of the polarization state mainly occurrs in the process of optical propagation in single mode fibers used in the system, the variation can be compensated by this polarization control. The continuous signal of backscattered light is down-converted and sent to the signal processor.

The fiber-based configuration mentioned above has some additional advantages. First, the pulse width corresponding to a certain range resolution can be varied easily from 0.2 μs to 2 μs corresponding to the gating pulse width of the AOM. Second, there is no difficulty in alignment of the transmitted and received beams since both beams use the same fiber. Furthermore, both beams are almost ideally diffraction limited because the fiber is a mature single mode fiber. The system efficiency of CDL systems is in general very sensitive to misalignment of beams and unideal beam diffraction. Therefore, these are distinct advantages of all-fiber CDL systems, especially in realizing high system efficiency.

The transmitted pulse has a peak power of about 10 W. The shift frequency of the AOM is 100 MHz and this is the IF frequency of the heterodyne-detected signal. The influence of baseband large Relative Intensity Noise (RIN) can be rejected by this frequency shift. The Pulse Repetition Frequency (PRF) is also variable up to 50 kHz. The reflectivity of the fiber end is adjusted to provide heterodyne-detected signal of internal reflection light having a suitable level for polarization control, and is set to -65dB using an anti-reflection coating. The effective aperture diameter, taken asthe $1/e^2$ intensity beam diameter, is 50 mm. The focal distance of the telescope can be varied from 150 m to infinity by changing the end position of the optical fiber. The receiving bandwidth is 100 MHz, corresponding to a measurement velocity range from -38 m/s to 38 m/s. The sampling frequency of the A/D converter is 200 MHz to reduce aliasing effect. In signal processing, the signals are gated by time gates corresponding to the ranges, and a periodogram for each range is calculated by Fast Fourier Transform (FFT). Incoherent accumulation is used to improve detection probability and velocity estimation accuracy. The accumulated periodograms are post-processed and the noise floor is rejected. The Line of Sight (LOS) velocity of each range is estimated from the first moment of post-processed periodogram around the peak frequency. In data acquisition, digital data for all range gates and accumulation numbers is first stored in the on-board memory of the A/D converter, and is then transferred to the PC memory *via* PCI-BUS. Signal processing is done with the PC hardware in series after this transfer. The acquisition time is 20 ms for a PRF of 50 kHz and the usual accumulation number of 1000. The data transfer time to PC memory is below 0.1 s because of the high data rate (more than a few tens of Mbytes/s) of the BUS. The calculation time is about 200 μs for a FFT of 256 points and this is the dominant factor which determines the refresh time. Consequently, in the case of a measurement of 15 ranges with the usual accumulation number, the refresh time becomes about 3 s. More detailed explanations of the key components of the system are denoted below.

5.2. Key Components

5.2.1. Laser Source

The laser source is a Distributed Feed-Back Erbium doped fiber laser (DFB-EDFL). The laser wavelength is 1545 nm and the output power is 20 mW. The linewidth of this DFB-EDFL was measured using the delayed self-heterodyne method [35] with a 20 km long fiber delay line. The spectrum of the delayed heterodyne-detected signal measured by a spectrum analyzer is shown in Fig. (9). The full width half maximum (FWHM) of this spectrum is 66 kHz. Since this includes the influence of the laser linewidth and frequency drift within the delay time, which is given by dL/c, where dL is the delay length (m), c is the speed of light (m/s), and the laser linewidth is less than 33 kHz. In general, the required velocity estimation accuracy is 1 m/s which corresponds to a Doppler frequency of 1.3 MHz, so the laser linewidth is enough for our specification.

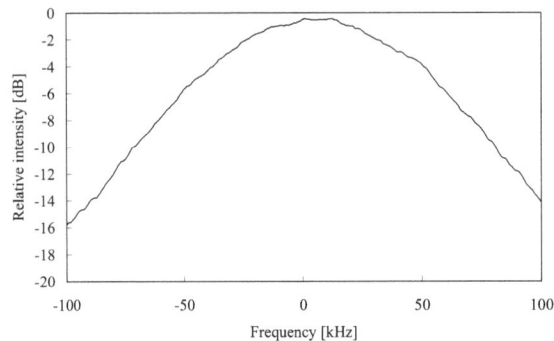

Fig. (9). Spectrum of self-heterodyne-detected signal obtained with DFB-FL.

The relation between local power and efficiency on power penalty, which is caused by the receiver thermal noise and the Relative Intensity Noise (RIN) of the laser source, is derived from the general heterodyne detection theory and expressed by

$$\eta_{PP} = \left(1 + \frac{2kTF_N h\nu}{\eta_q e^2 P_L R_L} + \frac{\eta_q P_L R_{in} R_B}{2h\nu} \right)^{-1}, \qquad (7)$$

where k is the Boltzmann constant (J/K), e is the electronic charge (C), T is the operation temperature (K), F_N is the noise figure, ν is the laser frequency (Hz), η_q is the quantum efficiency of the balanced receiver, P_L is the local power (W), R_L is the trans-impedance gain of the receiver (Ω), R_{in} is the RIN level of the laser source (/Hz) , and R_B is the RIN suppression ratio obtained by the balanced receiver. The specification of the balanced receiver used in this system are η_q=0.8, R_L=70 Ω, F_N=1, and R_B =-25 dB. The measured frequency dependence of RIN level is shown in Fig. **(10)**. The RIN level is about -160 dB/Hz for the frequency range of more than 50 MHz. Assuming these conditions and operation at room temperature T=295 (K), the relation between the local power and the power penalty is calculated using Eq. (7), and the result is shown in Fig. **(11)**. Calculated results obtained by setting RIN level at -150 dB/Hz and -140 dB/Hz are also shown in Fig. **(11)**. The local power of this system is 8.5 mW. This has been set by referring Fig. **(11)** and considering required input power for the fiber amplifier. A high efficiency of -0.4 dB is theoretically obtained by this local power, and it is shown that the level of -160 dB/Hz is enough to make the influence of RIN negligible. Noise spectra of the output from the balanced receiver were measured by turning the local light on and off, and the results are shown in Fig. **(12)**. The measured efficiency is -0.4 dB, and is in good agreement with the theoretical value.

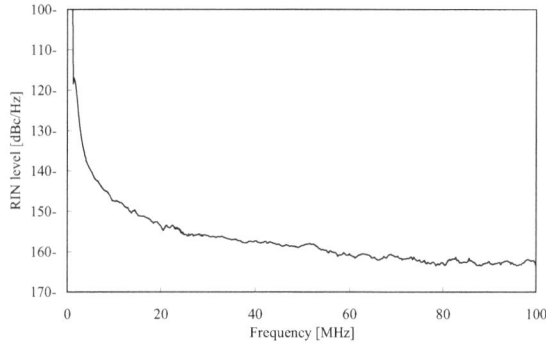

Fig. (10). Measured result of Relative Intensity Noise (RIN) level of DFB-FL.

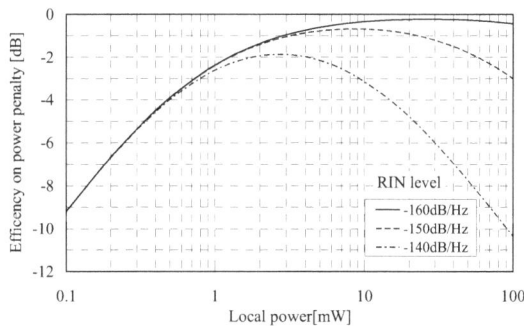

Fig. (11). Relation between local power and efficiency on power penalty in heterodyne detection.

5.2.2. Acousto-Optic Modulator

The AOM used in this system consists of two modulators connected in series to realize a very low extinction ratio. The shift frequency of each modulator is 50 MHz, and the resultant total shift frequency is 100 MHz.

In the AOM, an ultrasonic pulse is generated in a piezoelectric device by a driving signal. The input CW laser light is frequency-shifted and pulsed by the ultrasonic pulse.

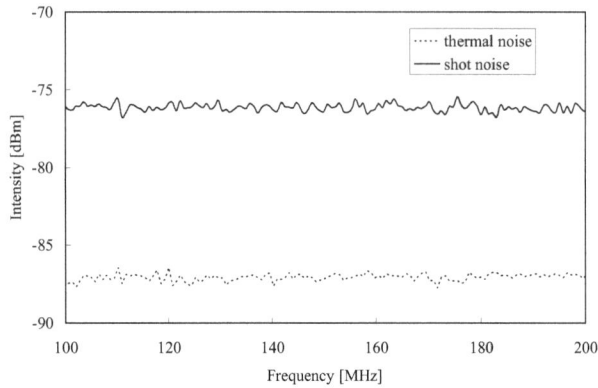

Fig. (12). Measured results of noise from balanced receiver.

If the extinction ratio is zero, only one modulator is needed. However, there are some leak pulses transmitted after the main pulse. They are supposedly caused by multiple reflections of the ultrasonic pulse in the AOM. These signals are amplified by the fiber amplifier, and act as internal reflection light. As a result of digital signal processing, these leak pulses generate inconvenient, artificial peaks in the periodgram at 0 m/s. Assuming that all of the pulse energy is included in one frequency bin of a periodgram by digital signal processing, the intensity ratio of the leak pulse to the shot-noise in the periodgram is given by

$$R_I = \frac{A_A A_F \eta_q P_T}{h\nu B},\qquad\qquad(8)$$

where P_T is the peak power of the transmitted pulse (W), A_A is the extinction ratio of the transmitted light, and A_F is the reflectivity of the fiber end. The extinction ratio should be sufficiently low to make R_I negligible. The relation between R_I and extinction ratio has been obtained using Eq. (8) and is shown in Fig. (13). In the calculation, the peak power is set to 10 W and the receiving bandwidth is 1 MHz corresponding to the case of 1 μs gate time and 150 m range resolution. The reflectivity of the fiber end is -65 dB as described above. It is shown that an extinction ratio of less than -73 dB is needed to make the signal of leak pulse below shot noise level ($R_I < 0$ dB).

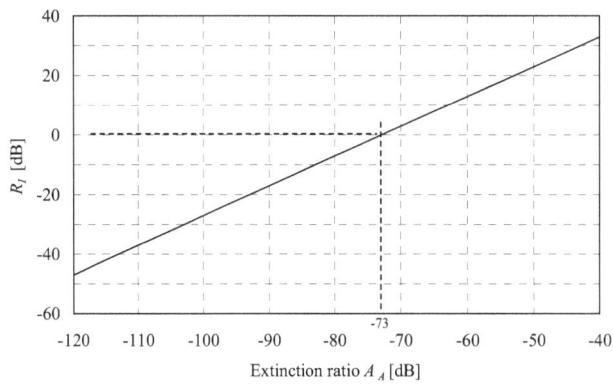

Fig. (13). Intensity ratio of leak pulse to shot noise in a frequency bin which corresponds to 0 m/s.

The extinction ratio for a single AOM was measured and the result is shown in Fig. (14). The horizontal axis shows the delay time after the peak of the main pulse, which corresponds to the time of flight of the leak pulse. It is shown that the ratios are higher by -50 dB to -60 dB than the required ratio described above for all delay times. Thus, the configuration in which two modulators are connected in series can realize an extinction ratio A_A of less than -100 dB and a shot noise level R_I of less than -20 dB.

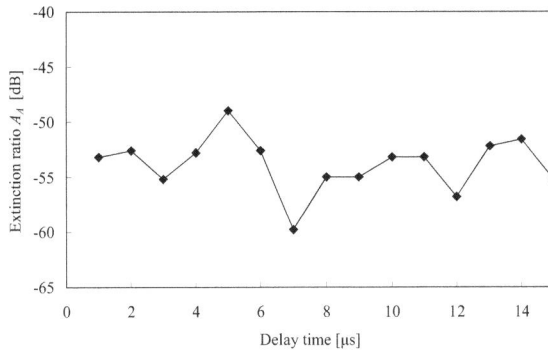

Fig. (14). Extinction ratio of single AOM versus delay time.

5.2.3. Amplifier

An Erbium Doped Fiber Amplifier (EDFA) which has an average output power of 1W is used. A two stage amplifier arrangement (preamplifier and power amplifier) limits the gain of power amplifier to less than 20 dB to prevent parasitic oscillation. A 1 nm bandpass filter is inserted between the two amplifier stages for reducing Amplified Spontaneous Emission (ASE) to maximize the extracted energy from the amplifier, and to minimize the ASE noise coupled to the balanced receiver through the internal reflection from the fiber end.

The peak power of the output pulse is limited to 10 W-level because of Stimulated Brillouin Scattering (SBS). The output pulse width is variable by setting the pulse width of the seed pulse and adjusting the pumping power of the power amplifier. The envelopes of output pulses are shown in Fig. **(15a)**, Fig. **(15b)**, and Fig. **(15c)**, in which the FWHM pulse widths are, (a) 0.2μs, (b) 0.3μs, and (c) 0.7μs, respectively. The relation between (1) pulse width (FWHM), (2) peak power, and (3) pulse energy, corresponding to the pulses shown in Fig. **(15)**, are summarized in Table **4**.

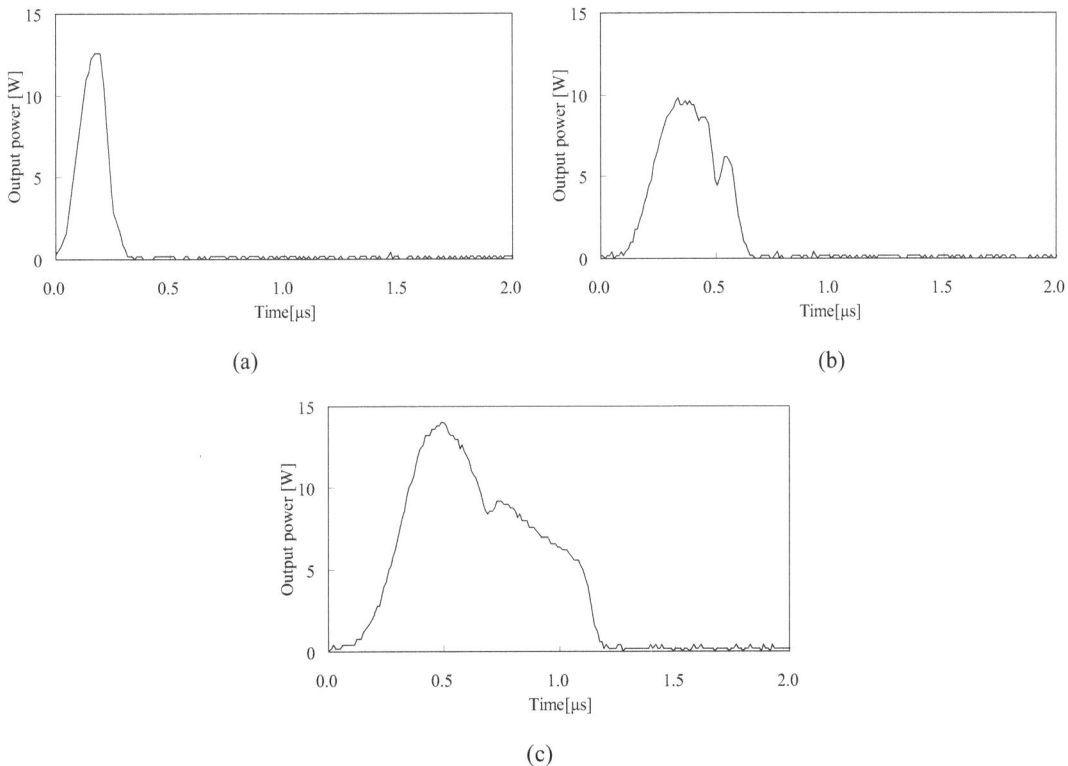

(a)

(b)

(c)

Fig. (15). Envelope of pulsed laser light transmitted from EDFA (the FWHM pulse width is (a) 0.2μs, (b) 0.4μs, and (c) 0.7μs).

Table 4: Specification of transmitting pulsed laser light from EDFA

Pulse envelopes	(a)	(b)	(c)
FWHM [μs]	0.2	0.4	0.7
Peak power [W]	12.5	9.8	14.0
Pulse energy [μJ]	1.8	3.2	8.3

5.2.4. Polarization Controller

The polarization controller controls the polarization state of the received light with piezoelectric fiber squeezers. The main components of the polarization control circuit are shown in Fig. **(16)**. The circuit consists of a peak-hold circuit and a CPU having an A/D and D/A conversion port. An automatic polarization control algorithm is installed on the CPU. The peak value of the heterodyne-detected internal reflection from the fiber end is detected by the peak-hold circuit. The CPU controls the polarization controller according to the control algorithm of the Hill-climbing method, and maximizes the peak value. This closed-loop feedback compensates the variation of the polarization state throughout the system. Since depolarization in the process of atmospheric propagation and reflection is small, heterodyne-detection efficiency of the signal of backscattered light is kept automatically high. The control frequency is about 100 Hz. A result of a wind sensing experiment to confirm the effect of polarization control is shown in Fig. **(17)**.

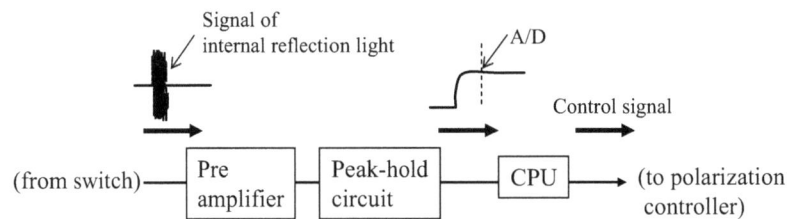

Fig. (16). Configuration of polarization control circuit.

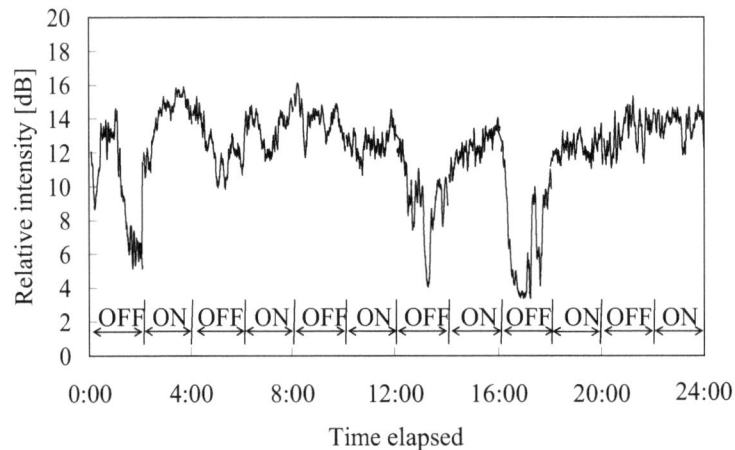

Fig. (17). Effect of automatic polarization control.

The transmitting pulse used in this experiment was the same as in Fig. **(15c)**. The vertical axis shows the detected relative signal intensity at a distance of 225m which is obtained from periodgrams. The polarization control was turned ON and OFF alternatively with intervals of 2 hours. When the polarization control was OFF, the signal intensity sometimes fell (for example, at time elapsed of 2:00, 13:00, 17:00). On the other hand, when the control was ON, the intensity was maintained at high values. From the result of this experiment, the effect of automatic polarization control was confirmed.

5.3. System Performance Evaluation

In this section, system performance is evaluated. The system efficiency is calibrated by a hard-target measurement. Performance in wind sensing is theoretically predicted based on the system specification and compared with experimental results.

The system efficiency has been calibrated by an experiment using a hard-target. A Spectralon® having ideal diffuse characteristic and known reflectivity of $\cong 1$ was used as the target and located at a range of 120 m. The transmitted beam was collimated. The transmitted pulsed laser light used in this experiment was the same as the one shown in Fig. (**15b**). This pulse was selected so that the pulse width became shorter than the target range. The receiving bandwidth was 2 MHz. The SNRs were averaged over 4000 pulses and compared with the calculated one.

Table 5: Summary of system efficiency

Factor	Value [dB]
Insertion loss of the optical components except for the telescope	-1.7
Reflection loss at the telescope (including both transmission and reception)	-0.1
Absorption loss at the telescope (including both transmission and reception)	-0.4
Power penalty	-0.4
Quantum efficiency of the balanced receiver	-1.0
System efficiency at focal range	-4.5
System efficiency (total)	-8.1

Table 6: Parameters in hard-target experiment

Description	Value
Laser frequency	194 THz
Pulse energy	3.2 μJ
Pulse width	0.3 μs
Effective aperture diameter	50 mm
Aperture correction factor	0.77
Focal range	∞
Receiving bandwidth	2 MHz
System efficiency	-10.7 dB
Target reflectivity	1.0
Target range	120 m

In this experiment, the target range was so short that we neglected the influence of C_n^2 and atmospheric transmission K. The measured narrowband SNR was 35.3 dB and in good agreement with the calculated one of 35.8 dB. The estimated system efficiency is -8.1 dB and summarized in Table **5**. Experimental parameters are listed in Table **6**.

The range dependence of SNR in wind sensing has been measured and compared with the calculated result. The transmitted pulsed laser light with a pulse width of 0.7 μs, shown in Fig. (**15c**), was used in this experiment. A total of 15 continuous range gates, each having a length of 1 μs, were obtained between a minimum range of 225 m and a maximum range of 2,325m. The range resolution in this condition was 150 m. A total of 20,000 periodgrams were incoherently accumulated for each range and wideband SNRs were obtained from the periodgrams. Fig. (**18**) shows the range dependence of wideband SNR for the system. The measured results are plotted and the line represents the calculated value based on Eq. (1) and Table **7**. The parameter values are shown only if they are different from Table **6**. The atmospheric backscatter coefficient was roughly predicted using the aerosol number distribution measured by a particle counter. C_n^2

and K were not measured, so in the calculation, these parameters were set to empirically reasonable values. There is good agreement between these two results and the measured range dependence of SNR is reasonable for the system. The dependence on the estimated LOS velocity is also shown in Fig. **(18)**, and periodgrams for range 225 m, 675 m, 975 m, 1575 m are shown in Fig. **(19a)**, Fig. **(19b)**, Fig. **(19c)**, Fig. **(19d)**, respectively. Range-dependent Doppler shifts corresponding to LOS velocities can be confirmed in these periodgrams. For the theoretical prediction of measurable range, the simulation result in [21] can be referred directly.

Table 7: Parameters in wind sensing experiment

Description	Value
Pulse energy	8.3 μJ
Receiving bandwidth	100 MHz
Atmospheric backscatter coefficient	8.3×10^{-7} /m/sr
Atmospheric refractive index structure constant	2.0×10^{-14} m$^{-2/3}$
Atmospheric extinction coefficient	0.95 /km

The required wideband SNR for the detection probability of more than 70% is -24.5dB for the system parameter shown in Table **7**. Here, the measurable range is defined as the maximum range at which wideband SNR exceeds the required value for this detection probability. By applying this required wideband SNR to the result of Fig. **(18)**, the measurable range is about 1,500 m. If our empirical value of $\beta = 3.1 \times 10^{-7}$ /m/sr is assumed, the measurable range is about 1,100 m.

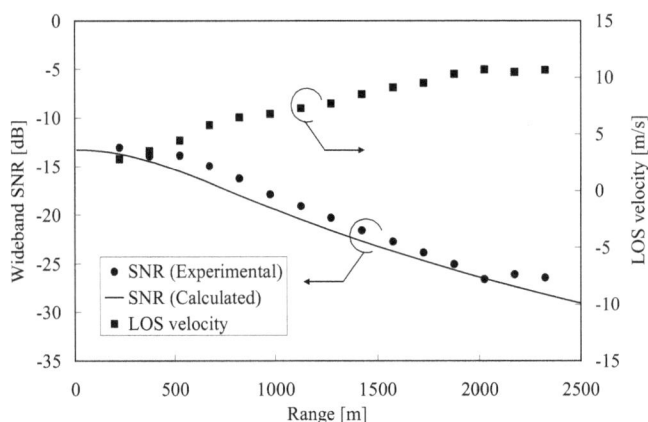

Fig. (18). Range dependence of SNR and LOS velocity in wind sensing (points: experimental, solid-line: theoretical).

The time record of measurable range obtained by an experiment is shown in Fig. **(20a)**. The minimum and maximum measurable ranges are about 375 m and 2,325 m, respectively, and the average measurable range is about 1,500 m. This is a reasonable value from the discussion based on the simulation. In this measurement, the measurable range was known to change as time progressed, and there was a region in which the measurable range dropped steeply. To check the reliability of hardware, aerosol number at the vicinity of the system was simultaneously measured with the particle counter. A time record of wideband SNR obtained at the nearest range (range of 225 m) is shown in Fig. **(20b)**. The vertical axis on the right hand side shows the aerosol number having a diameter of >0.3 μm. It is known that there was a clear correlation between the SNR and the particle count, and drops of SNR were caused by those of aerosol number, not by hardware performance.

The estimated LOS velocities obtained by the same experiment in Fig. **(20)**, are shown in Fig. **(21a)**, Fig. **(21b)**, Fig. **(21c)**, for target range 225 m, 975 m, 1575 m, respectively. Random estimation error occurred with high frequency when the measurable range was shorter than the target ranges.

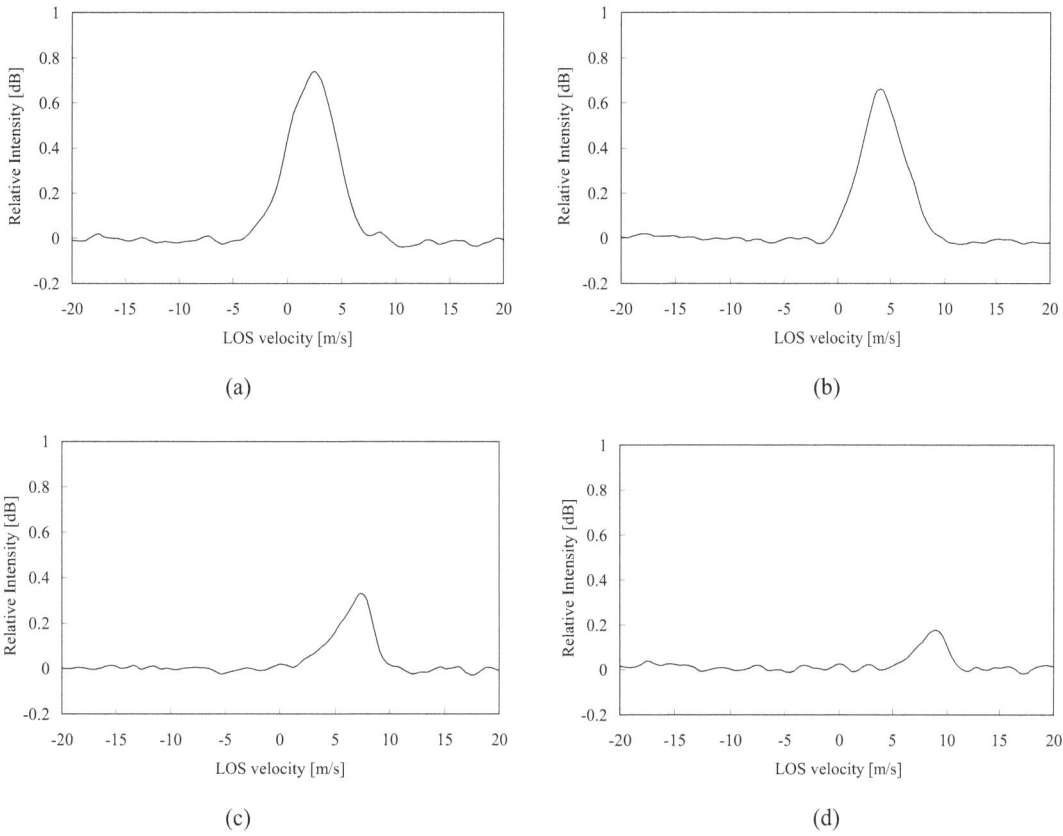

Fig. (19). Periodgrams obtained by the same experiment as in Fig. **(12)** for target range (a) 225m, (b) 675m, (c) 975m, (d) 1575m.

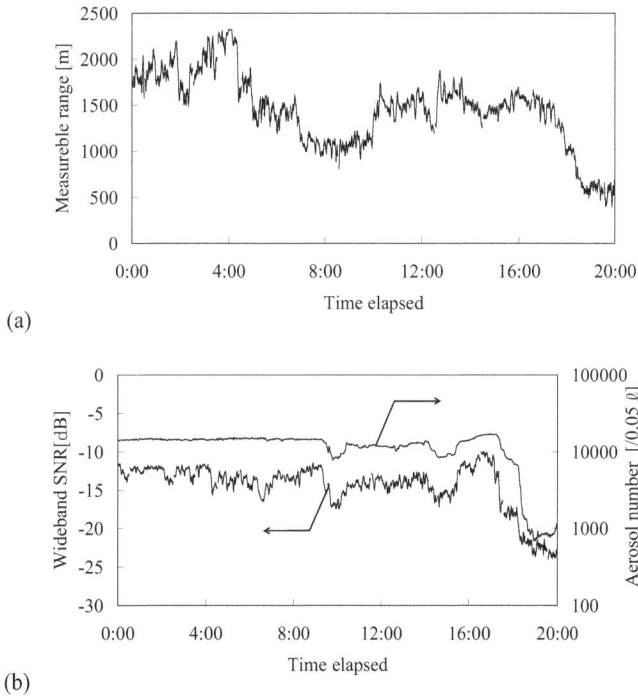

Fig. (20). Results of continuous operation test: (a) Time record of measurement range, (b) correlation of SNR and particle count.

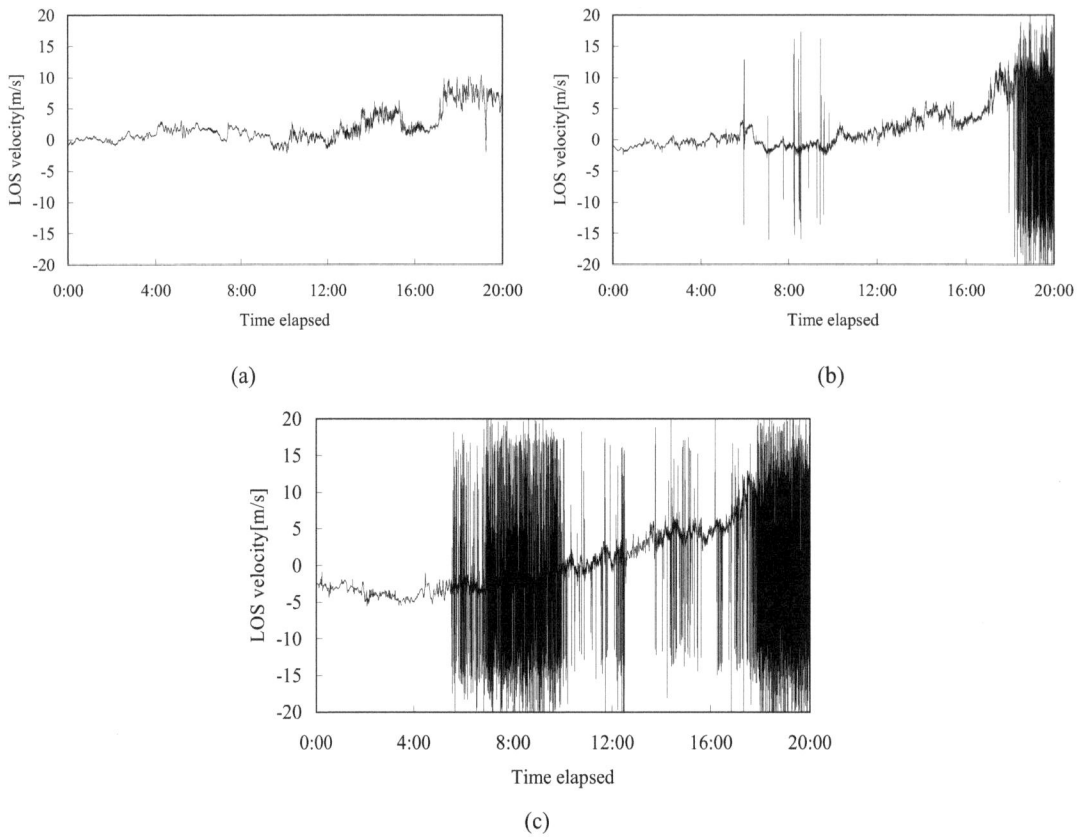

Fig. (21). Time record of estimated LOS velocity obtained by the same test as Fig. **(15)** for target range (a) 225m, (b) 975m, and (c) 1575m.

6. PRODUCT MODEL

6.1. System Configuration

Although the prototype model showed good performance in wind sensing, there had been some remaining issues. First, a beam scanner was needed to obtain the wind field and wind vector. Second, a real-time signal processor was needed to visualize the movement of wind field. Third, components should be reexamined by considering requirements for wind sensing CDL systems. We have developed the product model with solutions to these issues. The external appearance of the product model is shown in Fig. **(22a)**. The system consists of a main container of width 53 cm, height 65 cm, depth 56 cm, and weight 46 kg, an optical antenna of width 15 cm, height 15 cm, depth 30 cm, and weight 7 kg, connected with optical fiber cables, and a tripod of weight 2 kg. The main container contains a fiber-based optical transmitter/ receiver (T/R) unit, a PC-based signal processing unit, and its power supply unit in a 19-inch 10U rack case. Casters allow easy transport and quick setup (~5 min) at observation sites. The total power consumption is less than 400 VA, which allows operation using the power outlet of an automobile along with a commercially available DC-AC converter. By changing the scanning scheme, the system has four measuring modes as follows: (1) LOS wind velocity with a fixed beam direction, (2) a sectored (±20°) plan position indicator (PPI) mode with horizontal linear beam scanning, (3) a sectored range height indicator (RHI) mode with vertical linear beam scanning, and (4) velocity azimuth display (VAD) mode with conical beam scanning (±10°). The user interface is considerably improved by using simultaneously refreshed quick-look displays (LOS velocity, PPI, RHI, Doppler spectra, and wind vector) and touch-panel user interfaces. An example of the screen image (anemometer mode) is shown in Fig. **(22b)**, which indicates the direction of the horizontal wind speed and vertical wind speed in the measuring range. The refresh rate of these screens can have a value up to a few hertz depending on the scanning rate.

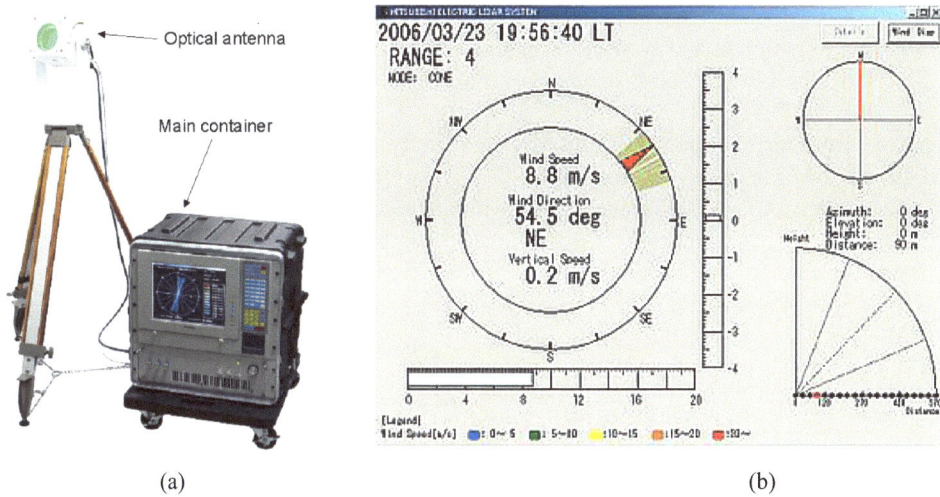

(a)   (b)

Fig. (22). External appearance of the product model of all-fiber pulsed CDL system. (a) Outer view and (b) an example of a quick-look display (anemometer mode).

The block diagram of the product model is shown in Fig. **(23)**. In this model, PRF is 4 kHz (16 kHz optional), and transmitted pulse energies are in the range of 1.8~4.6 μJ. The range resolution can be selected as 30, 75 or 150 m by changing the pulse width. The configuration of the optical T/R unit is basically the same as that of the prototype model except for the following points. In the product model, the polarization controller and the control circuit are removed by using polarization maintained fibers in all optical components. This substantially simplifies the system configuration. The laser source used in the prototype model was a DFB-FL which had a very narrow linewidth and was expensive. This is changed to a low-price DFB-Laser Diode (LD) which has a linewidth of about 1 MHz. Since the heterodyne-detected signal of the backscattered light has a spectral width of the same order, the DFB-LD can be used as the laser source. In general, AOM is one of the expensive components in the all-fiber CDL. Therefore, the number of AOM, which was two in the prototype model, is reduced to one by combining a Faraday Rotated Mirror (FRM).

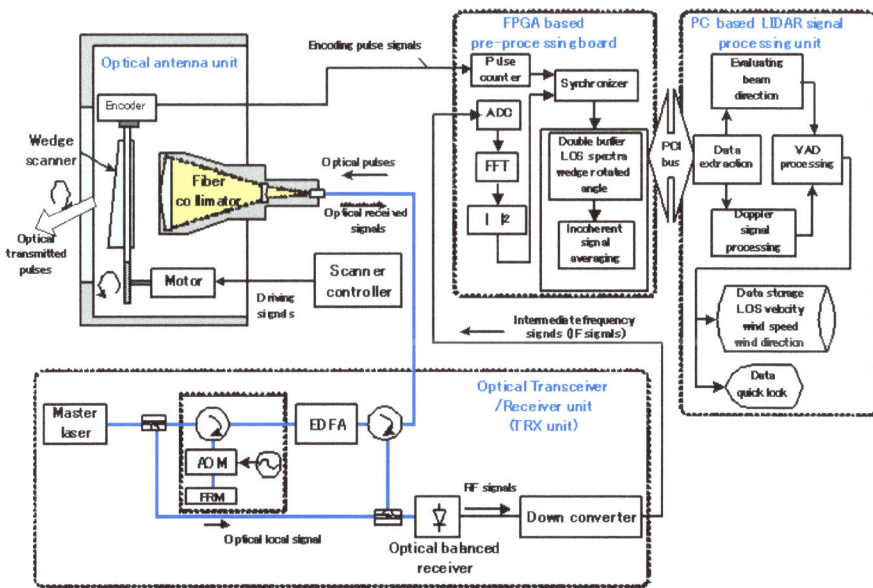

Fig. (23). Block diagram of the standard all-fiber pulsed CDL system.

The optical antenna unit comprises a fiber collimator and a compact scanner, both of which are covered by a waterproof case. The fiber collimator is a refractive-type fiber collimator with an effective aperture diameter of 60 mm, which can focus from 100 m to 1 km without any truncation. The passive-athermal design enables us to maintain a wavefront error less than $\lambda/14$ (Marechal criterion) over the temperature range of $20 \pm 30°C$; the antenna unit is operated outside the laboratory under ordinary conditions.

The compact scanner enabled output beams to be conically scanned by simply rotating a silicon wedge prism and directly monitoring its rotation angle. Angular information on the output beams can be obtained by combining an attainable wedge-rotation angle with a constant wedge-deflection angle. Furthermore, the optional mode has been readily designed for linear scanning by inserting an additional wedge prism that is reversely rotated to a primary prism.

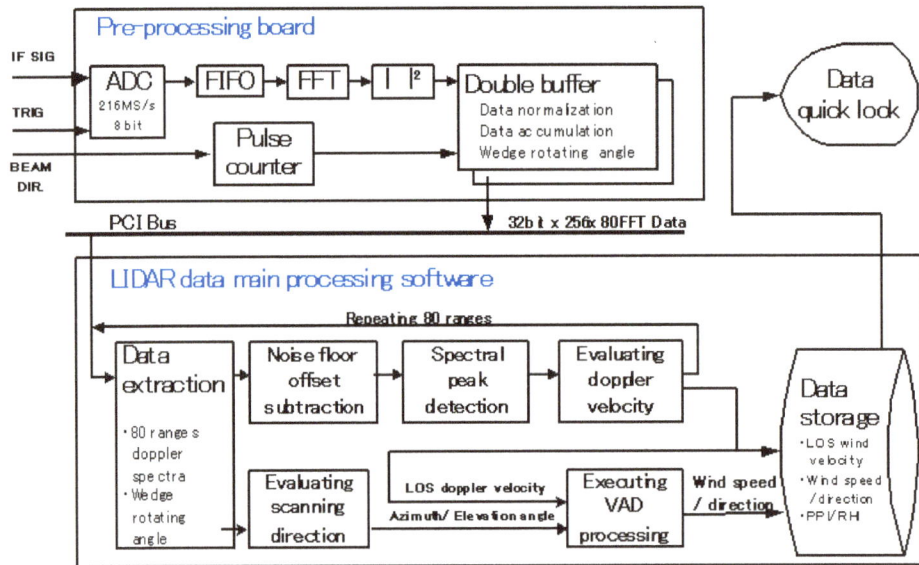

(a)

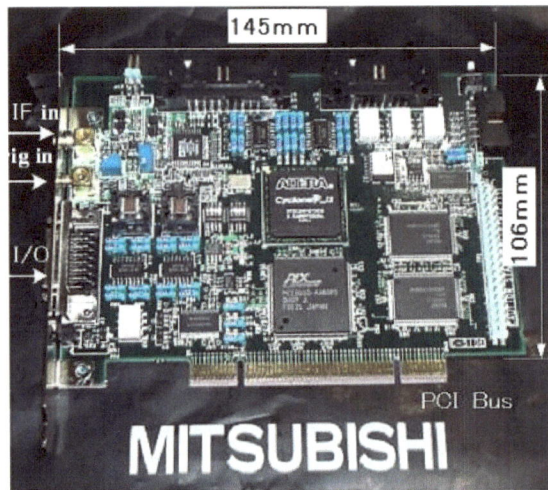

(b)

Fig. (24). Schematic block diagram of a PC-based signal processing unit: (a) schematic block diagram and (b) outer view of FPGA-based pre-processing board.

6.2. Signal Processing

A schematic block diagram of a PC-based signal processing unit is shown in Fig. **(24a)**. The field programmable gate array (FPGA) pre-processing board used in this unit has been newly developed for achieving high-speed Doppler signal processing after A/D acquisition, as shown in Fig. (**24b**). This pre-processing board comprises of a fast Fourier transform (FFT) unit, a data accumulation unit, and coupling with encoder data as beam direction from the scanner unit. In order to accommodate a bandwidth of 100 MHz corresponding to a Doppler velocity of ±38 m/s, input IF signals are sampled at a rate of 216 MS/s in the A/D converter with a resolution of 8 bit in the front-end of the pre-processing board. Then, the FFT of 256-point data, which includes appropriate zero padding, is performed simultaneously for each range as Doppler spectra in a single Pulse Repetition Interval (PRI), and also made incoherent accumulation up to 16,000. This design enables the use of a range gate up to 1.19 µs that corresponds to a range resolution of ~150 m, providing flexible operation for the range resolution by changing only the zero-padding length.

In real-time processing, 80 processing ranges for the Doppler spectra can be achieved at a PRF of 4 kHz. The processing ranges and PRFs have a trade-off relationship. Operation at a PRF of 16 kHz can also be sustained by reducing the number of ranges to 20 in a PRI. This satisfies the different PRF requirements for optical pulse sources.

The cumulative power spectra data for each beam direction are transferred to the double buffer as 32 bit × 256 × 20 data and are extracted from the main processing program in the PC *via* a PCI bus. The main program for spectral peak detection can be used to evaluate the Doppler velocity after subtracting the noise floor offset at every range gate. From this process, we can obtain LOS velocity, velocity variance, and detectability.

6.3. Data Display

The LOS velocity, velocity variance, and detectability (SNR) are displayed on the quick-look displays as shown in Fig. (**25a**). The Doppler spectra can also be monitored with the aid of images on the displays as shown in Fig. (**25b**). These displays enable us to quickly understand whether the observations are correctly performed on site. The spatial distribution of the LOS wind velocities can also displayed under different beam scanning conditions; such a display, enables us to easily determine the wind shear. The screen images for the sectored (±20°) PPI mode with horizontal linear beam scanning and the sectored RHI mode with vertical linear beam scanning are shown in Fig. (**26**). In order to obtain wind vector information, it is necessary to obtain more than three LOS velocities with different beam directions to determine the three-dimensional wind vector; knowledge of this vector enables the estimation of the true wind speed and direction. Provided the wind distribution is spatially uniform, LOS wind velocities vary periodically with the azimuthal angle of the conical-beam scanning. The velocity-azimuth display (VAD) uses the property that the horizontal wind speed, its direction, and the vertical wind speed can be evaluated from amplitude, phase, and bias components of the periodical variation of the LOS wind velocity. The images for the wind speed and direction on a quick-look display are shown in Fig. (**27a**) for the anemometer mode and in Fig. (**27b**) for the height distribution of wind-vector components.

In this example, the height distributions of the horizontal wind speed, wind direction, and vertical wind speed are obtained at a vertical resolution of 30 m for the height range between 30 m and 600 m; the data refresh rate is 1 Hz. The height range can be increased by increasing the resolution or by increasing the number of range segments. It is noted that our VAD-based calculations for wind-vector measurement are generalized to azimuth/elevation angles in order to be applicable to tilted-center-angles to the zenith in conical beam scanning. This leads to flexible deployment on site to avoid beam truncations with fixed obstacles, *e.g.*, trees, buildings, and electric power lines.

The refresh rate of the calculated wind vector is up to a few Hz depending on the scanning rate. The measured results are simultaneously stored in the disk drive as time-series data files in a spreadsheet format and are easily displayed by loading these files.

Additionally, horizontal wind-vector calculations are also performed for the horizontal linear beam scanning. This observation mode enables the horizontal distribution of the wind speed and wind direction near the ground surface to be determined.

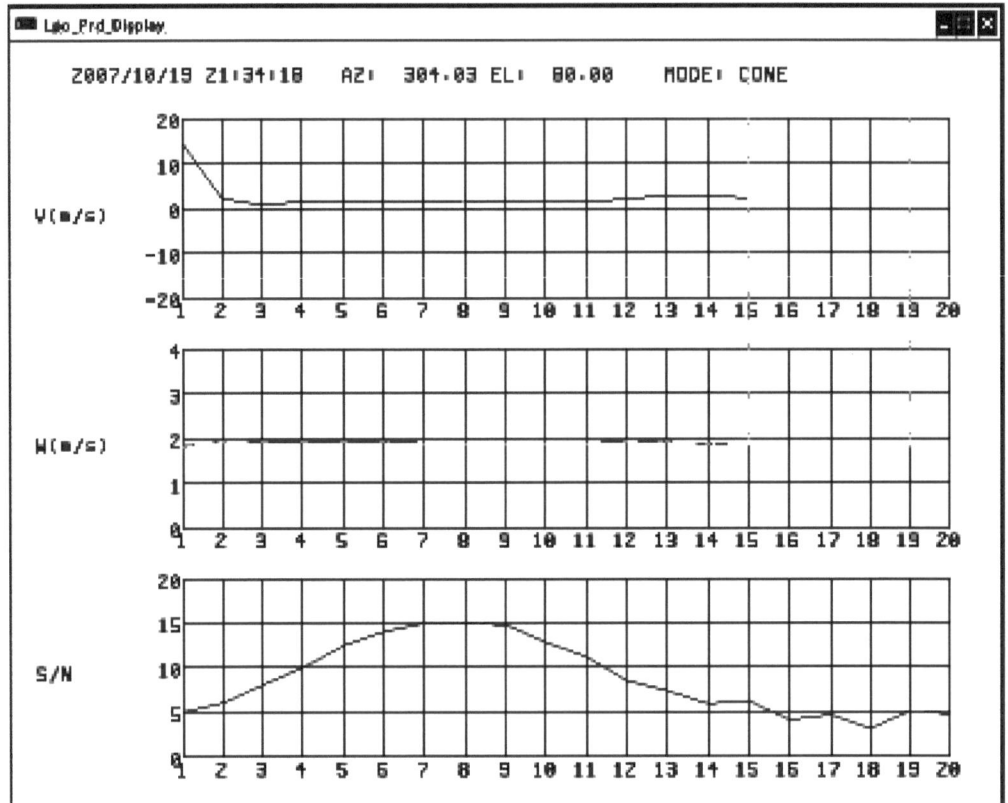

(a)

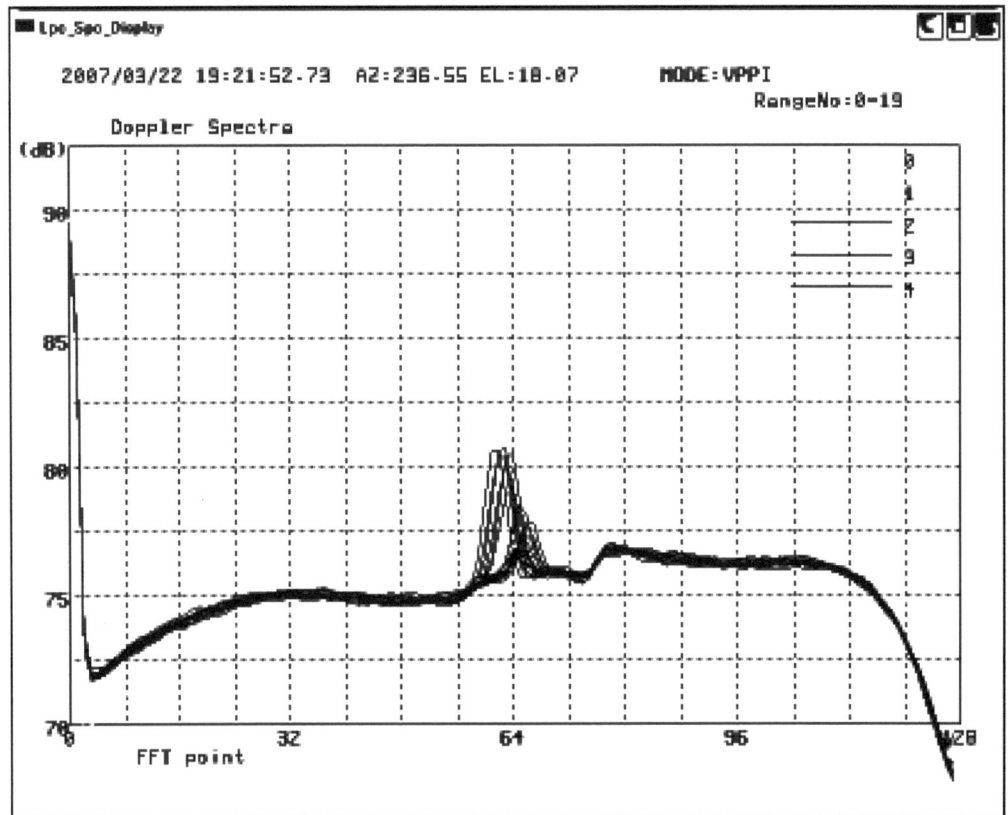

(b)

Fig. (25). Screen images of quick-look display for measured LOS data: (a) LOS wind velocity, velocity variance and detectability and (b) LOS Doppler spectra.

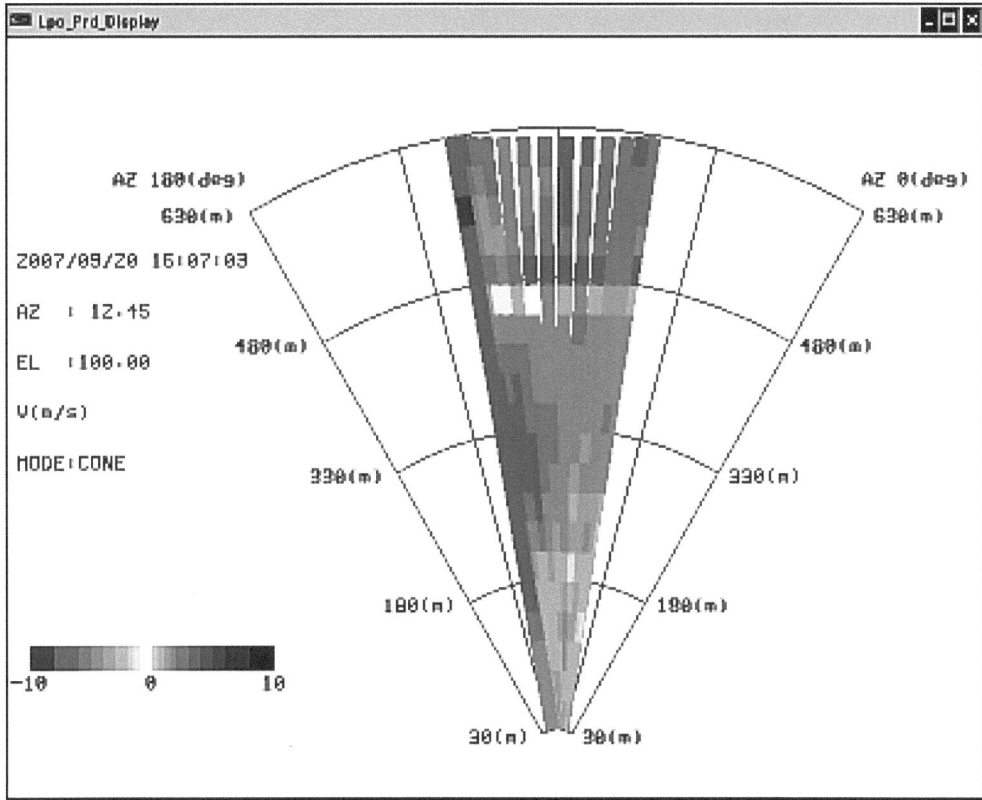

(a)

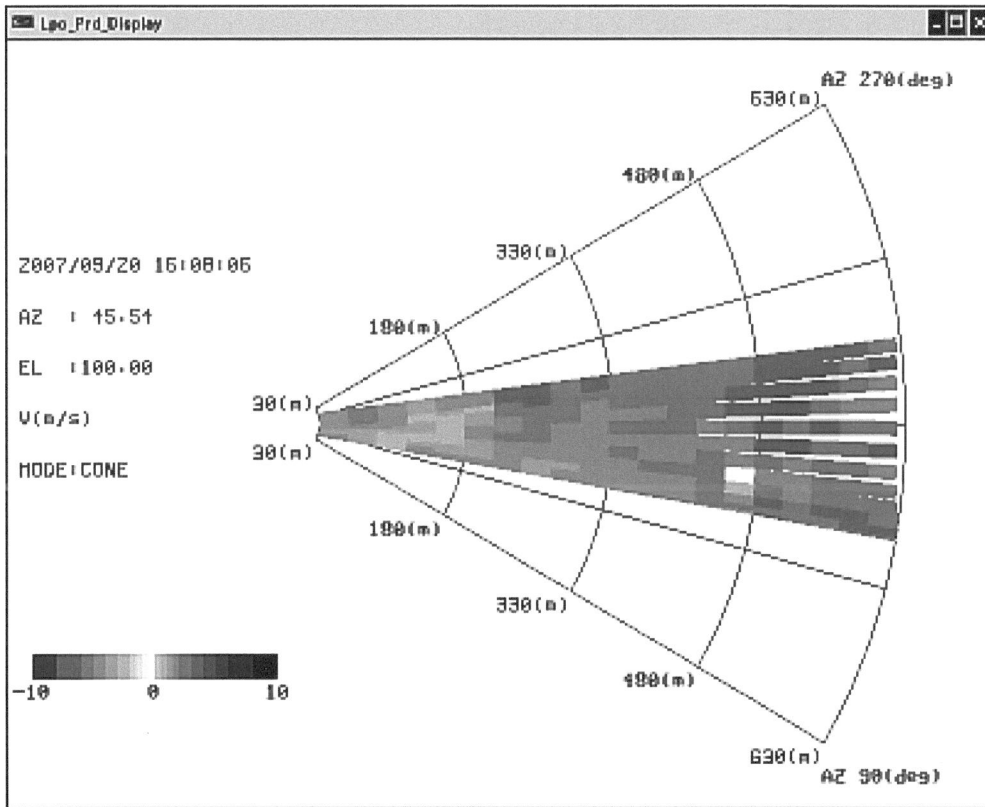

(b)

Fig. (26). Screen images of quick-look display for linear beam scanning: (a) sectored plan position indicator (PPI) mode and (b) sectored range height indicator (RHI) mode.

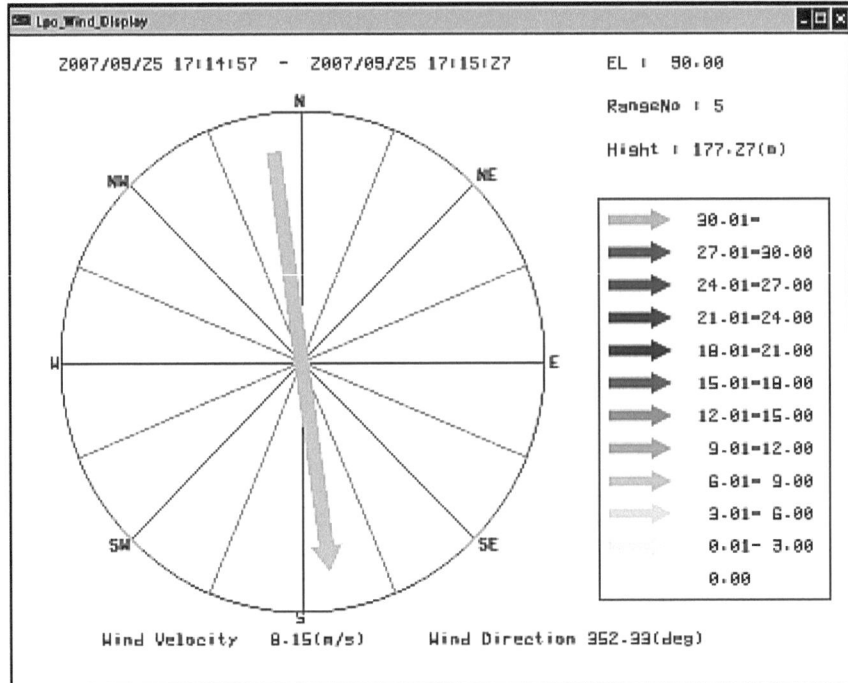

(a)

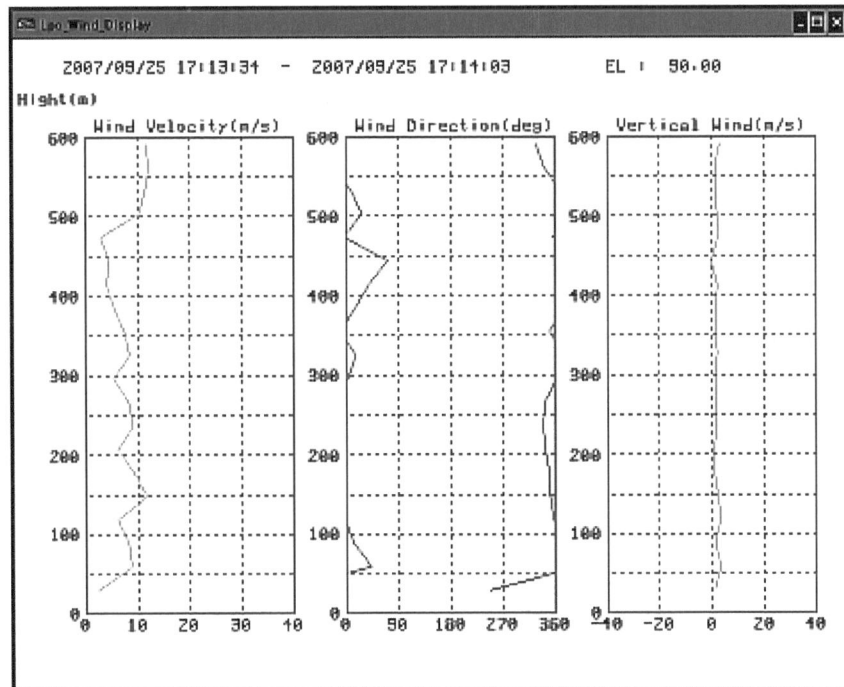

(b)

Fig. (27). Screen images of quick-look display for wind speed, wind direction, and vertical wind speed: (a) anemometry mode and (b) height distribution of wind-vector components.

An example of the screen image for horizontal wind speed and wind direction at an elevation angle of approximately 0° is shown in Fig. **(28)**. In this example, the horizontal wind speed and wind vectors are indicated as colored feathers segmented by 20 ranges in the radial range between 130 and 1650 m and a scanning range of ±20°. This measuring mode may be promising for wind-flow analysis in urban regions because of the measurement of the horizontal wind field at specific areas surrounding many buildings with no ground clutter.

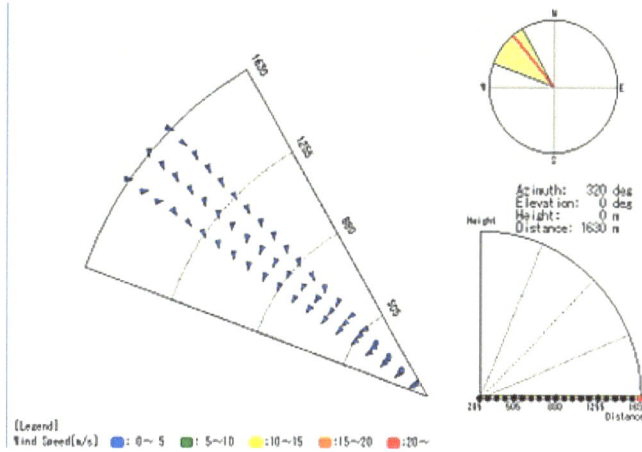

Fig. (28). Screen image of quick-look display for horizontal wind speed and wind direction under horizontal linear beam scanning.

7. WIND SENSING RESULTS

Measurements were performed with the CDL prototype system and an ultrasonic anemometer to evaluate the measuring accuracy of the wind velocity/wind direction for one week (from 17 September 2004) and one month (from 24 November 2004). The measurement was performed at the Tokyo Electric Power Corporation in Fukushima, which is located approximately 0.5 km from the Pacific coast. The measurement site was an open space surrounded by woods, and there was a 103-m-high tower where metrological observations were performed. An ultrasonic anemometer was mounted on the top of the tower; this anemometer provided continuous observations of the horizontal wind speed and wind direction every 6 s and also stored the data after averaging over 6 min. The measuring layout is shown in Fig. **(29)**. The CDL was positioned at approximately 50 m southeast of the tower, and was operated in a conical beam scanning mode with zenith angle of 15° to obtain the horizontal wind speed and wind direction. Twenty segments of LOS wind velocities were simultaneously measured from 50 m to 650 m with a range resolution of 30 m. Every 1 s, LOS wind velocities were obtained after 4000 incoherent integrations of each Doppler spectrum. The scanning rate was 5°/s along the azimuthal angle, and therefore, the wind-vector calculations using the VAD algorithm were performed using 72-direction of LOS data.

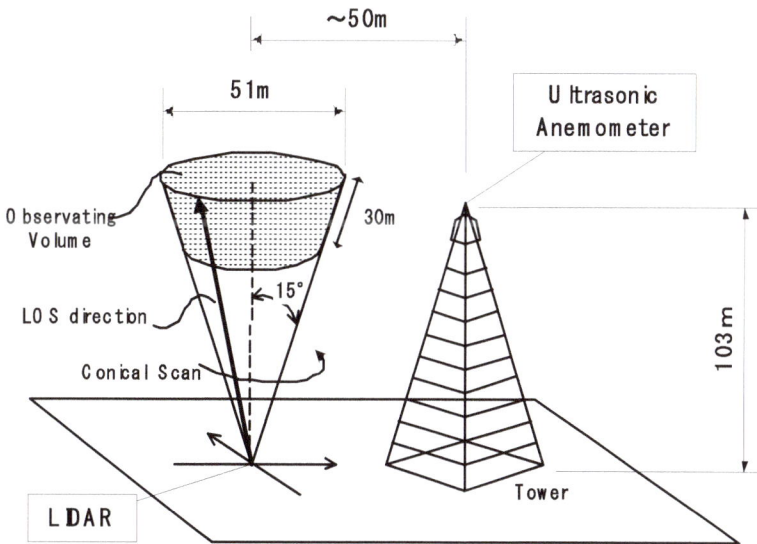

Fig. (29). Experimental layout for simultaneous measurement using the CDL prototype system and ultrasonic anemometer.

In order to compare these data with those from the ultrasonic anemometer, the specific segments at a height of 110 m were extracted from the calculated wind-vector data by the CDL and these data were averaged over 6 min. The results of simultaneous observation of the wind speed and wind direction by the ultrasonic anemometer and the CDL system for one week (10–17 September 2004) is shown in Fig. **(30)**.

The regression plot of the horizontal wind-speed data shows a good correlation with a correlation coefficient of 0.972, as shown in Fig. **(30a)**. A polar plot of horizontal wind direction also indicates good agreement with a residual error of less than 22.5°, as shown in Fig. **(30b)**. Note that there is no ambiguity of 180 degrees in wind direction of CLD data because of the bidirectional (upwind and downwind) measurement of the LOS wind velocity.

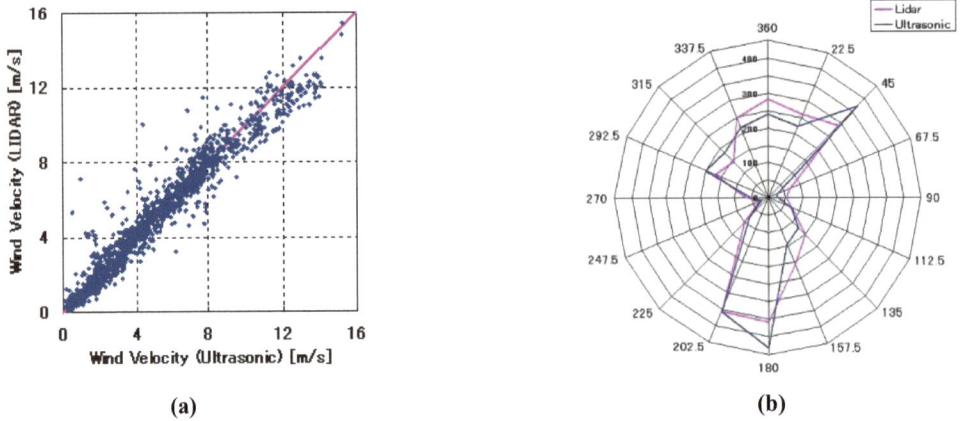

(a)   (b)

Fig. (30). Experimental results using the CDL and the ultrasonic anemometer for one week: (a) Regression plot of horizontal wind speed and (b) Polar plot of horizontal wind direction.

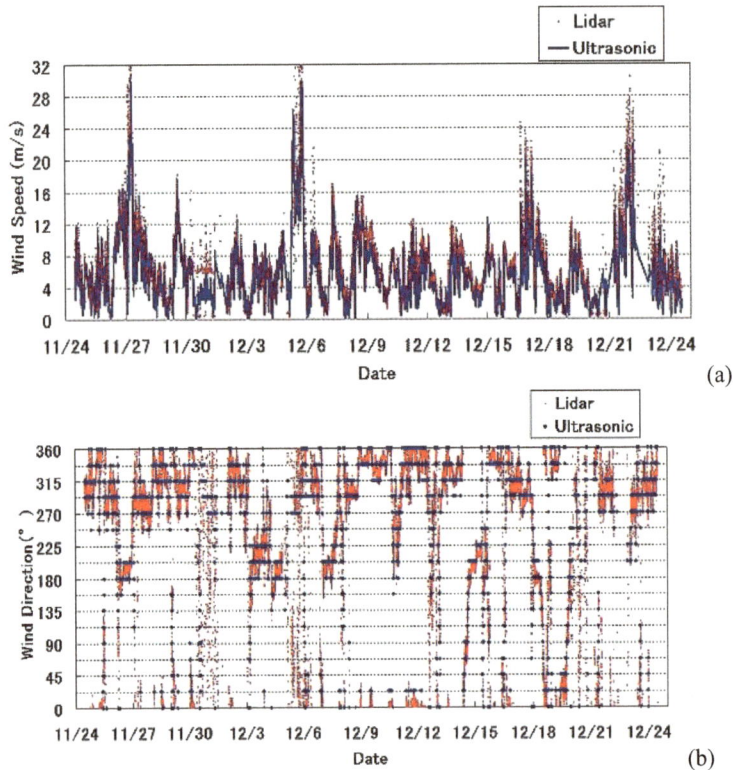

Fig. (31). Comparison between CDL and anemometer data for one month. Time series of (a) horizontal wind speed, (b) horizontal wind direction.

According to the long-term evaluation, the comparison of time series horizontal wind by the CDL and the ultrasonic anemometer data are shown in Fig. **(31)**. The measured data were averaged over 6 min for both the wind speed in Fig. **(31a),** and wind direction in Fig. **(31b),** and the averaged data are in fairly good agreement. The CDL system has continuously been operated without any problem.

The wind sensing results have shown that the fiber laser-based coherent Doppler lidar is capable of field measurement over a period of several weeks. This device is thought to be very useful in surveying wind fields for installation of wind power generation farms.

8. FUTURE PERSPECTIVE

In the future development, there are two directions in performance improvement for all-fiber CDL systems: higher range resolution, and longer range measurement. Since it is easy for all-fiber CDL to transmit shorter pulse of widths of few tens of nanoseconds, there is a possibility to realize a range resolution of less than 10 m with a minor system improvement. In this challenge, a required velocity estimation precision becomes much less than the Doppler frequency resolution. An important technical point is whether this precision can be realized. Concerning longer range measurement, SBS is now the bottleneck in improving the transmitting peak power. To overcome this issue, we have invented the configuration with a post optical amplifier near the optical antenna to transmit laser light to the atmosphere just after amplification to prevent the generation of SBS. By using a rare earth ion highly doped fiber with a short length, we have demonstrated a peak power of 90 W and measurement range of up to 8 km [36]. Another approach for longer range measurement is using a large mode area fiber amplifier [37]. By putting them to practical use, all-fiber CDL systems will be applied to much wider applications.

ACKNOWLEDGEMENTS

We express our great gratitude to the Fukushima second nuclear power plant of the Tokyo Electric Power corporation, which provided us with the measurement site shown in Fig. **(29)** and the ultrasonic anemometer data shown in Fig. **(30)** and Fig. **(31)**.

REFERENCES

[1]   R. Huffaker, R. Hardesty, "Remote sensing of atmospheric wind velocities using solid-state and CO_2 coherent laser systems", *Proceedings of IEEE*, Vol. 84, pp. 181-204 (1996).

[2]   J. Vaughhan, K. Steinvall, C. Werner, P. Flamant, "Coherent laser radar in Europe", *Proceedings of IEEE*, Vol. 84, pp. 205-226 (1996).

[3]   R. Huffaker, A. Jelalian, J. Thompson, "Laser-Doppler system for detection of aircraft trailing vortices", *Proceedings of IEEE*, Vol. 58, pp. 322-326 (1970).

[4]   J. Bilbro, G. Fichtl, D. Fitzjarrald, M. Krause, "Airborne Doppler lidar wind field measurements", *Bulletin of the American Meteorological Society*, Vol. 65, pp. 348-359 (1984).

[5]   J. Bilbro, C. Dimarzio, D. Fitzjarrald, S. Johnson, W. Jones, "Airborne Doppler lidar measurements", *Applied Optics*, Vol. 25, pp. 3952-3960 (1986).

[6]   F. Hall, R. Huffaker, R. Hardesty, M. Jackson, T. Lawrence, M. Post, R. Ritcher, B. Weber, "Wind measurement accuracy of the NOAA pulsed infrared Doppler lidar", *Applied Optics*, Vol. 23: 2503-2506 (1984).

[7]   M. Post, W. Neff, "Doppler lidar measurements of winds in a narrow mountain valley", *Bulletin of the American Meteorological Society*, Vol. 67, pp. 274-281 (1986).

[8]   M. Kavaya, S. Henderson, J. Magee, C. Hale, R. Huffaker, "Remote wind profiling with a solid-state Nd:YAG coherent lidar system", *Optics Letters*, Vol. 14, pp. 776-778 (1989).

[9]   T. Kane, W. Kozlovsky, R. Byer, C, Byvik, "Coherent laser radar at 1.06 µm using Nd:YAG lasers", *Optics Letters*, Vol. 12, pp. 239-241 (1987).

[10]   J. Hawley, R. Targ, S. Henderson, C. Hale, M. Kavaya, D. Moerder, "Coherent launch-site atmospheric wind sounder: theory and experiment", *Applied Optics*, Vol. 32, pp. 4557-4568 (1993).

[11]   S. Henderson, C. Hale, J. Magee, M. Kavaya, A, Huffaker, "Eye-safe coherent laser radar system at 2.1 µm using Tm, Ho:YAG lasers", *Optics Letters*, Vol. 16, pp. 773-775 (1991).

[12]   S. Henderson, P. Suni, C. Hale, S. Hannon, J. Magee, D. Bruns, E. Yuen, "Coherent laser radar at 2 µm using solid-state lasers", *IEEE Transactions on Geoscience and Remote Sensing*, Vol. 31, pp. 4-15 (1993).

[13]   R. Targ, M. Kavaya, R. Huffaker, R. Bowles, "Coherent lidar airborne windshear sensor: performance evaluation", *Applied Optics*, Vol. 30, pp. 2013-2026 (1991).

[14]   R. Targ, B. Steakley, J. Hawley, L. Ames, P. Forney, D. Swanson, R. Stone, R. Otto, V. Zarifis, P. Brockman, R. Calloway, S. Klein, P. Robinson , "Coherent lidar airborne wind sensor II: flight-test results at 2 and 10 μm", *Applied Optics*, Vol. 35, pp. 7117-7127 (1996).

[15]   K. Asaka, T. Yanagisawa, Y. Ooga, K. Hamazu, T. Tajime, Y. Hirano, "Development of 1.5-μm eye-safe coherent Doppler lidar system", *Proceedings of the 11th Coherent Laser Radar Conference*, pp. 147-150 (2001).

[16]   T. Yanagisawa, K. Asaka, K. Hamazu, Y. Hirano , "11mJ, 15Hz single-frequency diode-pumped Q-switched Er, Yb:phosphate glass laser", *Optics Letters*, Vol. 26, pp. 1262-1264 (2001).

[17]   C. Carlsson, F. Olsson, D. Letalick, M. Harris, "All-fiber multifunction continuous-wave coherent laser radar at 1.55 μm for range, speed, vibration, and wind measurements", *Applied Optics*, Vol. 39, pp. 3716-3726 (2000).

[18]   G. Pearson, P. Roberts, J. Eacock, M. Harris, "Analysis of the performance of a coherent pulsed fiber lidar for aerosol backscatter applications", *Applied Optics*, Vol. 41, pp. 6442-6450 (2002).

[19]   S. Kameyama, T. Ando, K. Asaka, Y. Hirano, S. Wadaka, "Compact all-fiber pulsed coherent Doppler lidar system for wind sensing", *Applied Optics*, Vol. 46, pp. 1953-1962 (2007).

[20]   T. Ando, S. Kameyama, Y. Hirano, "All-fiber coherent Doppler LIDAR technologies at Mitsubishi Electric Corporation", *IOP Conference Series: Earth and Environmental Science*, 1: 012011 (2008).

[21]   S. Kameyama, T. Ando, K. Asaka, Y. Hirano, "Semianalytic pulsed coherent laser radar equation for coaxial and apertured systems using nearest Gaussian approximation", *Applied Optics*, Vol. 49, pp. 5169-5174 (2010).

[22]   S. Kameyama, T. Ando, K. Asaka, Y. Hirano, S. Wadaka, "Performance of discrete-fourier-transform-based velocity estimators for a wind-sensing coherent Doppler lidar system in the Kolmogorov turbulence regime", *IEEE Transactions on Geoscience and Remote Sensing*, Vol. 47, pp. 3560-3569 (2009).

[23]   C. Sonnenschein, F. Horrigan, "Signal-to noise relationships for coaxial systems that heterodyne backscatter from the atmosphere", *Applied Optics*, Vol. 10, pp. 1600–1604 (1971).

[24]   R. Frehlich, M. Kavaya, "Coherent laser radar performance for general atmospheric refractive turbulence", *Applied Optics*, Vol. 30, pp. 5325–5352 (1991).

[25]   P. Salamitou, F. Darde, P. Flamant, "A semi-analytic approach for coherent lase radar system efficiency", *Journal of Modern Optics*, Vol. 41, pp. 2101–2113 (1994).

[26]   Y. Zhao, M. Post, R. Hardesty, "Receiving efficiency of monostatic pulsed coherent lidars. 1: Theory", *Applied Optics*, Vol. 29:, pp. 4111–4119 (1990).

[27]   R. Hardesty, "Performance of a spectral peak frequency estimator for Doppler wind velocity measurement", *IEEE Transactions on Geoscience and Remote Sensing*, Vol. 24, pp. 777–783 (1986).

[28]   B. Rye, R. Hardesty, "Discrete spectral peak estimation in incoherent backscatter heterodyne lidar. I: Spectral accumulation and the Cramer-Rao lower bound", *IEEE Transactions on Geoscience and Remote Sensing*, Vol. 31, pp. 16–27 (1993).

[29]   B. Rye, R. Hardesty, "Discrete spectral peak estimation in incoherent backscatter heterodyne lidar. II: Correlogram accumulation", *IEEE Transactions on Geoscience and Remote Sensing*, Vol. 31, pp. 28–35 (1993).

[30]   R. Frehlich, "Crame-Rao bound for Gaussian random process and applications to radar processing of atmospheric signals", *IEEE Transactions on Geoscience and Remote Sensing*, Vol. 31, pp. 1123–1131 (1993).

[31]   R. Frehlich, M. Yadlowsky, "Performance of mean-frequency estimators for Doppler radar and lidar", *Journal of Atmospheric and Oceanic Technology*, Vol. 11, pp. 1217–1230 (1994).

[32]   R. Frehlich, "Simulation of coherent Doppler lidar performance in the weak-signal regime", *Journal of Atmospheric and Oceanic Technology*, Vol. 13, pp. 646–658 (1996).

[33]   R. Frehlich, "Effects of wind turbulence on coherent Doppler lidar performance. *Journal of Atmospheric and Oceanic Technology*, Vol. 14, pp. 54–75 (1997).

[34]   R. Frehlich, "Velocity error for coherent Doppler lidar with pulse accumulation", *Journal of Atmospheric and Oceanic Technology*, Vol. 21, pp. 905–920 (2004).

[35]   T. Okoshi, K. Kikuchi, A. Nakamura, "Novel method for high resolution measurement of laser output spectrum", *Electronics Letters*, Vol. 16, pp. 630-631 (1980).

[36]   T. Ando, S. Kameyama, K. Asaka, Y. Hirano, H. Tanaka, H. Inokuchi, "All fiber coherent Doppler LIDAR for wind sensing", *Proceedings Of Material Research Society Symposium*, 1076-K04-05 (2008).

[37]   A, Dolfi-Bouteyre, G. Canat, M. Vall, B. Augère, C. Besson, D. Goular, L. Lombard, J. Cariou, A. Durecu, D. Fleury, L. Bricteux, S. Brousmiche, S. Lugan, B. Macq, "Pulsed 1.5-μm LIDAR for Axial Aircraft Wake Vortex Detection Based on High-Brightness Large-Core Fiber Amplifier", *IEEE Journals on Selected Topics of Quantum Electronics*, Vol. 15, pp. 441-450 (2009).

CHAPTER 8

3D Laser Radar for Traffic Safety System

Kiyohide Sekimoto[1*], Kouichirou Nagata[1] and Yutaka Hisamitsu[2]

[1]Security Project Department, Infrastructure Operations, IHI Corporation, 3-1-1 Toyosu, Koto-ku, Tokyo 135-8710 Japan and [2]Electrical System Department, Corporate Research & Development, IHI Corporation, 1 Shinnakahara-cho, Isogo-ku, Yokohama-city 235-8501 Japan

Abstract: 3D laser radar is a high-speed figure recognition system that measures the position and speed of objects in real time. It can function despite poor weather conditions as it works by measuring the distance to an object with a laser pulse. By applying 3D laser radar for use in road and rail traffic safety, IHI has realized a system that can measure the position and speed of vehicles and pedestrians. Currently, more than 800 of these systems are in operation in Japan and this number will highly likely increase in future, contributing to improvements in traffic safety.

Keywords: Laser radar, obstacle detection, three-dimensional scanning, traffic safety, level crossing, vehicle, train, pedestrian, gondola, collision avoidance.

1. INTRODUCTION

Three-dimensional (3D) laser radar is a device which measures the position and speed of vehicles or persons in real time. It must maintain stable performance under all weather conditions, including heavy rain, fog, dust, *etc.*, and be resistant to external disturbances such as lightning strikes.

The 3D laser radar is applied to a traffic condition monitoring device for ITS (Intelligent Transport System) and a level crossing obstacle detector. The device was first developed to improve the cargo work efficiency of transporting machines, as the driving support system which recognized the shape of the object. The detection speed and reliability was improved and the device was used to detect pedestrians and vehicles on a road or in a pedestrian crossing. We are first among the world to put 3D laser radar to practical use on railroads and roads [1-3].

2. OUTLINE OF 3D LASER RADAR

2.1. Principle of Measurement

3D laser radar emits a laser light pulse from a laser diode. It detects the reflected light pulse from an object, measures the time of flight of laser pulse, and calculates the distance to the object. Furthermore, it measures the three-dimensional shape by scanning the laser beam in two dimensions.

Fig. **(1)** shows the detection method of pedestrians and vehicles. At first, laser beam pulses are scanned in two dimensions throughout the measurement area and measures a three-dimensional coordinate of each point. The height information with respect to the road surface is extracted, and a neighboring point group is treated as one object and the position and shape are found. By repeating this process, the position and speed of the objects are calculated.

2.2. System Configuration

The 3D laser radar system consists of a laser radar head and controller, as shown in Fig. **(2)**. The laser radar head processes pulses emitted by a laser diode through polygon and swing mirrors to scan the entire monitoring area in horizontal and vertical directions twice every second. The laser light reflected off

---

*Address correspondence to Kiyohide Sekimoto: Security Project Department, Infrastructure Operations, IHI Corporation, 3-1-1 Toyosu, Koto-ku, Tokyo 135-8710 Japan; Tel: +81-3-6204-7235; E-mail: Kiyohide_sekimoto@ihi.co.jp

Tetsuo Fukuchi and Tatsuo Shiina (Eds)

objects is guided detected by high-speed, high-sensitivity photodiodes, and converted to electrical signals. The time interval counter measures the distance to an object based on the time interval between the trigger signal (which indicates the emission of a laser pulse) and the output signal of the phtodiode (which indicates the reception of the laser pulse reflection). The signal processing board controls the motion of polygon and swing mirrors, and measures the 3-D space data in which the origin is the position of the laser radar head, based on the time interval (distance) between the trigger pulse and photodiode signal, the rotation angle of the polygon mirror, and the angle of the swing mirror. These data are sent to the detection data processing section in the controller. In the detection data processing section, an object recognition process is executed, and a recognized object is instantaneously judged to be an obstacle based on crossing alarm conditions and other conditions. If a recognized object is judged to be an obstacle, an alarm is generated.

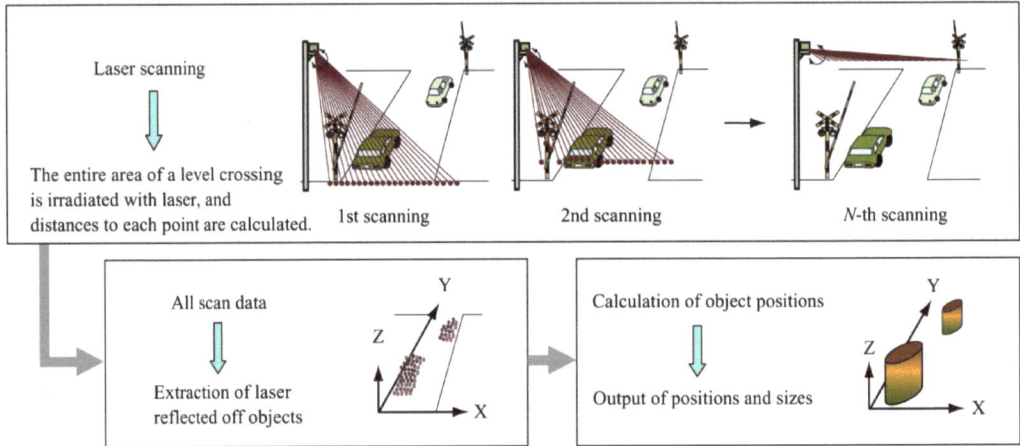

Fig. (1). Detection method of objects.

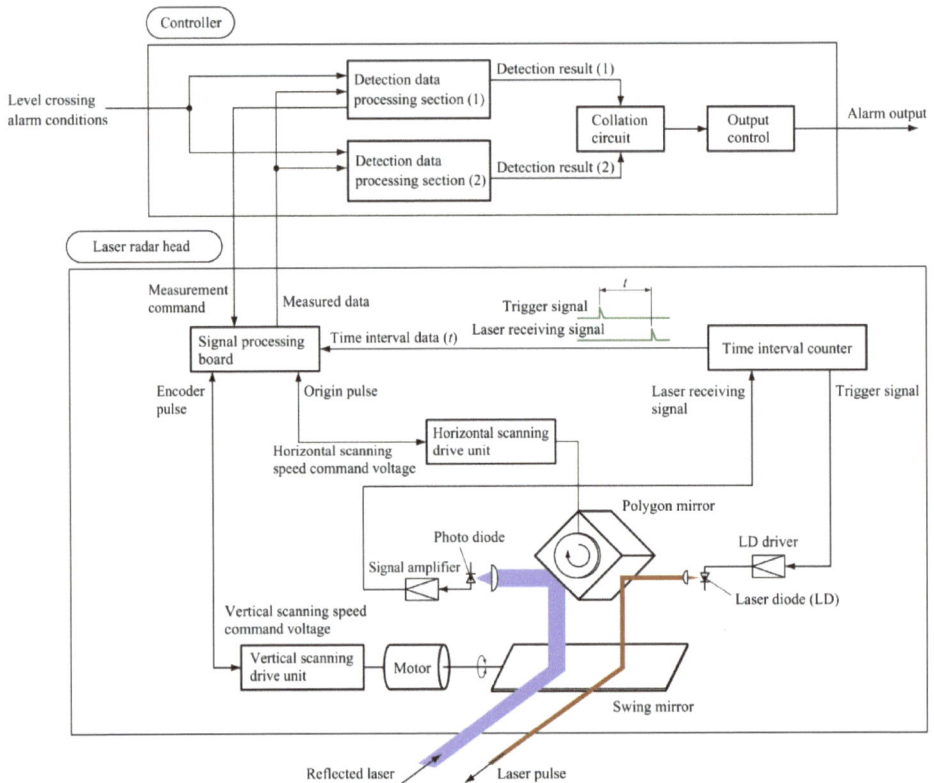

Fig. (2). System configuration.

In this system, the hardware that executes the detection processing tasks is made duplex. That is, detection tasks are performed by two sets of hardware, and the result of a detection task performed by one set of hardware is correlated with that performed by the other set of hardware in order to prevent a detection error or false alarm that may occur in the event of system abnormality or failure. This acts as a failsafe function. Furthermore, this system is provided with a fault diagnosis function to monitor changes in the system conditions, including the occurrence of abnormality or malfunction, blocking of laser, postural change of the laser radar head, *etc.*

2.3. Detection Algorithm

As shown in Fig.**(1)**, the laser radar head scans in two dimensions, converts the measured data (distance, horizontal angle, vertical angle) into three-dimensional plane coordinates, and detects data that has a height with respect to the road surface. Within the detected data, the laser radar recognizes nearby data groups as belonging to a single object.

Since the data is virtually processed in real time, detection processing is performed for each single-axis horizontal scan (data from one scan) as shown in Fig. **(1)**. Fig. **(3)** shows the status of measurements from one scan (side view and plane view). The details of the processing are as follows:

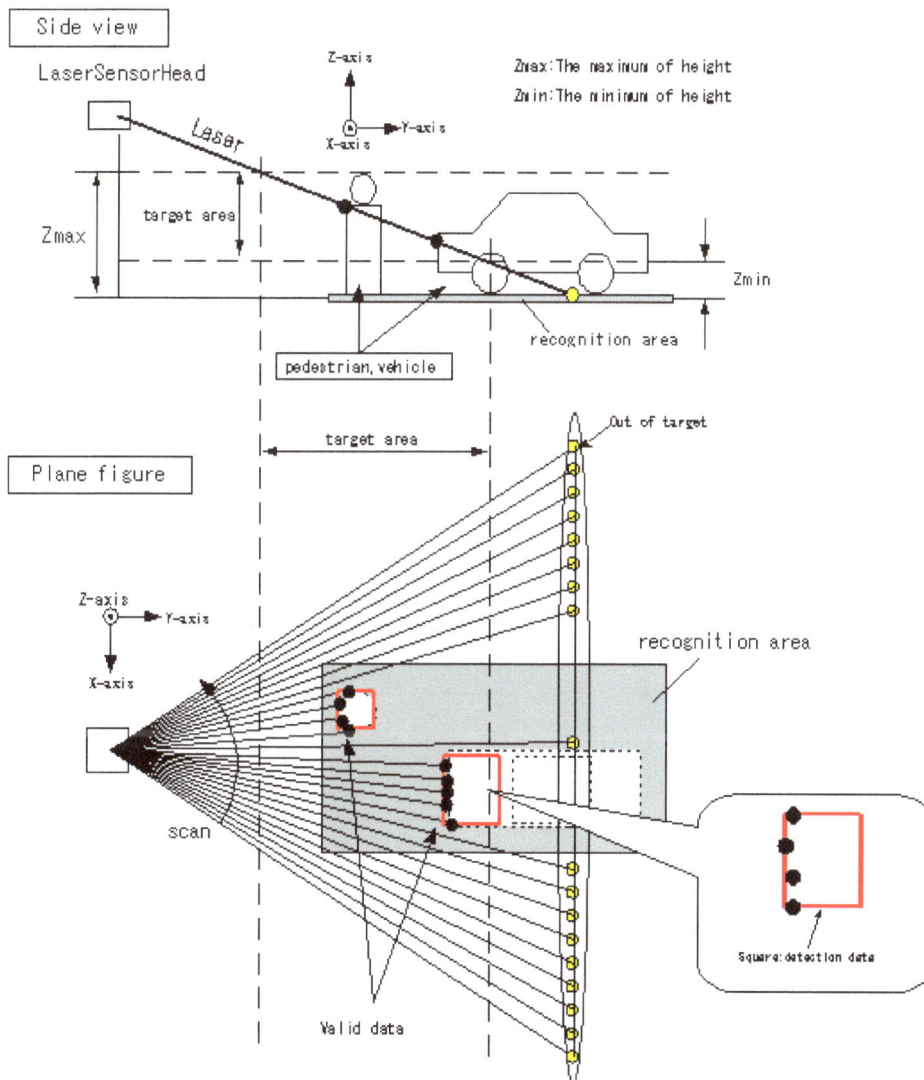

Fig. (3). Detection algorithm.

1. Measurement data from one scan (polar coordinate system: distance, horizontal angle, vertical angle) is converted to XYZ three-dimensional plane coordinates with an arbitrary point serving as the origin (for the tests, the center of the intersection was set as the origin).

2. Measurement data within a certain height range is detectedm, with the height threshold set to Z_{min} to Z_{max}. In Fig. **(3)**, solid circles indicate in-range data; empty circles indicate out of range data.

3. Detected data that is close together is grouped and considered as an object candidate.

4. The detection results are displayed as the smallest possible square that incorporates all of the measured data for each group.

5. The timetable produced from the detection results are recorded as a detection time record.

In this manner, detection processing is repeated from the data obtained from each scan.

In addition, the speed of objects is calculated from the differences between the position and time data from each scan.

2.4. Specifications

Table **1** shows the specifications of 3D laser radar. The detection range is within 90 degrees in the horizontal direction and 60 degrees in the vertical direction and is mechanically determined by the rotation angle of the polygon mirror and the swing angle of the swing mirror. For a level crossing, obstacles of dimension 1 m x 1 m x 1 m or larger, which includes all vehicles, are detected with a time resolution of 0.1 s. The maximum range is about 30 m.

Table 1: Specifications of the 3D laser radar

Item		Specification
Object to be detected		For level crossing: Obstacle whose dimensions are 1 m (width) × 1 m (height) × 1 m (depth) or larger For ITS: Vehicle, pedestrian, bicycle, obstacle, *etc.*
Detection range	Horizontal angle of view	Max 90 degrees
	Vertical angle of view	Max 60 degrees
	Distance	For level crossing: ∼30m For ITS: ∼200m
Detection time		Min 0.1sec
Accuracy	Position detection	±0.1m
	Speed detection	±0.1km/h
Dimensions and weight	Laser radar head	W570×H336×D300mm, 17kg
	Controller	W500×H350×D350mm, 24kg

3. APPLICATION EXAMPLE

3.1. Application for Level Crossing Obstacle Detection System

3.1.1. General Description

Since conventional level crossing obstacle detection systems are designed with the laser beam crossing type, pairs of light receiving and emitting devices must be installed on both sides of a railroad track. If there are turnout switches or station platforms close to a crossing, there are cases where a level crossing obstacle detection system cannot be installed. For a level crossing obstacle detection system designed with the loop coil type, large-scale installation work must be done since detecting coils must be laid underground along the railroad track inside a crossing.

Features of this system are that the laser radar head is installed on a special concrete pillar erected outside a railroad track and that the entire area of a crossing (up to 20 m long and 10 m wide) can be monitored by a single piece of equipment. This makes the layout design and installation work much easier compared with conventional level crossing obstacle detection systems and, therefore, enables considerable space-saving and shortening of the work period possible. Level crossing obstacle detection systems designed with the laser beam crossing type are accompanied by various problems, including tampering with light receiving or emitting devices by passers-by, displacements of optical axes, fouling of light receiving or emitting devices as trains pass during rainfall, and so forth. Since this system is installed at a height of 4 m or higher from the road surface, it is free of these problems. In the case of a level crossing obstacle detection system designed with the loop coil type, maintenance work must be done with considerable toil if detection coils are broken. A great advantage of this system is that no breaking of wires occurs and maintenance work is simplified.

3.1.2. Case of Measurement of Level Crossing Obstacle

3D laser radar detects vehicles and obstacles left on level crossings in any weather conditions including snow and rain, contributing to prevention of accidents at railroad crossings. (Note: The 3D laser radar for level crossing obstacle detection in this paper for railroad operators in Japan limits detection targets to only vehicles so that it securely detects targets along railroad lines even during harsh weather including snow and rain. Therefore, this system has different specifications compared to 3D laser radars for other uses. The device is installed outside the railway tracks, enabling installation and maintenance easy [4, 5]. Fig. **(4)** shows the installation of 3D laser radar and Fig. **(5)** shows a result of measurement of level crossing obstacle. The device is installed outside the railway, enabling installation and maintenance easy.

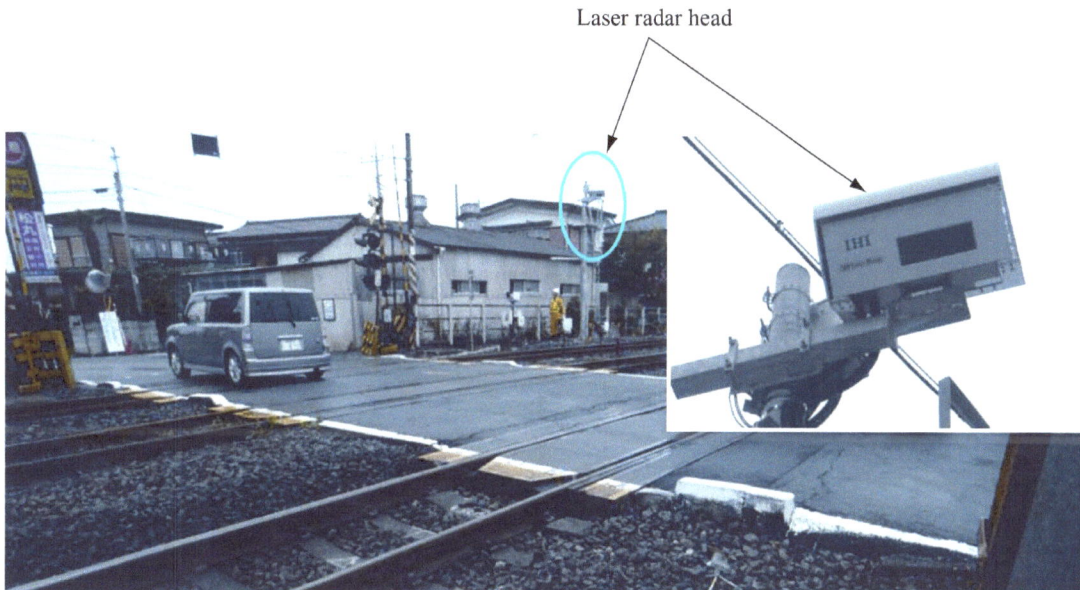

Fig. (4). Installation of 3D laser radar level crossing obstacle detection system.

Furthermore, a maintenance operation terminal (notebook PC) can be connected to this system to input individual detection parameters, such as the shape of a crossing (length, width, road surface height, number of railroad tracks, *etc.*), and the alarm and obstacle detection conditions specified by each railroad company. Therefore, this system can be flexibly used with various different types of crossings.

Condition of level crossing
(Detected results are displayed in
red frames in camera image)

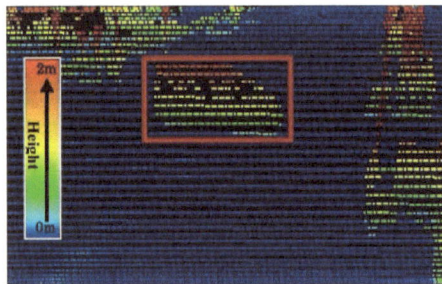

Detected image with 3D laser radar

Fig. (5). Result of measurement of level crossing obstacle. Upper image: obstacle detection result in red frame superimposed on the visual camera image, lower image: height distribution.

3.2. Application to Road Traffic Safety System

Japan's "IT New Reform Strategy (2006 formulation)" aims to make the roads the safest in the world, currently in public-private partnership to develop a safe driving support system through cooperation with road vehicles.

The strategy includes pedestrian collision avoidance system, rear-end collisions avoidance system and during turning collision avoidance system in the safety driving support system. And to measure the situation of invisible range of vehicles and pedestrians from vehicles, the system provides information to alert the driver. IHI is applied to 3D laser radar for these systems in fiscal 2008; ITS-Safety 2010 participated in large-scale field trials.

Fig. **(6)** shows this system installed at a road in the Tokyo area. The 3D laser radar head is mounted on a pole and detects vehicle traffic. Fig. **(7)** shows a measurement example, in which positions of detected vehicles are displayed in red frames and positions of motorcycles and bicycles are displayed in yellow frames. Since the system can estimate the size of the vehicle, it can distinguish large trucks and buses from passenger vehicles.

3.3. Application to Pedestrian Detection and Tracking System

The 3D laser radar can detect and track pedestrians on a crosswalk to count the number of pedestrians and measure their velocity. We are seeking applications for a traffic accident prevention system and a signal control system based on traffic conditions. Fig. **(8)** shows an example of the result of pedestrian detection. In the upper figure, the detected pedestrians are shown in red frames. The lower figure shows the distribution of the height of objects. A group of points exceeding a threshold height is classified as a pedestrian. Note that, even if two pedestrians overlap in the visual image, the system can classify them as two distinct pedestrians.

Fig. (6). Installation of road traffic safety system.

Fig. (7). Measurement example.

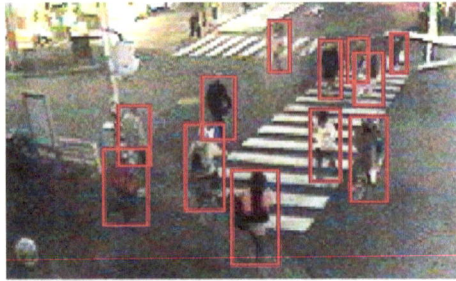

Detection of pedestrians
(Detected results are displayed in
red frames in camera image)

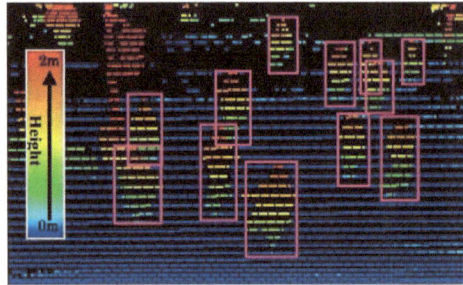

Detected image with 3D laser radar

Fig. (8). Result of pedestrian detection. Upper image: detection results in red frames superimposed on the visual camera image, lower image: distribution of height of objects.

3.4. Application to Flow Detection of People

The 3D laser radar can also detect the flow of people at entrance gates in railroad stations or other places in which congestion might occur. We are developing a congestion relief system for rush hour by recognizing the state of congestion from detected results. Fig. **(9)** shows the results of flow detection of people. The upper figure shows the detected people in red frames. The lower figure shows the distribution of the height of objects.

Detection of pedestrians
(Detected results are displayed in
red frames in camera image)

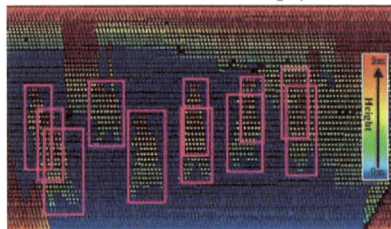

Detected image with 3D laser radar

Fig. (9). Results of flow detection of people.

3.5 Application to Gondola Swing Detection

Gondolas (suspended trams) are used in ski areas and sightseeing spots. The 3D laser radar can be used to detect gondola swing by sensing and tracking the movement of gondolas. Accidents can be prevented by generating an alarm if the swing level is too large. Fig. **(10)** shows the results of gondola swing detection.

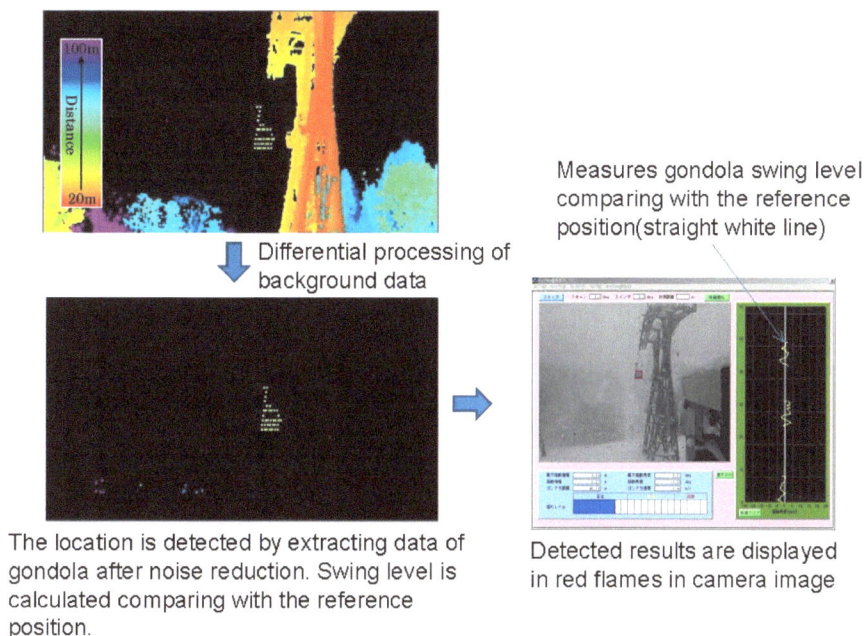

Differential processing of background data

Measures gondola swing level comparing with the reference position(straight white line)

The location is detected by extracting data of gondola after noise reduction. Swing level is calculated comparing with the reference position.

Detected results are displayed in red flames in camera image

Fig. (10). Results of Gondola swing detection.

4. SUMMARY

After the Tohoku earthquake (March 11, 2011) and the Fukushima nuclear accident in Japan, safety and security for unexpected impact has gained stronger attention throughout the world. The need for safety and security is found in every corner of social life (*e.g.* every person has a risk of being hit by a car when crossing a road), in addition to natural disasters. We paid attention to safety of road and railroad traffic, which are everyday means of transportation. For prevention of traffic accidents, we developed a traffic flow recognition system by using the 3D laser radar.

3D laser radar has the ability to realize a safe and secure traffic society. We plan to improve the 3D laser radar technology for a safer future.

As shown in the application to gondola swing detection, the 3D laser radar has potential applications in addition to traffic safety.

DISCLOSURES

Part of information included in this chapter has been previously published in *IHI Engineering Review*, Volume 41, Number 2, pages 51-57, 2008.

REFERENCES

[1]   K. Sekimoto, Y. Hisamitsu, M. Ishii, "Laser-Based Shape Recognition System", *Electronic Engineering*, Vol. 40, pp. 9-12 (1998).
[2]   K. Sekimoto, N. Kamagami, Y. Hisamitsu, K. Ono and K. Nagata, "Development of 3D Laser Radar for Traffic Monitoring", *Ishikawajima-Harima Engineering Review* Vol.43, pp. 114-117 (2003).

[3]   Y. Nakajima, "Development and Implementation of Newest Detection System (Stereo Camera Type Falling Object Detection System and 3D Laser Radar Level Crossing Obstacle Detection System)", *Signal Seminar, Text of 2006* (2006).

[4]   E. Ota, N. Yamaguchi, K. Sekimoto, K. Okajima, "Verification of the Ability for Obstacle detection with a 3D Laser Radar", *Technical Papers of The Institute of Electrical Engineers of Japan*, TER-05-25 (2005).

[5]   Y. Hisamitsu, K. Sekimoto, K. Nagata, M. Uehara, E. Ota, "3-D Laser Radar Level Crossing Obstacle Detection System", *IHI Engineering Review*, Vol. 41, No. 2, pp. 51-57 (2008).

CHAPTER 9

Remote Sensing of Concrete Structures Using Laser Sonic Waves

Yoshinori Shimada* and Oleg Kotyaev

Institute for Laser Technology, 2-6, Yamada-oka, Suita, Osaka 565-0871, Japan

Abstract: A laser-based remote sensing system (LRSS) for detecting defects in concrete has been developed. The diffraction efficiency of a photorefractive crystal (PRC) in the LRSS was increased by an applied electric field. A stabilization system to stabilize the interference pattern in the PRC using the running hologram technique was constructed. Defects in concrete can be located using initiation and detection of impact echo and standing Lamb waves (or natural mode of vibration). The prototype of the LRSS was assembled and set on a small truck. Field experiments were carried out to investigate real concrete defects of a bridge in bullet-train line in Japan. The LRSS scan concrete surfaces to produce a two-dimensional map of real concrete defects. The observed predominant frequency of concrete vibration was consistent with data from impact hammering method.

Keywords: Concrete, photorefractive effect, laser interferometer, homodyne detection, hologram, vibration detection, Lamb wave, ultrasound, impact echo detection, defect detection.

1. INTRODUCTION

Nondestructive locating of potentially dangerous defects of civil structures is a key technique for a safe society. This chapter describes laser-based sensing as a promising technique for inspection of concrete structures, such as bridges and tunnels. At the present time, hammering is still the most frequently used technique for concrete inspection. However, quantitative hammering inspection is difficult due to human subjectivity of inspecting personnel. Moreover, the inspection procedure is not only labor intensive but also requires numerous staff and a long time for periodic inspection of long tunnels and bridges. Other inspection methods use ultrasound [1], active infrared imaging [2], and electromagnetic waves [3]. Table 1 shows a summary of general inspection methods for tunnel walls. All these methods have their advantages and disadvantages. The main advantages of laser-based sensing using laser sonic waves [4] are realization of remote inspection, insensibility to surface roughness, and high operation rate. Therefore, laser-based sensing is suitable technique as an alternative inspection method.

The operating principles of the LRSS are described in Sections 2 and 3. Some results of locating defects both in laboratory samples and a real bridge in bullet-train line in Japan (Shinkansen train) are described in Sections 4 and 5.

2. LASER PHOTOREFRACTIVE INTERFEROMETER

2.1. Principle of Vibration Detection

In the laser-based remote sensing approach, a pulsed laser is used to generate elastic vibration in the inspected object and a CW laser is used for detection of initiated vibration [4, 5]. The vibration mode depends on the inner structure of the inspected object. The presence of a defect affects the vibration parameters. The criterion of defect location is the difference in the vibration mode in areas with and without defects.

To increase the sensitivity of vibration detection, the most commonly used detection device is a laser interferometer, where interference between a reference beam and radiation scattered by a vibrating concrete surface is analyzed. However, traditional interferometers cannot operate effectively with the signal which occurs after scattering of laser radiation by a concrete surface. Significant wavefront mismatching takes

---

*Address correspondence to Yoshinori Shimada: Institute for Laser Technology, 2-6, Yamada-oka, Suita, Osaka 565-0871, Japan; Tel: 081-6-6879-8737; E-mail: shimada@ile.osaka-u.ac.jp

Tetsuo Fukuchi and Tatsuo Shiina (Eds)

place between the scattered signal, which is characterized by speckle structure (very complex wave front) and the high quality reference beam (flat wave front). This phenomenon affects the interferometer performance, decreasing detection efficiency.

Table 1: Comparison of non-destructive inspection methods.

	Laser system	Ultrasound system	Infrared system	EM-wave system	Hammering
Advantage	· Non-contact, Remote sensing · High speed scanning · Wide band frequency response · Curved surface scanning · Detail scanning	· Small system · Detail scanning	· Possible to recognize defect size	· Possible to investigate deeper defect · Recognize defect size	· Easy for defect inspection
Disadvantage	· Influence of outside vibration · Large system	· Need of surface contact · Difficulty of curved face inspection	· Long time for inspection due to heating	· Need contact inspection	· Non-recording system for defect location · Long time for inspection

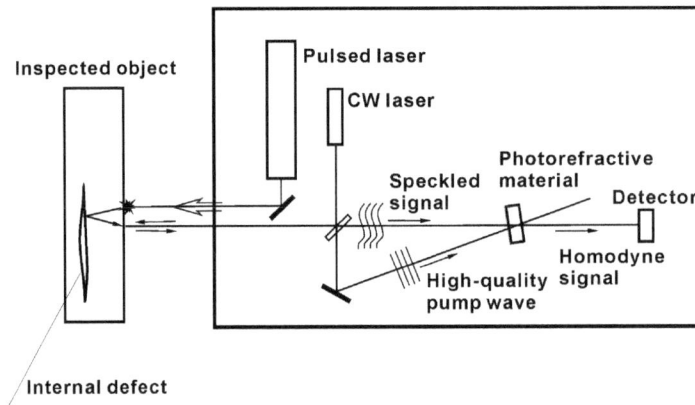

Fig. (1). Schematic diagram of the detection technique.

The alternative technique which is free of this problem is based on homodyne detection *via* the two-wave mixing process in a photorefractive material [6, 7].

A Schematic diagram of the detection technique is shown in Fig. **(1)**. The principle of detection is as follows. The CW laser beam is split into two beams: a high quality pump beam, and a probe beam which illuminates concrete surface. The same system collects radiation scattered by the concrete surface. This scattered signal and the pump beam intersect each other in the photorefractive crystal (PRC). The crystal is used as a nonlinear medium for recording a dynamic hologram. In the presented system, $Bi_{12}SiO_{20}$ (BSO) crystal is chosen. The dynamic hologram is recorded *via* interference between the pump and signal beams. Because of the roughness of the surface, the signal scattered by the concrete surface has a speckle structure (very complex wavefront). However, the pump beam diffracted by the recorded dynamic hologram has exactly the same wavefront and propagates exactly in the same direction as the signal beam. This phenomenon is the main benefit of using the nonlinear interferometer with dynamic hologram, since there is no wavefront mismatch between the signal and diffracted pump beams. As a result, the interferometer is capable of effectively processing speckled signals.

Because of the surface displacement of vibration, the phase of the signal scattered by vibrating surface is modulated at the frequency of vibration. The resulting interference pattern inside the PRC oscillates at the frequency of phase modulation, but the dynamic hologram recorded in the crystal does not oscillate, because the response time of the photorefractive nonlinearity is much longer than period of phase modulation. The recorded dynamic hologram adapts itself to low frequency variations of the signal beam while it is transparent to the high frequency phase modulation to be detected.

Pump (or reference) wave
$E_{ref} = A_{ref} \sin(t)$

Phase modulated wave
$E_{pm} = A_{pm} \sin(t + \sin(0.1t) + \pi/2)$
$T_{pm} = T_v,$ here $T_{pm} = 10\ T_{osc}$

Interference between pump and
phase modulated waves
$E_{Interf} = E_{ref} + E_{pm}$

Detected intensity
$I_{HS} = (E_{Interf})^2$
$T_I = 10\ T_{osc} = T_{pm} = T_v$

time

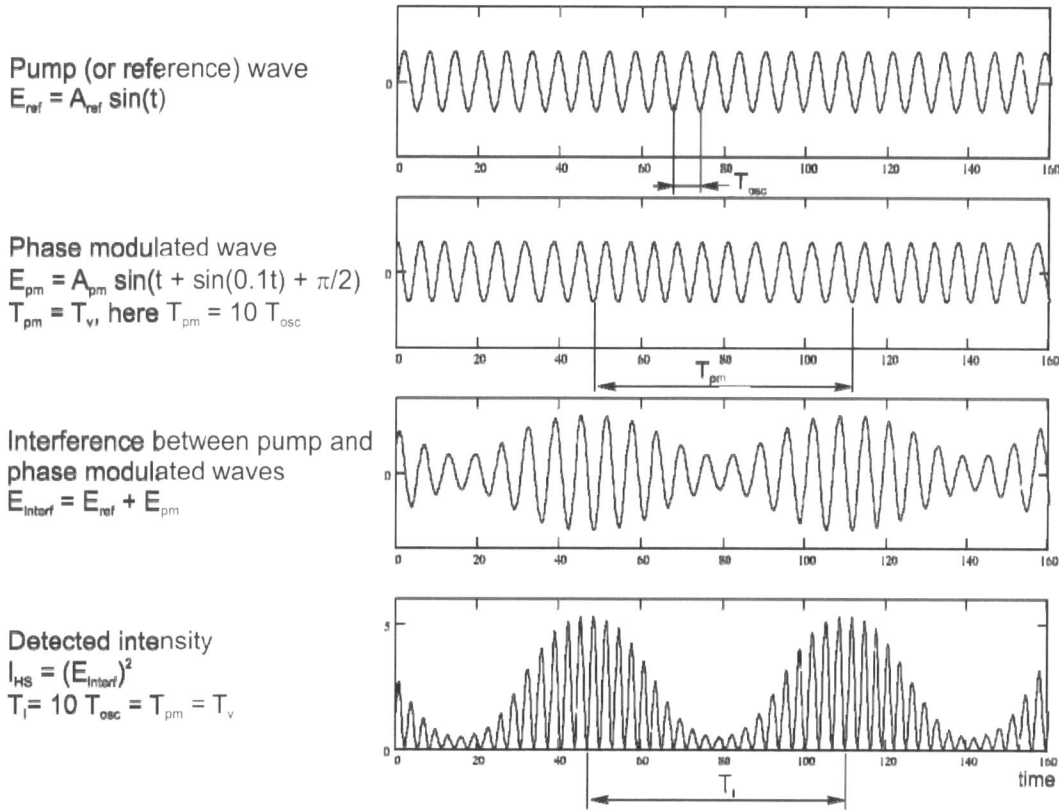

Fig. (2). Simulation of homodyne detection.

When the pump beam encounters the hologram, a diffracted beam appears in the direction of the transmitted signal beam. The diffracted beam is not phase modulated like the transmitted signal from concrete, because the hologram is not oscillating. Interference between these two beams results in the conversion of phase modulation into an amplitude modulation, which is the basis of homodyne detection.

Fig. **(2)** shows a simulation of interference between the phase-modulated signal passed through the hologram and the monochromatic pump wave diffracted by the hologram. The signal phase is modulated at the frequency of vibration initiated in the inspected object. In this simulation, the period of vibration TV and the corresponding period of phase modulation TPM is 10 times larger than the period of carrier oscillation TOSC. Interference between the signals produces an amplitude-modulated homodyne signal IHS. The period of the amplitude modulation is equal to the period of phase modulation and corresponds to the period of vibration of the inspected object. The resulting signal is detected by a photodetector.

The main vibration parameters to be analyzed are the waveform and its spectrum. Analysis of these parameters provides information regarding the inner structure of the inspected object.

Various types of vibration can be initiated by laser impact: ultrasonic (elastic) wave (P and S waves), surface waves, and natural vibration (or standing Lamb waves). Ultrasonic waves can be used in the well-known impact echo technique. Surface waves are useful for investigation of surface cracks, and standing Lamb waves can be used for express analysis providing real-time information on the presence or absence of inner defects. The current design of the LRSS can be used for initiation and detection of any types of vibration.

2.2. Photorefractive Effect and Two-Wave Mixing

The main component of the homodyne detection system is the photorefractive crystal (PRC) with recorded dynamic hologram. Fig. **(3)** illustrates the spatial variation of intensity inside the PRC, space-charge density, space charge field, and the induced refraction index [8].

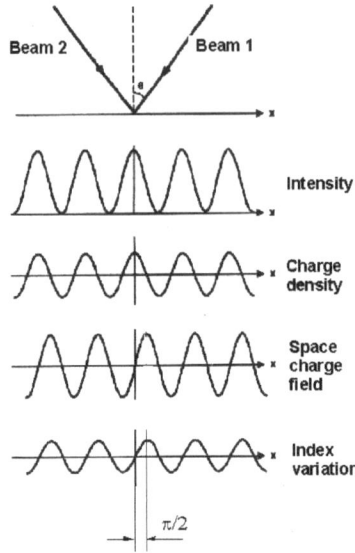

Fig. (3). Creation of dynamic hologram in a photorefractive medium.

Processes of recording and reading-out the dynamic hologram are as follows. Two coherent laser beams, beam 1 and beam 2, intersect each other inside PRC. Due to interference between the two beams, an array of dark and bright regions appears. In the bright regions, photoionized charges are generated by the absorption of photons. These charge carriers can diffuse away from the bright regions leaving behind positively charged ionized donors. If these charge carriers are trapped in the dark regions, they will remain there because there is no light to re-excite them. This leads to a charge separation, in which dark regions are charged negatively and bright regions are charged positively. The build up of space charge separation continues until the diffusion current is counterbalanced by the drift current.

Space charge variation induces a space charge field inside the PRC. The space charge field in turn induces a variation of the refraction index *via* the electro-optic effect. This index variation can be considered as a dynamic grating or hologram.

It should be noted that the space charge field and refraction index variation are shifted in space by $\pi/2$ relative to the interference pattern. The direction of the grating shift is determined by the sign of the electro-optic coefficient and crystal orientation. This shift is very important in the energy exchange between the interacting beams. As it is known, after diffraction by a phase grating, a phase shift of $-\pi/2$ appears in the diffracted beam. That means that diffraction by the photorefractive grating will result in total phase shift $\Delta\phi$ equal to 0 or π depending on crystal orientation, as in Fig. (4). Beam 1 diffracted by the grating has a phase shift of 0, constructively interferes with beam 2 and amplifies it. Beam 2 diffracted by the grating has a phase shift of π, destructively interferes with beam 1, and attenuates it.

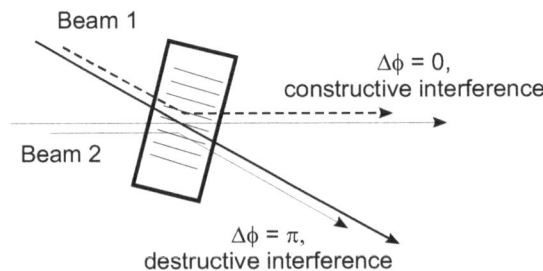

Fig. (4). Two-wave mixing in a photorefractive medium.

Both constructive and destructive interference can be used for vibration detection. The depth of amplitude modulation in the homodyne signal is proportional to the power of each interacting beam. In the present experiments, the constructive interference configuration is used for amplification of the signal scattered by the surface of a concrete object.

3. ENHANCEMENT OF DETECTION SENSITIVITY

If there is no electric field applied on the PRC, then the diffusion mode of the photorefractive effect is realized [9]. To enhance detection sensitivity of the photorefractive interferometer, drift mode of the photorefractive effect can be used. Moreover, additional enhancement can be achieved by realization of a moving hologram in the crystal. Experimental results of using the photorefractive crystal in the drift mode with a moving hologram are presented.

The efficiency of homodyne interferometry depends on the power of the beams forming the homodyne signal. Increasing beam power (within the dynamic range) leads to an increase in sensitivity. However, the signal to be analyzed usually has very low power; which cannot be significantly increased. Therefore, sensitivity can be enhanced by increasing the power of the pump beam diffracted by the recorded grating. This means that the diffraction efficiency of the grating must be increased.

The diffusion mode of the photorefractive effect is characterized by a very low diffraction efficiency, of the order of 10-5. However, the diffraction efficiency can be increased by applying an electric field to the PRC. In this case, the photorefractive effect is realized in the drift mode, resulting in increased mobility of charge carriers and deeper modulation of the electric field. The resulting modulation of refraction index is much higher, producing a dynamic grating with higher diffraction efficiency.

For the present experiments, a BSO crystal, shown in Fig. **(5)**, is used as a photorefractive material. The dimensions of the crystal is 5 x 5 x 10 mm. The aperture is 5 x 5 mm with antireflection coating for 532 nm. Golden electrodes are coated on sides of the crystal to provide an uniform electric field inside. The two-wave mixing process in the crystal has been examined in terms of diffraction efficiency and signal gain [10], which is defined as ratio of homodyne signal power (signal plus diffracted pump) to initial signal power (signal only).

Fig. (5). $Bi_{12}SiO_{20}$ (BSO) crystals.

Fig. **(6)** shows the influence of the applied field strength on signal gain. In this experiment, the signal power is 7 mW and the pump power is 25 mW. As one can see, the BSO crystal provides a signal gain of 3.5 when it is under an applied field of 6 kV/cm. This means that the power of the signal from concrete is increased by 3.5 times in the two-wave mixing process at the expense of the pump beam.

The field strength dependence does not appear to be saturating. Applying a stronger field may provide stronger signal gain. However, we did not increase the applied field strength beyond 6 kV/cm to avoid strong corona discharge and air breakdown. The use of a vacuum chamber to prevent such breakdown may allow the use of higher field strength.

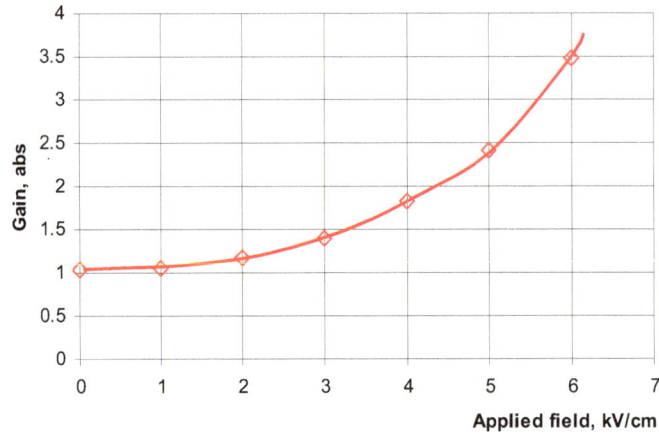

Fig. (6). Field strength dependence of signal gain in absolute value.

Fig. **(7)** shows the dependence of resulting signal gain on the intersection angle. One can see that under an applied field, the crystal provides maximum amplification at smaller intersection angles. In the experiments, minimum possible angle was technically limited to 3 degrees.

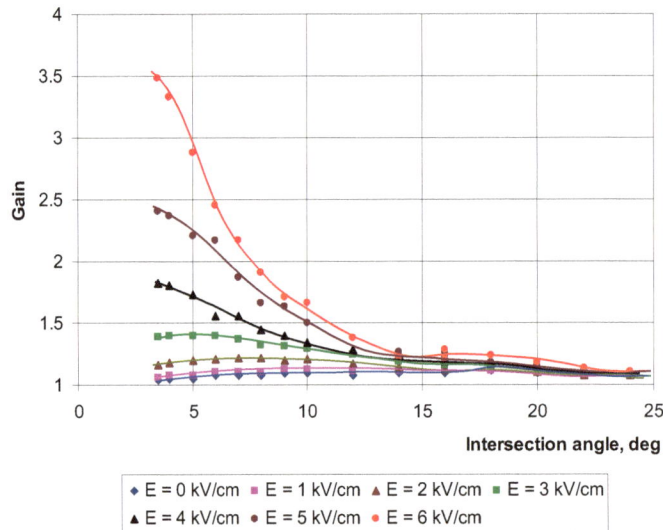

Fig. (7). Intersection angle dependence of signal gain.

3.1. Drift Mode with Moving Hologram

As mentioned above, the dynamic hologram recorded in a PRC in the diffusion mode of photorefractive effect is shifted by $\pi/2$ relative to the interference pattern. In this case, the optimum phase shift condition for realization of the most effective energy transfer between the interacting beams is $-\pi/2$, which is met automatically. However, in the drift mode of photorefractive effect, the presence of the electric field affects the recorded hologram location. As a result, the phase shift is in general almost zero in this case, and in spite of higher diffraction efficiency in the drift mode, the energy transfer is not optimum.

However, it is possible to introduce the necessary phase shift to optimize the energy transfer. For that purpose, the interference pattern recording the hologram should run with an appropriate velocity. The velocity should not exceed the value at which the phase change period in the interference pattern is close to the photorefractive non-linearity response time. For BSO crystals, the photorefractive non-linearity response time is about 20 ms. That means, frequency of phase change in the interference pattern should not exceed $(1/20 \text{ ms})^{-1} = 50$ Hz.

If the intersection angle between the interacting beams, of wavelength l = 0.532 mm, is about 5 degrees (0.087 rad), then the hologram recorded in a BSO crystal (refraction index n = 2.56) will have a period of L = 0.532[mm]/(0.087/2.56) = 16.2 mm. That means that the movement velocity of this hologram should not exceed v_m = 16.2[μm]/20[ms] ≈ 800 μm/s.

To realize the moving interference pattern, a motorized mount for one of the mirrors in the pump beam path was used. Fig. **(8)** shows the experimental setup for optimization of the interference pattern velocity. Output radiation of a CW Nd:YVO$_4$ laser is split in two beams: probe and pump. The probe beam illuminates the surface of the inspected object, which is a small piece of concrete. To avoid signal instability, the oscillator is installed on the same optical table on which the interferometer is assembled. Scattered radiation is collected and used as a working signal. The pump beam reflected by a mirror on the motorized mount is directed to the PRC. The solid line in the figure corresponds to realization of a small intersection angle appropriate to the drift mode; the dashed line shows the beam path in the diffusion mode. Part of the pump and signal beams (dotted lines) is taken to monitor the phase change in the interference pattern and corresponding velocity of running hologram.

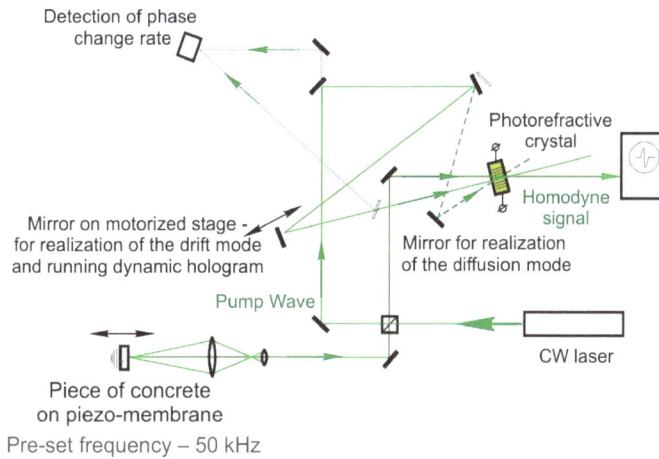

Fig. (8). Experimental setup for investigation of moving hologram.

Fig. **(9)** shows the resulting homodyne signal (lower trace in the left figures), phase change in interference pattern (upper trace in the left figures), and frequency spectrum of the phase change rate (right figures). Each record was obtained according to the following procedure: during 0-0.5 s, the zero level was recorded (signal and pump beams were blocked); during 0.5-1.5 s, only the signal power was recorded (signal beam was open, pump beam was blocked); after 1.5 s, the signal and pump beams were both open and the resulting homodyne signal was recorded. Results for motorized mirror movement speeds of 0, 0.05, 0.10, 0.15, 0.20, 0.25, 0.30, 0.35 mm/s are shown in F Fig. **(9a)**, Fig. **(9b)**, Fig. **(9c)**, Fig. **(9d)**, Fig. **(9e)**, Fig. **(9f)**, Fig. **(9g)** and Fig. **(9h)** respectively.

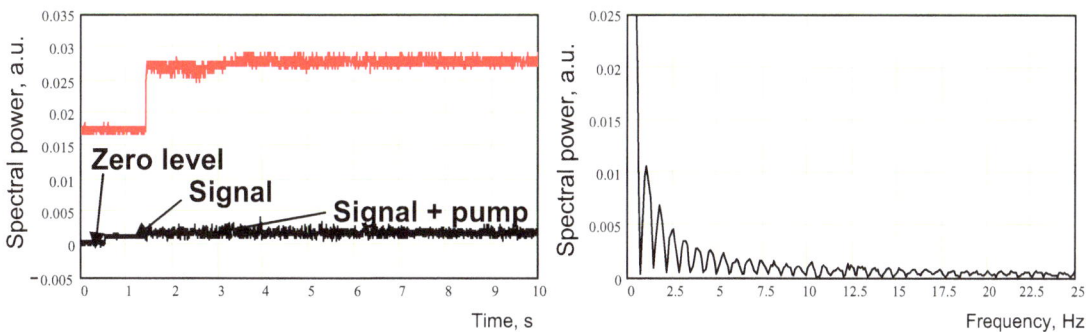

(a) Motorized mount movement speed: 0, phase change frequency: 0, signal gain: 1.5

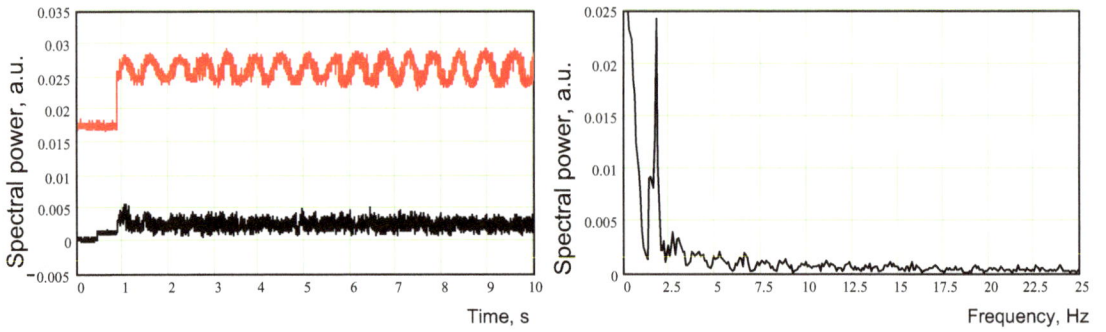

(b) Motorized mount movement speed: 0.05 mm/s, phase change frequency: 2 Hz, signal gain: 2.1

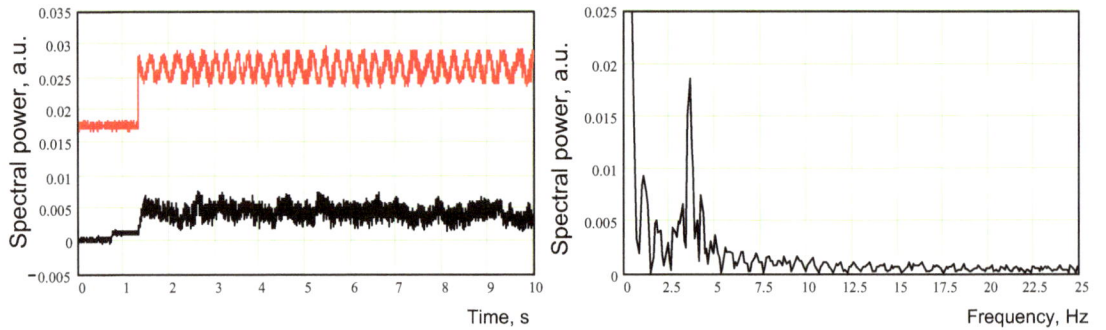

(c) Motorized mount movement speed: 0.1 mm/s, phase change frequency: 3.7 Hz, signal gain: 3.9

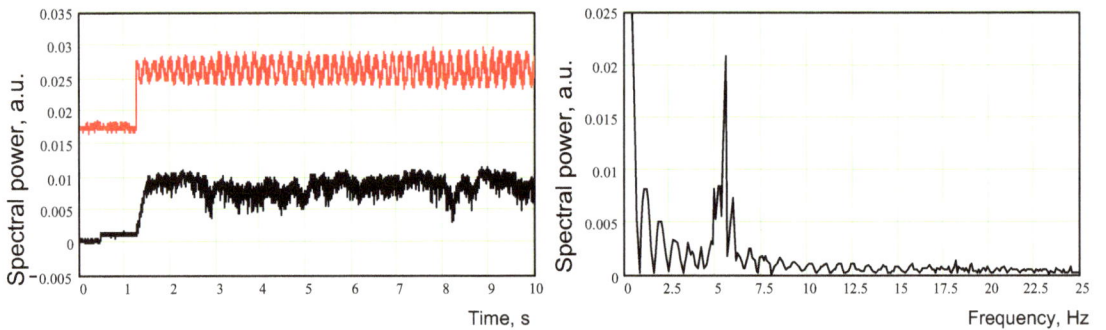

(d) Motorized mount movement speed: 0.15 mm/s, phase change frequency: 5.5 Hz, signal gain: 8.4

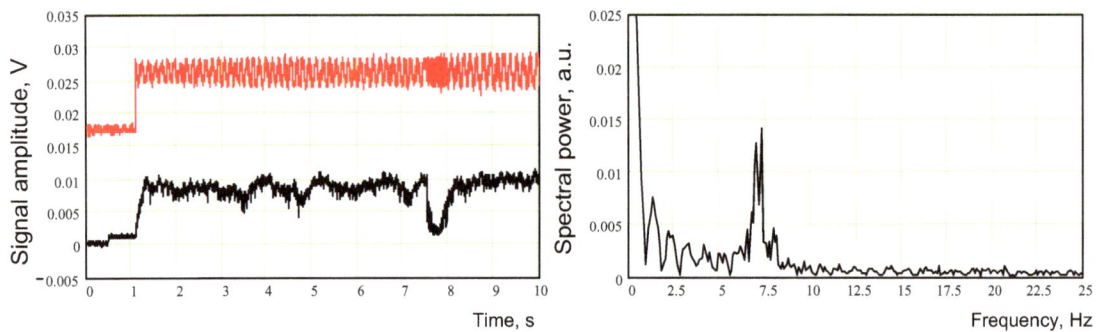

(e) Motorized mount movement speed: 0.2 mm/s, phase change frequency: 7.4 Hz, signal gain: 7.4

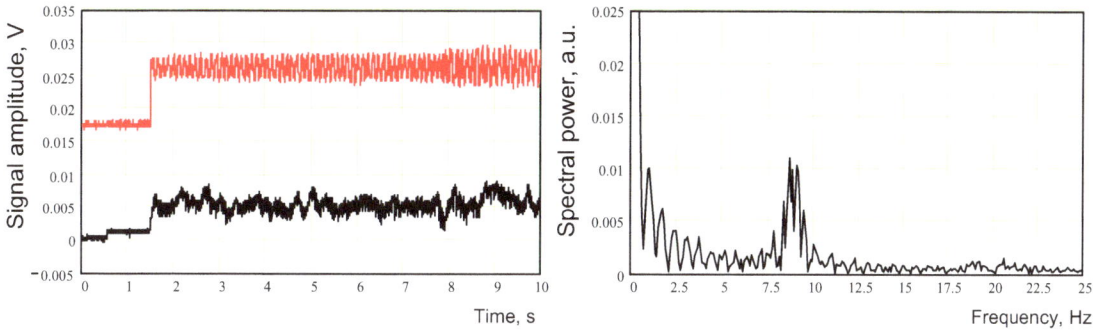

(f) Motorized mount movement speed: 0.25 mm/s, phase change frequency: 8.8 Hz, signal gain: 4.5

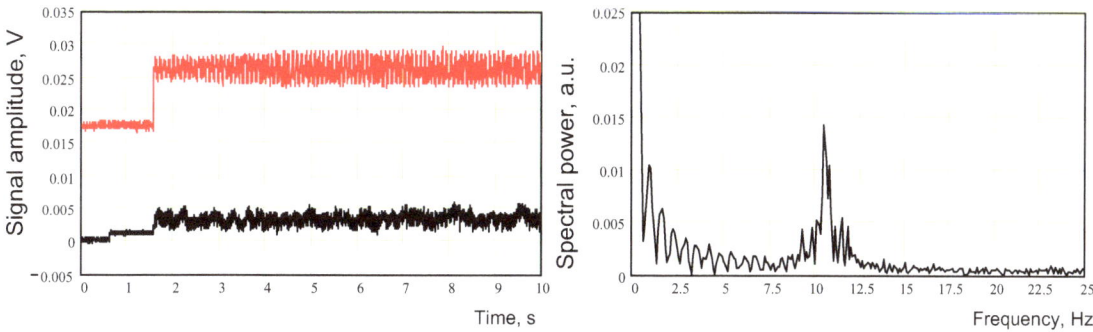

(g) Motorized mount movement speed: 0.3 mm/s, phase change frequency: 10.5 Hz, signal gain: 2.8

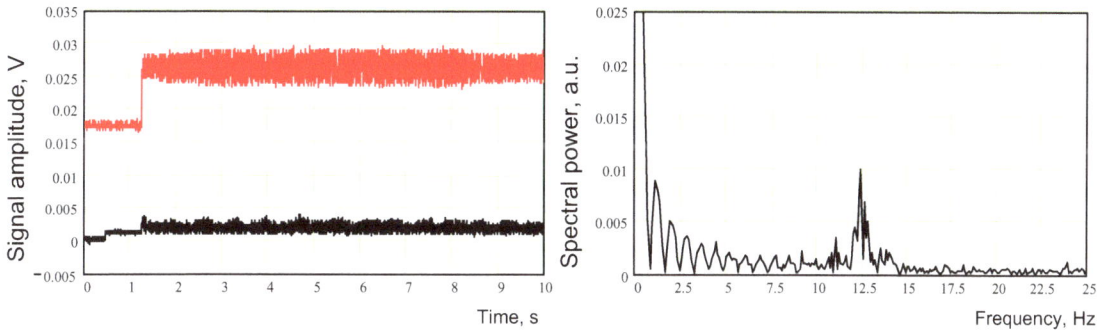

(h) Motorized mount movement speed: 0.35 mm/s, phase change frequency: 12.5 Hz, signal gain: 1.4

Fig. (9). Influence of hologram movement on the signal gain.

When the motorized mirror is moving, the optical path of the pump beam is changing, resulting in synchronized movement of the interference pattern in the PRC and the in front of the detector monitoring the phase change rate. The detected waveform looks like a harmonic function with frequency equal to the phase change rate inside the PRC. Corresponding spectra are shown in Fig. **(9)**. The spectra are presented in linear scale for better visualization of the spectral peaks.

In these experiments, the signal from concrete has low power (0.75 μW), which corresponds to the actual power of scattered and collected radiation in the real inspection procedure.

In this experiment, signal gain is defined as ratio of the averaged power level of amplified signal (after opening the pump beam) to the initial level (before opening the pump beam).

When the hologram is still, the signal gain does not exceed 1.5. However, when the motorized mount moves and drives the hologram, the signal gain changes. One can see that there is some optimum in the movement speed. Fig. **(10)** shows a maximum around 6 Hz. In this experiment, the intersection angle was 5 degrees, and the corresponding period of interference pattern and the hologram was 16.2 μm. That means, for the current conditions, the optimum running hologram velocity is 16.2[μm]x6[Hz] ≈ 100 μm/s. In this case, the signal gain reaches value around 8.5, and the recorded signal is amplified up to 0.01 V, as in Fig. **(9d)** and Fig. **(9e)**.There is a problem of using the running hologram technique for real inspection. As it was mentioned above, in the experiments, the inspected object was installed on the same optical table on which the interferometer is assembled. The table is equipped with an air suspension system isolating the interferometer from external vibration.

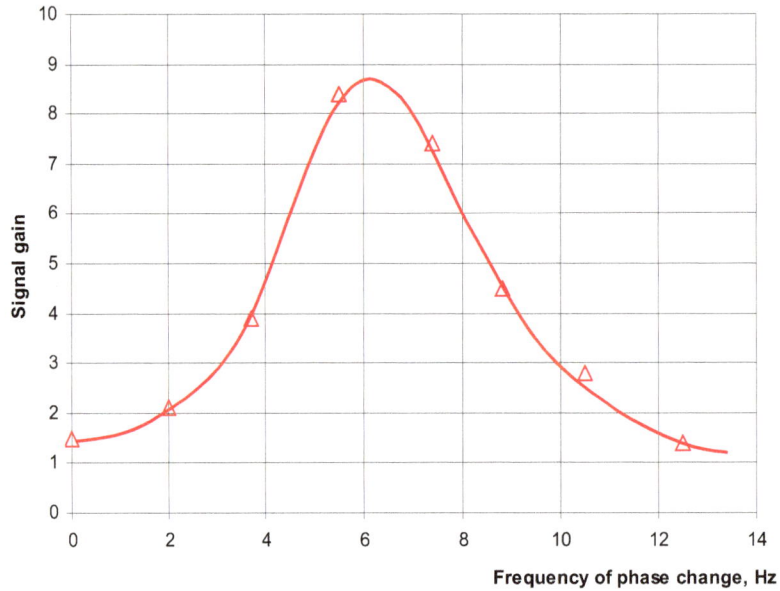

Fig. (10). Phase change rate dependence of the signal gain.

In this case, the interference pattern inside the photorefractive crystal is comparatively stable and its movement driven by the motorized mount is smooth. However, when the inspected object is placed on a separate table, the signal becomes very unstable. The resulting interference pattern is bouncing irregularly, and using the running hologram technique makes no sense.

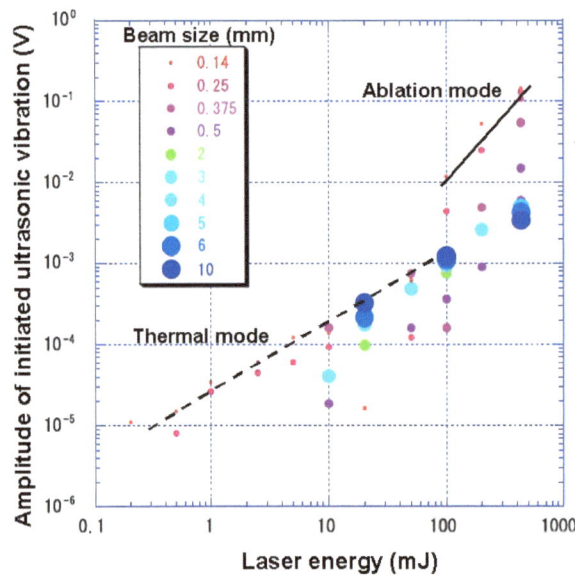

Fig. (11). Relative strength dependence on impact laser energy and beam size.

However, according to the experimental results, this technique is an attractive method to increase the signal gain. To make it this technique useful for real inspection, it is necessary to stabilize the interference pattern in the PRC. For that purpose, special stabilization system should be designed and introduced. This is described in Section 4.2.

4. PRINCIPLE OF LOCATING DEFECTS IN CONCRETE

4.1. Laser-Initiated Vibration in Concrete

 In this section, we describe the dependence of initiated ultrasonic vibration (P-wave) amplitude on the laser impact energy. In this experiment, the impact laser beam initiates a vibration in the concrete sample, which is a 100 x 200 x 200 mm concrete block. A Nd:YAG laser of pulse energy 0.4 J, pulse duration 10 ns, and beam diameter varying from 0.14 to 10 mm is used as the source of laser impact. The P-wave is initiated by laser impact, passes through the 10-cm thick concrete layer, and is detected by the homodyne interferometer. Transmission configuration is used to avoid influence of laser initiated R-wave and acoustic noise in air. A CW Nd:YAG laser (second harmonic) is used for vibration detection. The power of the homodyne signal and that of the pump are 10 µW and 20 mW, respectively.

Depending on the impact laser beam size and energy, thermal or ablation mode of laser impact is realized. A 50-shot average is used to increase the signal-to-noise ratio. Fig. **(11)** depicts the amplitude of initiated ultrasonic vibration versus laser impact energy. When the laser energy is less than 100 mJ, impact mode is thermal regardless of the impact beam size. In these conditions, the vibration amplitude is proportional to the impact laser energy.

If the laser energy is higher than 100 mJ and the beam size is in the range from 0.14 mm to 0.375 mm, ablation takes place on the concrete surface. In the ablation mode, the dependence slope is approximately 3-4 times steeper than that in thermal mode.

4.2. Two Methods of Vibration

The LRSS can initiate and detect both impact echo and standing Lamb waves for inner-defect location. Both methods are illustrated in Fig. **(12)**. The impact echo detection in Fig. **(12a)** can find the defect depth *via* measurement of delay time of the impact echo. However, vibration detection is difficult in this case because the surface displacement of vibration is comparatively small. In contrast, a standing Lamb wave (or natural vibration mode) has a larger surface displacement. As a result, detection of a standing Lamb wave, shown in Fig. **(12b)**, is easier and actually it is similar to the conventional hammering method. According to this similarity, laser-based inspection can be a good alternative technique to hammering method.

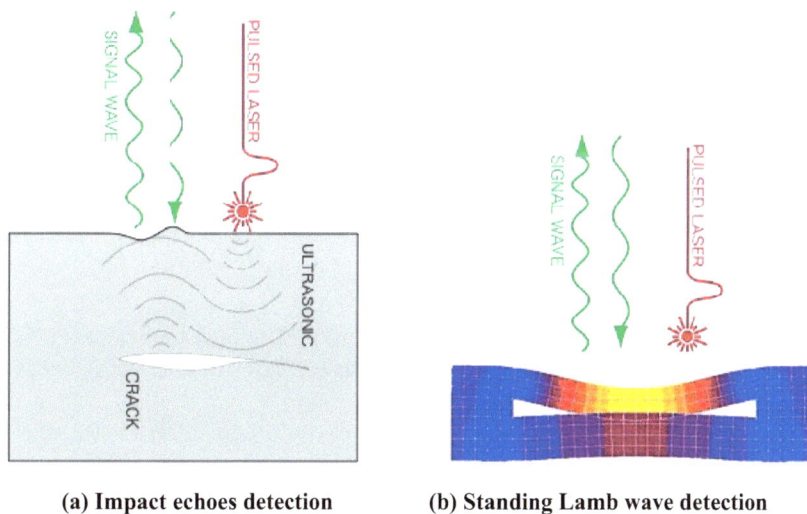

(a) Impact echoes detection   (b) Standing Lamb wave detection

Fig. (12). Inner-defect detection methods.

Fig. **(13)** shows a schematic diagram of the experimental setup used for examination of the two inspection methods. A pulsed Nd:YAG laser generates laser ablation impact. The impact energy is 400 mJ and the impact beam size is about 4 mm, producing a laser peak intensity of 300 MW/cm^2. A CW second harmonic Nd:YVO$_4$ laser is used for the detection of impact echo and standing Lamb waves. Output radiation (up to 1.8 W) is split into two beams: the pump beam with a power of 50 mW within the PRC aperture and the probing beam using the rest of the output power. The probing beam is focused on the sample surface. The pump beam and scattered signal record a dynamic hologram in the PRC and produce the homodyne signal which is detected by a photomultiplier tube and analyzed by an oscilloscope. The PRC is used in the drift mode, and the applied electric field strength is 5 kV/cm.

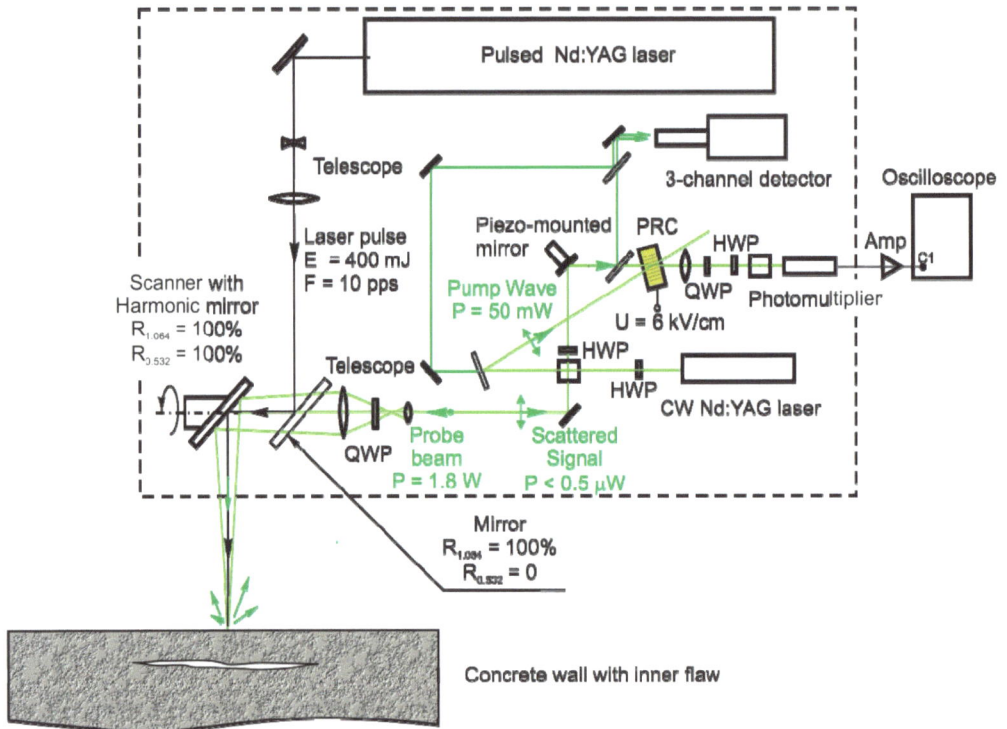

Fig. (13). Schematic diagram of experimental setup.

A small fraction of the pump and signal beams is used for analysis of the interference pattern behavior in the interference pattern stabilization system. The main components of the stabilization system are a 3-channel detector for pattern analysis, and piezo-mounted mirror for correction of pattern location.

The fraction of the pump beam is delivered to the 3-channel detector in the stabilization system. The fraction of the signal beam is taken from the signal beam reflected by the piezo-mounted mirror. As a result, moving the piezo-mounted mirror leads to synchronized motion of interference pattern in both the 3-channel detector and inside the photorefractive crystal.

4.3. Detection of Standing Lamb Wave Modes

Fig. **(14)** shows initiation of different standing Lamb wave modes in a sample with an inner defect. When the area above the defect center is tested, the first harmonic of the Lamb wave is initiated. When the area in the middle between the defect center and its edge is tested, the second harmonic of the Lamb wave appears.

When the impact/detection position is located over an area with no defect, no detectable Lamb wave is initiated. Therefore, the criterion for evaluation of the presence or absence of an inner defect is the appearance of an initiated Lamb wave. The mobile LRSS prototype assembled for the field experiments will use this criterion for the defect location in field conditions.

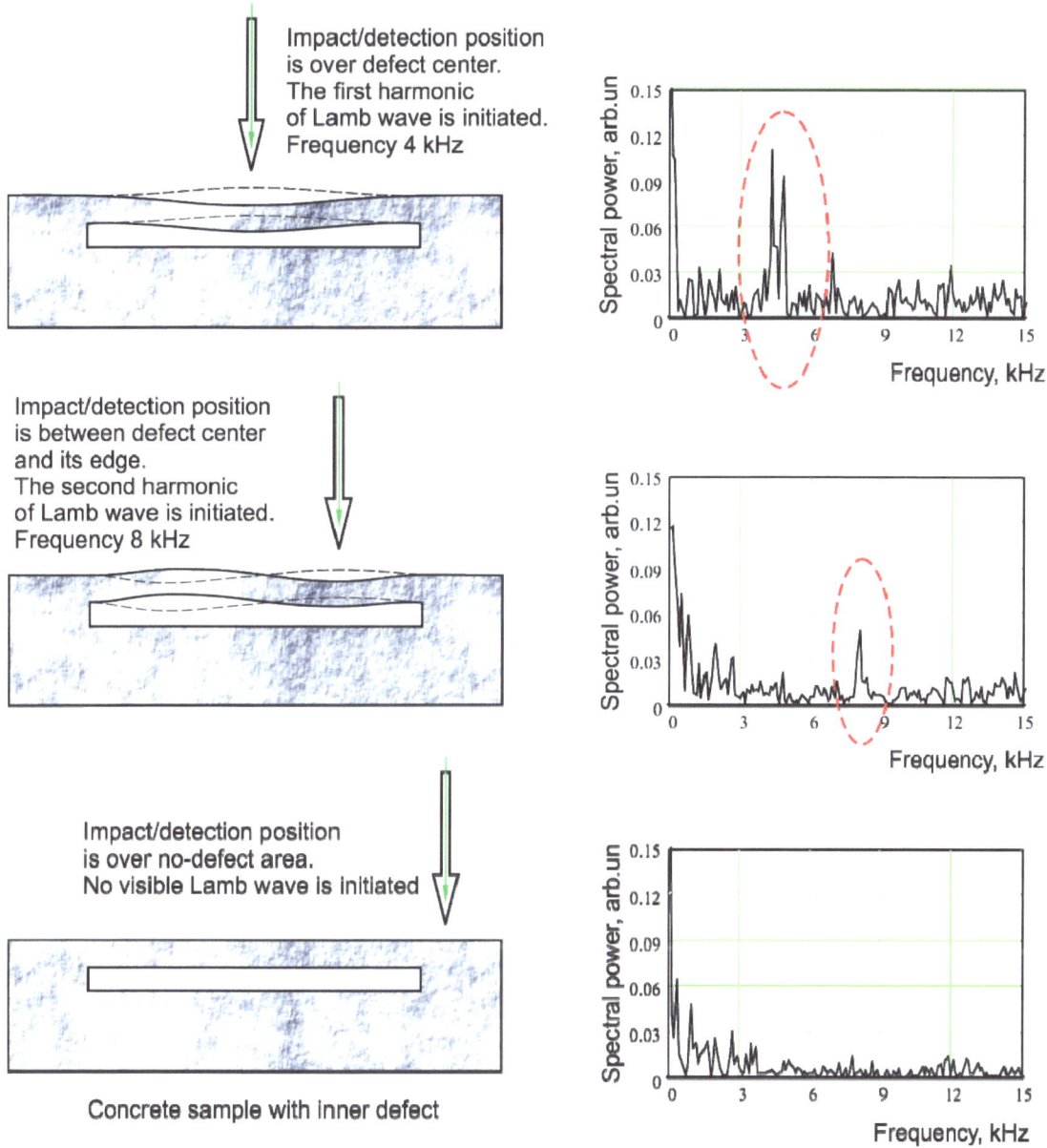

Fig. (14). Standing Lamb wave modes initiated by laser impact in a concrete sample with inner defect. When the area above the defect center is tested, the first harmonic of the Lamb wave (~4 kHz) is initiated (upper figure). When the area midway between the defect center and its edge is tested, the second harmonic of the Lamb wave (~8 kHz) is initiated (middle figure). When the area with no defect is tested, no detectable Lamb wave is initiated (lower figure).

4.4. Impact Echo Detection

Standing Lamb waves can be used for defect location and approximate evaluation of the transverse dimensions of the inner defect. Comparatively low frequency of standing Lamb wave modes in concrete (1-10 kHz) does not provide an evaluation accuracy better than several cm. Moreover, it is impossible to evaluate the depth of the defect location or the thickness of the concrete layer between the outer surface and the inner defect.

For more accurate measurement of the defect dimensions, initiation and detection of impact echo (reflection of P- and S-waves) can be used. The frequency of P-waves in concrete (10-100 kHz) is much higher than the frequency of standing Lamb modes. Evaluation of defect dimensions may be more accurate. Moreover, the time delay between impact and approaching echo allows calculation of the thickness of concrete layers.

(a)

(b)

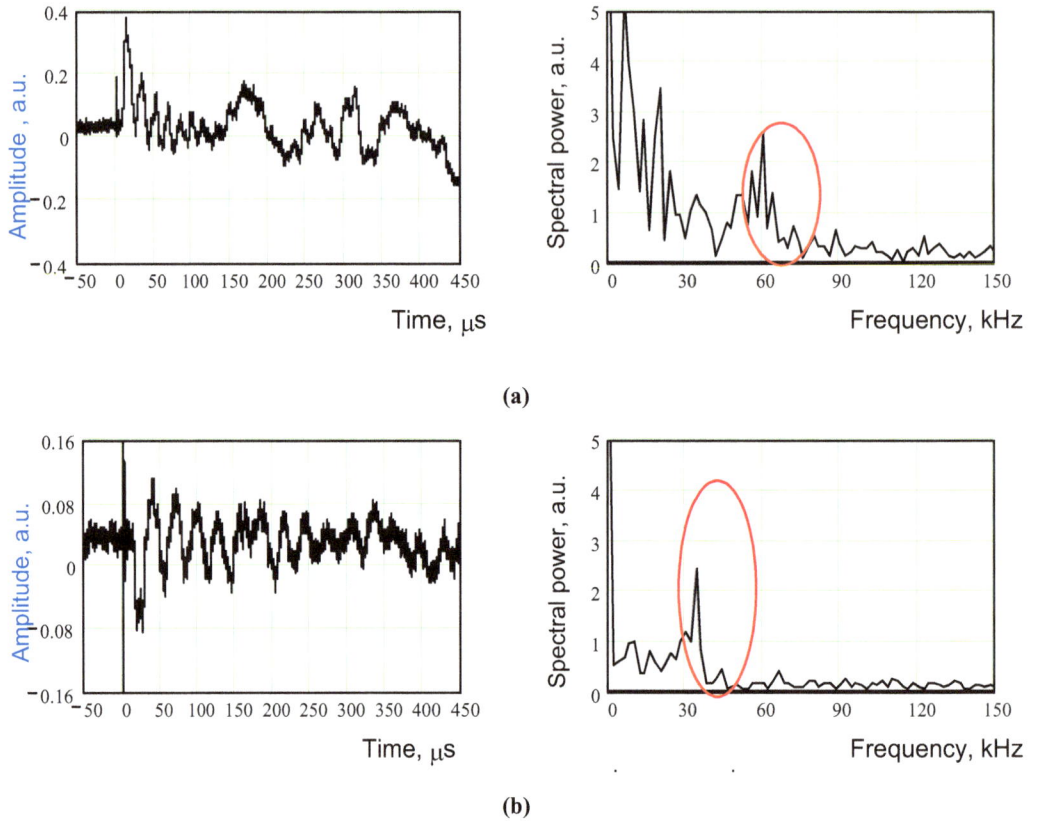

Fig. (15). Waveforms (left) and spectra (right) of impact echo initiated and detected in concrete plates with thickness of **(a)** 30 mm and **(b)** 50 mm. Reflection configuration, ablation mode.

However, sometimes it is difficult to observe the approaching echo of P-waves. It may be overlapped by some noise signals, for instance the noise caused by ablation. Fortunately, in this case, multiple echoes can be detected and analyzed. Fig. **(15a)** and **(15b)** show waveforms and spectra of multiple impact echoes detected in concrete plates of thickness 30 mm and 50 mm. The waveforms of multiple echoes look like quasi-harmonic oscillation. The amplitude of the oscillation decreases gradually due to attenuation in concrete. The oscillation lifetime is almost 0.5 ms, longer than typical duration of noise signal, which is about 10 µs in Fig. **(15)**. The corresponding spectrum has a comparatively high peak at the echo frequency. This frequency allows evaluation of the thickness of concrete layer where the echo takes place.

The plates used in the experiments are made of concrete characterized by a P-wave velocity of 3.6 km/s. Echo frequency detected in the concrete plates have peaks around 60 kHz in Fig. **(15a)** and 35 kHz in Fig. **(15b)**. That means that the time interval between successive echoes is 16.7 and 27.8 µs, respectively. This corresponds to the round trip time taken by P-waves between the two surfaces of a concrete plate (or between the outer surface of inspected object and an inner defect in the real inspection procedure). The thickness of the plates (or defect depth) can be calculated by (0.5)(3.6 km/s)(16.7 µs) = 30 mm and (0.5) (3.6 km/s) (27.8 µs) = 50 mm, respectively, and agrees exactly with the actual thickness of the concrete plates. The same procedure can be used for evaluation of real defect depth.

5. FIELD EXPERIMENT

5.1. Experiment Description

To examine the capability of LRSS to locate a real concrete defect in field conditions, a specially designed mobile LRSS prototype was assembled and transported to a bridge of the bullet-train line in Japan. Fig. **(16a)** shows the experiment site, and Fig. **(16b)** shows the system in operation, with the green beam of the vibration detection laser. The distance from the system to concrete surface is 4.5 m.

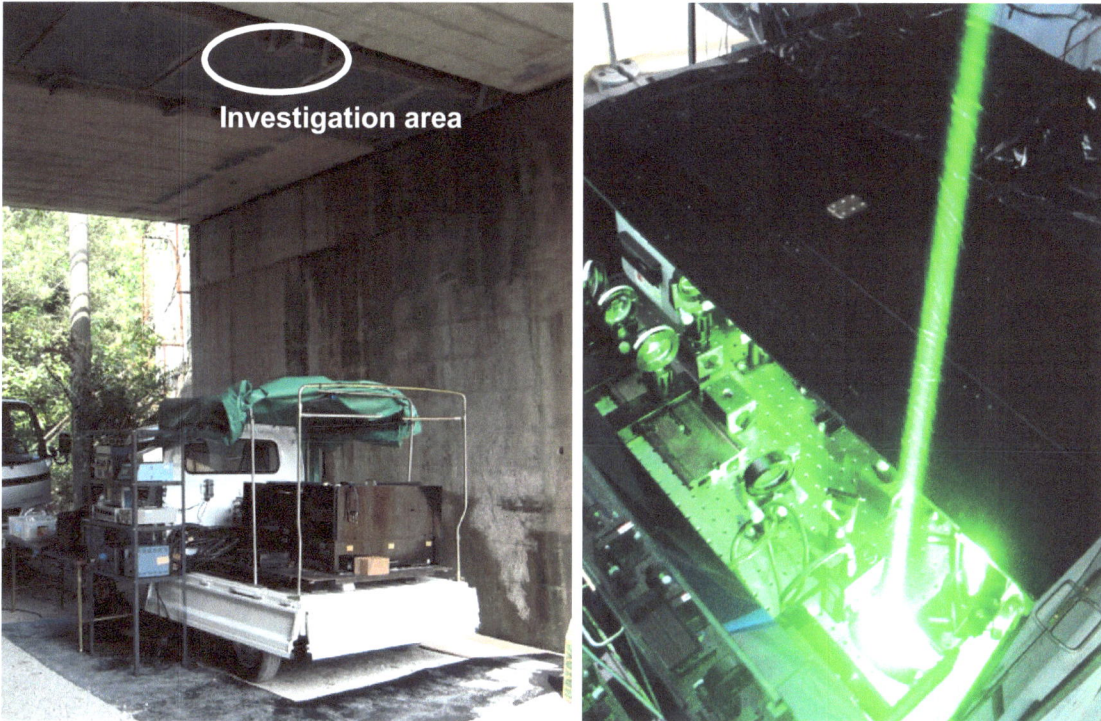

(a) Circle shows investigation area   (b) Green light indicates vibration detecting laser

Fig. (16). Laser-based remote sensing system for a field experiment.

The Direction of the impact and detection beams is controlled by a scanning mirror, to change the beam position over the inspected area. The beams scan the concrete surface in two dimensions (x and y directions) with a step size of 2 cm.

5.2. Experimental Results

The scanning result is the two-dimensional map of an inner defect in concrete shown in Fig. (**17**). The bright gray area corresponds to no-defect situation. The dark gray area indicates the presence of an inner defect in concrete. This area has a predominant vibration frequency of 0.9-1.2 kHz. The frequency spectrum obtained at point A in Fig. (**17**) is shown in Fig. (**18**), which shows a predominant peak at 1.144 kHz. This is in good agreement with the data obtained by the hammering method at point C in Fig. (**17**), which is shown in Fig. (**19**). The hammering method shows a predominant peak at 1.038 kHz, which is reasonably close to the value obtained using laser. The predominant frequency does not appear in the frequency spectrum obtained at point B in Fig. (**17**). The defect size was evaluated to be approximately 20 x 10 cm, with an accuracy of better than 2 cm.

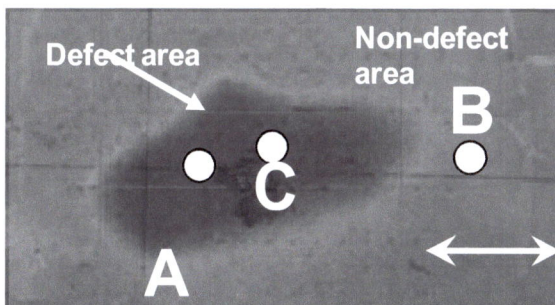

Fig. (17). Scanned result with a two-dimensional map. Dark area corresponds to area with defect.

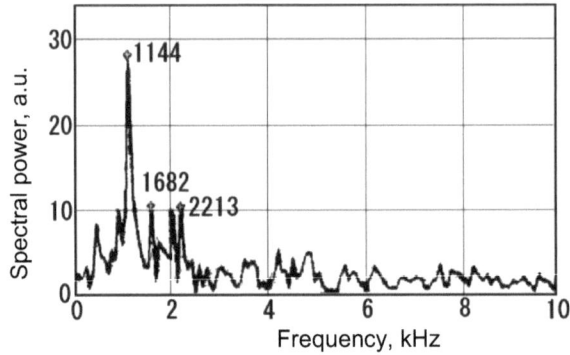

Fig. (18). Frequency spectrum of vibration at point A in Fig. **(17)** measured by laser.

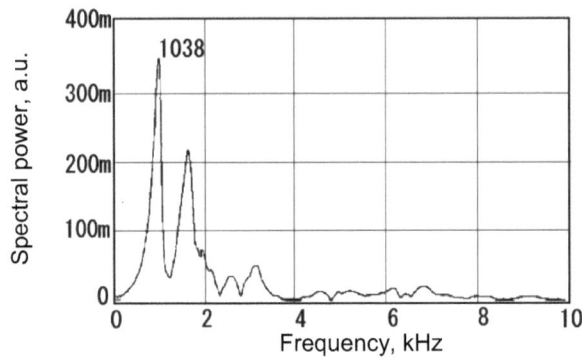

Fig. (19). Spectrum of vibration at point C in Fig. **(17)** measured by hammering method.

6. CONCLUSION

The laser-based remote sensing system (LRSS) and the principles of vibration detection with the use of photorefractive interferometry have been described. The drift mode of photorefractive effect is more effective than diffusion mode in terms of diffraction efficiency, energy exchange and resulting sensitivity of photorefractive interferometry. Realization of a running hologram results in additional increase of detection sensitivity.

The experimental results presented in this chapter demonstrate initiation and detection of both standing Lamb wave modes and impact echo in concrete. In principle, both phenomena can be used for laser-based remote nondestructive inspection of concrete structures.

The mobile LRSS prototype has been designed, assembled and tested in field conditions under a bridge of the bullet-train line (Shinkansen train). The LRSS prototype is designed for initiation and detection of standing Lamb waves. During the scanning of the inspected area, it generates two-dimensional map of the defect structure in real time. The prototype has demonstrated reliable capability of the real defect location with acceptable accuracy (better than 2 cm).

The application of inspection technology based on initiation and detection of impact echo, which will allow more accurate evaluating parameters of various types of concrete defects, is of further study.

REFERENCES

[1]   P. Mclntire, *Nondestructive testing handbook, Vol. 7: Ultrasonic testing*, American Society for Nondestructive Testing (1991).

[2]   H. Kanada, Y. Ishikawa, T. Uomoto, "Utilization of near-infrared spectral imaging system for inspection of concrete structures", *4th International Symposium on New Technologies for Urban Safety of Mega Cities in Asia*, 2005.

[3]   S. Tanaka, M. Kadowaki, "Non-Destructive Inspection of Reinforced Concrete by Electromagnetic Wave (Radar) Based on a Peak Pattern Recognition of Received Signals", *IEEJ Transactions on Electronics, Information and Systems*, Vol. 125, pp. 50-56 (2005). (in Japanese)

[4]   T. Uomoto, K. Kobayashi, "Measurement of Fiber Content of Steel Fiber Reinforced Concrete by Electra-Magnetic Method", *Journal of the American Concrete Institute*, Vol. 81, pp. 233-246 (1984).

[5]   C. Scruby, L. Drain, *Laser Ultrasonics, Techniques and applications*, Adam Hilger.

[6]   O. Kotyaev, S. Uchida, "Nondestructive inspection of concrete structures with the use of photorefractive two-wave mixing", *Proceedings of SPIE*, Vol. 4702, pp. 241-249 (2002).

[7]   P. Delaye, A. Blouin, L. Montmorillon, I. Biaggio, D. Drolet, J. Monchalin, G. Roosen, "Photorefractive detection of ultrasound", *Proceedings of SPIE*, Vol. 3137, pp. 171-182 (1997).

[8]   P. Yeh, *Introduction to photorefractive nonlinear optics*, John Wiley & Sons, Chapter 3 (1993).

[9]   S. Stepanov, V. Kulikov, M. Petrov, "Running holograms inphotorefractive $Bi_{12}SiO_{20}$ crystals", *Optics Communications*, Vol. 44, pp. 19-23 (1982).

[10]   A. Marrakchi, J. Huignard, "Diffraction efficiency and energy transfer in two-wave mixing experiments with $Bi_{12}SiO_{20}$ crystals", *Applied Physics*, Vol. 24, pp. 131-138 (1981).

CHAPTER 10

Minor Constituent Detection and Electric Field Measurement Using Remote Laser-Induced Breakdown Spectroscopy

Takashi Fujii[*]

Electric Power Engineering Research Laboratory, Central Research Institute of Electric Power Industry, 2-6-1 Nagasaka, Yokosuka, Kanagawa 240-0196, Japan

Abstract: Laser-induced breakdown spectroscopy (LIBS) is attractive for fast, on-site, and remote measurement of trace elements with high spatial resolution. The measurement of chlorine concentration in concrete, which can be useful for the evaluation of durability of reinforced concrete structures, was performed with a sensitivity of better than 0.18 kg/m^3, which is below the threshold chlorine concentration of 0.6 kg/m^3 at which the reinforcing bars in concrete structures start to corrode. In addition, ultrashort laser pulses have several advantages for application to LIBS. The propagation of an ultrashort high-intensity laser pulse in the atmosphere produces a bundle of filaments, which can be generated for a distance of more than several hundreds of meters and have sufficient intensity for producing plasmas at various targets. LIBS using filaments, called filament-induced breakdown spectroscopy (FIBS), is very useful for remote measurement of trace elements. Remote detection and identification of microparticles in air by FIBS at a distance of 16 m was demonstrated. In addition, as a new application of LIBS, remote measurement of the electric field is presented.

Keywords: Laser-induced breakdown spectroscopy, chlorine, concrete, filament, emission spectroscopy, ultrashort pulse laser, plasma, saltwater aerosols, remote measurement, electric field.

1. INTRODUCTION

Remote detection and identification of trace elements is useful in industrial applications, such as diagnosis of materials and buildings, and atmospheric measurement. Laser-induced breakdown spectroscopy (LIBS) is attractive for fast, on-site, and remote measurement of trace elements [1, 2]. In addition, LIBS using ultrashort pulse, high intensity lasers have several advantages, such as the reduction of the breakdown energy threshold [3, 4] and in white light noise. The propagation of an ultrashort high-intensity laser pulse in the atmosphere produces a bundle of filaments owing to the equilibrium between Kerr lens focusing and plasma defocusing [5-9]. Filaments can be generated for a distance of more than several hundred meters, and they have high intensity, of the order of up to 10^{14} W/cm^2, which is higher than the threshold for producing plasmas for various targets. Therefore, LIBS using filaments, called filament-induced breakdown spectroscopy (FIBS), is useful for remote measurement. In addition, in the measurement of the constituents of microparticles in air, the use of multiple filaments has a great advantage, because a large number of microparticles can be ionized along a bundle of filaments in air over a long distance. This can be used for lidar measurement of microparticles. In addition, the application of LIBS to remote electric field measurement has been proposed and demonstrated [10, 11].

2. LASER-INDUCED BREAKDOWN SPECTROSCOPY

2.1. Characteristics

LIBS can detect and identify trace elements in a target by spectral analysis of the emission from a plasma induced by irradiation of laser pulses on the target. Quantitative measurement is also possible using the emision intensity. Fig. **(1)** shows a energy diagram showing the mechanism of emission from a laser-induced plasma. The atoms constituting the target are ionized and/or excited by the laser plasma. The excited atoms decay to the ground level *via* radiative transition, which accompanies optical emission, or by collisional quenching, which does not.

[*]Address correspondence to Takashi Fujii: Electric Power Engineering Research Laboratory, Central Research Institute of Electric Power Industry, 2-6-1 Nagasaka, Yokosuka, Kanagawa 240-0196, Japan; Tel: +81-46-856-2121; E-mail: fujii@criepi.denken.or.jp

Tetsuo Fukuchi and Tatsuo Shiina (Eds)

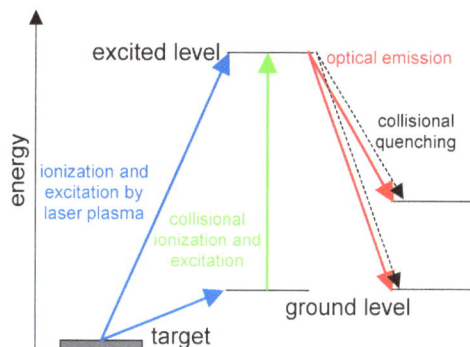

Fig. (1). Energy diagram of emission from a laser-induced plasma.

LIBS has advantages over other analytical methods, such as suitability for real time, on-site measurement, measurement in hazardous areas, simultaneous measurement of multiple species, and high spatial resolution. Real time, on-site measurement is possible because no pre-treatment of samples is necessary. Measurement in hazardous areas is possible because of the remote and noncontact nature of the method. Simultaneous multiple species measurement is possible becuase emission lines corresponding to different elements can be measured simultaneously. High spatial resolution is due to the fact that the plasma is generated in a small region determined by the focus spot of the laser beam, which is generally less than several 100 μm in diameter.

Characteristics of laser-induced plasmas, which are important to understand the performance of LIBS, can also be measured from the emission spectra. The electron density can be measured by Stark broadening or Stark shift; the temperature can be measured from Boltzmann plots of multiple emission lines and by the Saha equation [12].

2.2. Industrial Applications of LIBS

LIBS has been applied to remote material analysis of metallic samples, such as measurement of Cu content of 316H austenic stainless steel superheater bifurcation tubing in a nuclear power station [13].

Laser-induced breakdown can be combined with other spectroscopic methods for analysis of samples. For example, the isotope ratio $^{235}U/^{238}U$ in a laser-induced plasma using UO_2 samples has been measured by atomic laser fluorescence spectroscopy or laser atomic absorption spectrometry [14,15].

Laser-induced breakdown can also be combined with laser-induced fluorescence (LIF) for higher sensitivity and selectivity. By the use of this combination, trace concentrations of Cr and Si in certified steel samples can be measured down to detection limits of 105 ppm and 95 ppm, respectively [16]. The detection limit of phosphorus (P) in steel has also been reported to be 5.6 μg/g [17]. This technique is also effective for analysis of heavy metals such as As, Cd, Cr, Cu, Hg, Ni, Pb, Tl, Zn in soils, with detection limits in the order of μg/g [18]. The technique is also applicable to elements in water solution. The detection sensitivity of Fe in water solution is estimated to be 10 ppb [19], and sub-ppb levels are possible for alkali metals, such as Na and K.

2.3. Measurement of Chlorine Concentration in Concrete

The measurement of chlorine concentration is important in evaluating the durability of reinforced concrete structures, particularly at coastal locations, because reinforcing bars in concrete are corroded by chloride ions [20]. So far, chlorine concentration in concrete has been measured by standard chemical methods using a core sample with a diameter of about 100 mm extracted from the concrete structure. However, chemical methods are time-consuming and normally require a chemical laboratory for quantitative measurement. Moreover, the spatial resolution of chemical methods is limited by the thickness of slicing the core sample, which is generally 5-10 mm.

LIBS measurement of chlorine concentration in concrete is difficult, because chlorine atoms are difficult to excite owing to the high excitation energy of over 10 eV [21,22], and also because of the interference from emission of other elements. The required sensitivity is 0.6 kg/m^3, the threshold at which the reinforcing bars in concrete structures start to corrode.

2.3.1. Experimental Setup

Ground and pressed samples of concrete were used as targets in LIBS measurement. The volume concentrations of chlorine in the concrete samples before grounding, converted from the weight concentrations in the pressed samples used for the experiments, are shown in Table **1**.

Table 1: Chlorine concentration of concrete samples used for LIBS measurement

Sample	Chlorine concentration	
	kg/m^3	wt%
1	5.40	0.238
2	5.21	0.229
3	2.65	0.117
4	1.04	0.046
5	0.65	0.029
6	0.18	0.008

The experimental setup is shown in Fig. **(2)**. Laser 1 and laser 2 are Nd: YAG lasers (Hoya Continuum, Powerlite 8010, second harmonic) operating at a repetition rate of 10 Hz. The output pulse of laser 1 was focused by a lens of 250 mm focal length and irradiated on the sample surface to produce an ablation plasma. Measurement using only laser 1 is referred to as single pulse measurement. The output pulse of laser 2 was focused on the ablation plasma generated by laser 1 using a lens of focal length 250 mm and irradiated parallel to the sample surface. The focal point of the pulse of laser 2 was adjusted by changing the position of the focusing lens and/or a mirror located on a movable stage. Measurement using both laser 1 and laser 2 is referred to as double pulse measurement. Helium, as a buffer gas, was blown onto the sample surface. The pressed samples were set on a rotating holder, and each laser pulse was irradiated onto a fresh surface of the target. A photograph of laser irradiation on the target is shown in Fig. **(3)**.

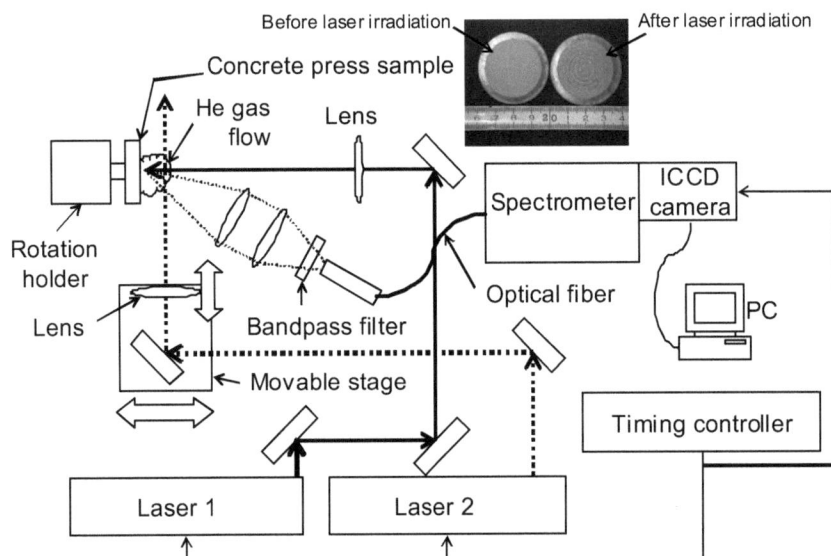

Fig. (2). Experimental setup for LIBS measurement of chlorine concentration in concrete [20].

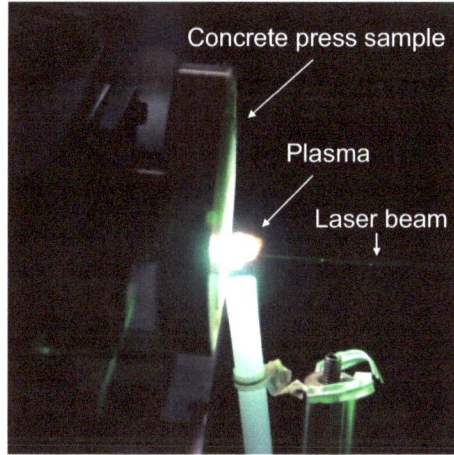

Fig. (3). Photograph of laser irradiation on the target.

Chlorine emission at wavelength 837.59 nm was used for LIBS measurements, because it has a strong intensity and is not overlapped by spectral lines of other elements contained in the concrete sample [23]. The emission was focused onto the entrance of an bundled optical fiber through a bandpass filter (center wavelength 838 nm, bandwidth 14 nm) to reduce interference from emission lines of elements such as calcium, carbon, and nitrogen.

The output emission from the bundled fiber was directed into a spectrometer, and the spectrum was detected using an intensified CCD (ICCD) camera. The gate width, gate delay of the ICCD camera from the laser shot, and interpulse delay were controlled by a signal synchronized with the laser shot, as shown in Fig. (**4**) and Fig. (**5**). For each measurement, data analysis was performed using 3-10 emission spectra obtained under the same conditions, and each spectrum was averaged over 50 laser pulses.

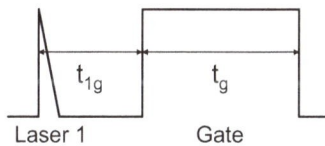

Fig. (4). Time delay between laser pulse and ICCD gate in single pulse measurement.

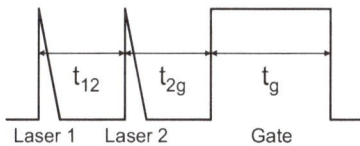

Fig. (5). Time delay between laser pulses and ICCD gate in double pulse measurement.

2.3.2. Experimental Results

In LIBS measurement, the white-light noise, which is due to bremsstrahlung in the plasma, is the major source of noise. It reaches a maximum just after laser irradiation and decays faster than the spectral emission of individual elements. Therefore, in order to reduce the white-light noise and increase the signal-to-noise ratio (SNR), the ICCD camera gate delay from laser irradiation (t_{1g} in single pulse measurement, t_{2g} from irradiation by laser 2 in double pulse measurement), interpulse delay between the two lasers (t_{12}), and gate width (t_g) should be optimized. Fig. (**6**) shows the emission spectra when t_{1g} was changed from 50 ns to 5 μs in single pulse measurement. The chlorine emission peak is not clearly observed at t_{1g}=50 ns, owing to the large white-light noise, but is clearly observed at t_{1g}=0.5 μs, after the white-light noise has sufficiently decreased. The chlorine emission intensity decreases along with the background intensity when t_{1g} is increased up to 5 μs.

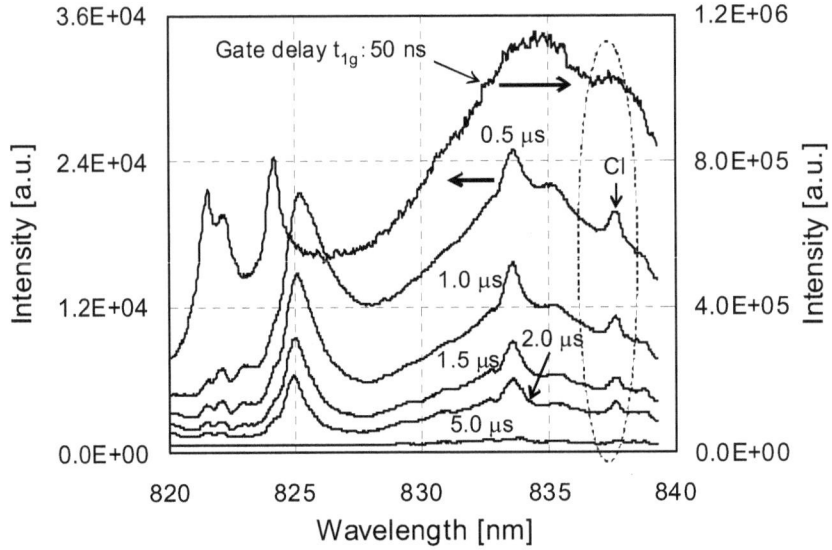

Fig. (6). Variation of emission spectra with gate delay time from laser irradiation (t_{lg}) in single pulse measurement [20].

Fig. **(7)** shows the chlorine emission intensity and SNR plotted against t_{lg}, based on from the results shown in Fig. **(6)**. The chlorine emission intensity decreases for $t_{lg} > 0.5$ µs, and becomes almost zero at 5 µs. However, SNR is constant for $t_{lg} = 0.5 \sim 2.5$ µs, owing to the reduction of noise intensity, and decreases with increasing delay time for $t_{lg} > 2.5$ µs. From these results, the optimum gate delay time from laser irradiation for single pulse measurement was determined to be $t_{lg} = 0.5$ µs.

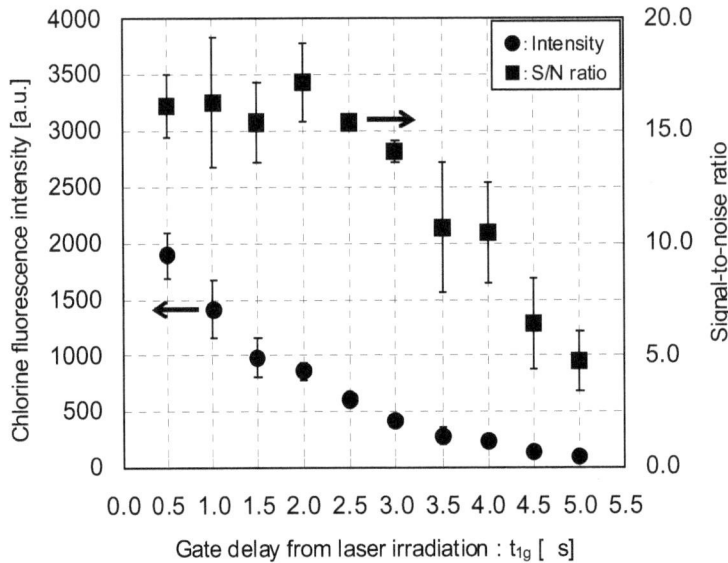

Fig. (7). Chlorine emission intensity and SNR versus gate delay time from laser irradiation (t_{lg}) in single pulse measurement [20].

LIBS spectra for single and double pulse measurements obtained after optimization of the delay and gate times are shown in Fig. **(8)**. In addition to chlorine, emission peaks of Ca (the main element in concrete), C, Fe, and N are observed. The emission intensities of these elements, particularly Cl and C, are higher for double pulse measurement than for single pulse measurement, although the background intensity does not exhibit a large difference.

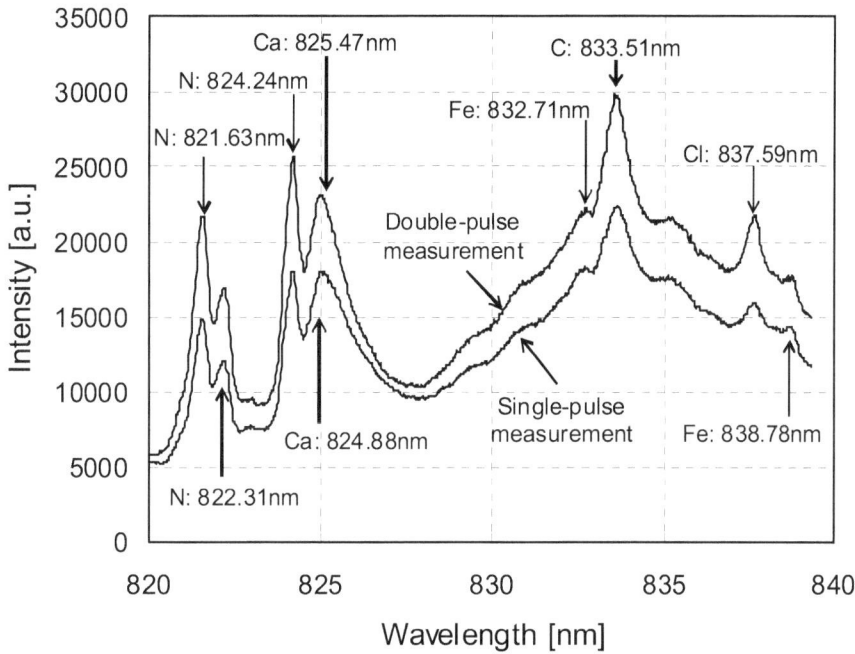

Fig. (8). Emission spectra obtained under optimized conditions in single and double pulse measurements [20].

Emission spectra for samples with chlorine concentrations of 0.18–5.4 kg/m^3 were obtained in single and double pulse measurements. The results are shown in Fig. **(9a)** and Fig. **(9b)**, respectively. The chlorine emission peak was detected with a SNR of more than 2, even for the sample with a chlorine concentration of 0.18 kg/m^3, for both single and double pulse measurements. These results show that the detection limit is below 0.6 kg/m^3, which is the threshold chlorine concentration at which the reinforcing bars in concrete structures start to corrode. Fig. **(10)** shows the dependence of the chlorine emission intensity on the chlorine concentration of samples for single and double pulse measurements. The linear relationship between chlorine emission intensity and chlorine concentration was verified for samples with chlorine concentrations from 0.18 kg/m^3 to 5.4 kg/m^3. These results show that LIBS is effective for the quantitative measurement of the chlorine concentration in concrete.

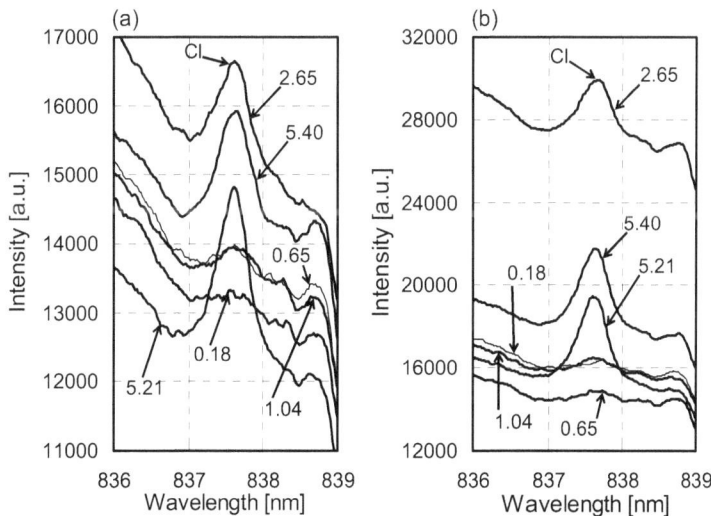

Fig. (9). Laser-induced emission spectra versus chlorine concentration for (a) single pulse and (b) double pulse measurement. The numbers in the figure denote chlorine concentrations in kg/m^3 [20].

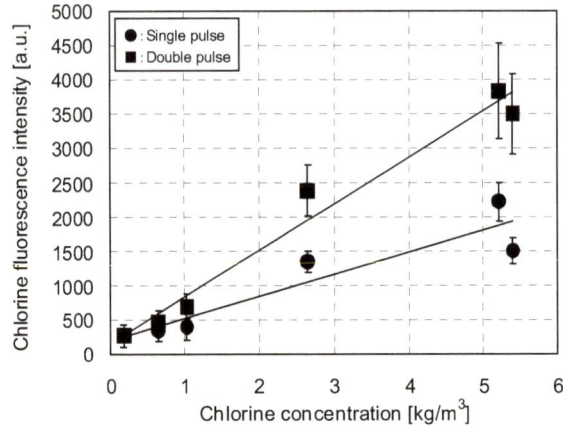

Fig. (10). Chlorine emission intensity versus chlorine concentration in single and double pulse measurements [20].

The double pulse measurement with an orthogonal laser beam arrangement is advantageous for increasing the chlorine emission intensity without damaging the target surface.

2.3.3. Application Method

Fig. **(11)** shows a schematic diagram of the application of LIBS for the measurement of chlorine concentration profile of a core sample extracted from a concrete structure. The chlorine concentration profile can be measured along the axial direction of the core sample, which corresponds to the depth into the concrete structure. The laser beam is focused on the core sample, and the emission from plasma is collected by an optical fiber and analyzed by a spectrometer. By moving the focal point, fast on-site measurement of chlorine penetration profile in concrete becomes possible. In addition, since LIBS can measure the chlorine concentration in the region determined by the focal spot of the laser beam, the concentration profile can be measured with high spatial resolution. These results can lead to improved accuracy of prediction of the chlorine penetration into concrete structures. Moreover, the diameter of the core sample can be reduced (compared to chemical methods), resulting in the reduction of the load on the concrete structure.

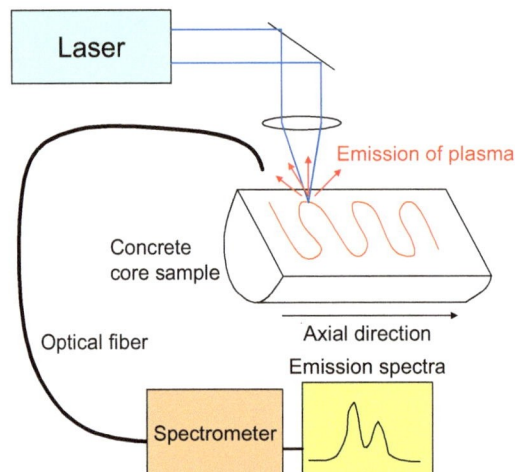

Fig. (11). Schematic diagram of LIBS measurement of chlorine concentration profile in a core sample obtained from a concrete structure.

The chlorine concentration on the surface of concrete structure can also be measured remotely. Fig. **(12)** shows a schematic diagram of remote measurement of chlorine concentration on the surface of a concrete structure. The laser beam is focused on the surface of the concrete structure, the emission from plasma is collected by a telescope and analyzed by a spectrometer. By moving the focal point, the chlorine concentration profile on the surface of the concrete structure can be measured remotely in real time. This method is useful for screening of locations which may suffer from strong salt damage.

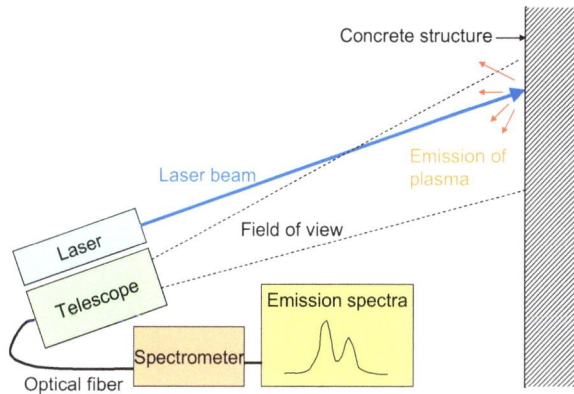

Fig. (12). Schematic diagram of remote LIBS measurement of a chlorine concentration profile on the surface of a concrete structure.

2.4. Long-Distance Remote LIBS

Remote LIBS using nanosecond lasers has been demonstrated for distances of more than tens of meters. Groenlund *et al.* demonstrated a remote imaging LIBS for metal targets such as copper, aluminum and iron at a distance of 60 m using the third harmonic of a Nd:YAG laser with an energy of 170 mJ [24]. The wavelength of 355 nm has an advantage for eye safety compared to visible wavelengths.

The identification of explosives such as RDX and TNT was demonstrated using LIBS [25,26]. By using a double-pulse configuration, the interference from atmospheric oxygen and nitrogen was diminished. The explosives were identified by measuring the ratios of several elements such as oxygen, nitrogen, hydrogen, and carbon. Using laser pulses with a energy of 275 mJ at wavelength 1064 nm, remote measurement of explosives and bio-materials using LIBS at 20 m was demonstrated [26]. Bulk explosives RDX, explosive residues, biological species, and chemical warfare simulates were detected using spectra in the wavelength region 200-860 nm.

3. FILAMENT-INDUCED BREAKDOWN SPECTROSCOPY

3.1. Filamentation Induced by High-Intensity Femtosecond Laser Pulses

Filamentation generated by high-intensity femtosecond laser pulses is a recently discovered phenomenon that has been extensively studied over the last decade [5-9]. Filaments are laser-induced plasma structures, which are formed by the equilibrium between Kerr lens focusing and plasma defocusing.

Kerr lens focusing is caused by the variation in the refractive index of air due to the high laser intensity. The refractive index n is given by $n=n_0+n_2I$, where n_0 is the refractive index in the absence of the laser pulse, n_2 is a constant, and I is the incident laser intensity. The laser intensity is usually higher at the center than at the edge. Since $n_2>0$, n becomes larger at the center than at the edge, and as a result, the filament acts as a convex lens. When the laser intensity exceeds the critical power P_{cr} (several GW in the atmosphere), the Kerr lens effect dominates over the divergence of the laser beam, and induces laser beam focusing.

Plasma defocusing is caused by the variation in the refractive index of a plasma. When the laser intensity reaches 10^{13}-10^{14} W/cm^2, multiphoton ionization occurs, creating a plasma. The electron density ρ in the plasma reduces the refractive index according to $n=n_0-\rho/2\rho_c$, where $\rho_c \sim 1.7\times10^{21}$ cm^{-3} for wavelength 800 nm. Since the refractive index of the plasma is less than that of the surrounding air, the plasma acts as a concave lens.

However, filament-like structures without plasma have recently been observed [27]. In order to explain these phenomena, a new model in which the laser beam divergence after self focusing is caused by high order Kerr effects is presented [28]. For laser intensity drastically exceeding the critical value P_{cr}, several filaments are generated in the laser beam (multi-filamentation). The intensity of each filament is cramped to 10^{13}-10^{14} W/cm^2 [29,30]; the number of filaments increases along with the laser intensity.

Fig. **(13)** shows a plasma generated by laser beam propagation, when the laser beam was focused by a spherical mirror with a focal length of 20 m. The inset in Fig. **(13)** shows a typical laser beam profile. Multiple filaments were observed as bright spots.

Fig. (13). Multiple filamentation produced by ultrashort laser pulses.

The profile of the laser beam, when focused by a spherical mirror with a focal length of 10 m, changed during propagation. An image of the laser beam cross section after 5.8 m propagation from the spherical mirror is shown in Fig. **(14a)**. At this point, filamentation was not observed. Multiple filamentation was clearly observed at 6.5 m propagation from the spherical mirror, as shown in Fig. **(14b)**. A number of bright filaments were observed after 7.9 m propagation, as shown in Fig. **(14c)**. The filaments coalesced after 9.2 m propagation, as shown in Fig. **(14d)** [31]. These results show that the filament plasma can be maintained over a distance of several m.

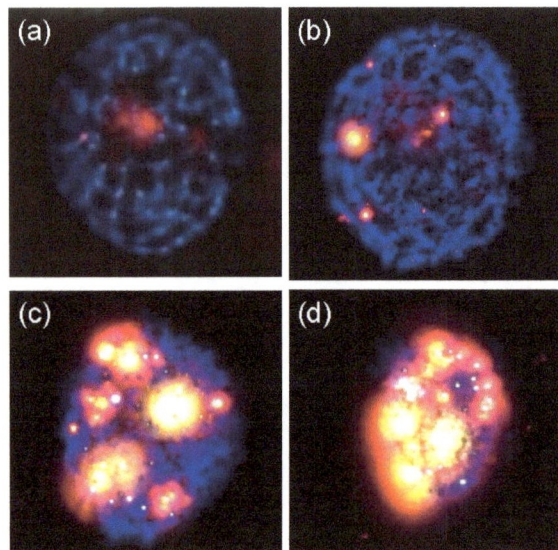

Fig. (14). Images of the laser beam cross section after (a) 5.8 m, (b) 6.5 m, (c) 7.9 m, and (d) 9.2 m propagation from the spherical mirror.

The filaments are reported as having diameter 0.1-1 mm, laser intensity $\sim 10^{14}$ W/cm^2, and plasma electron density 10^{16}-10^{18} cm^{-3} [32-37]. The electron density strongly depends on the laser intensity and/or focusing condition [36]. The electron temperature has been estimated to be 0.5 eV [37]. Although the filament loses its energy due to plasma generation, it can continue for a long distance because the laser beam around the filaments acts as an energy bath to supply energy to the filaments [38-40]. Generation of filaments for over 200 m [34], white light generation caused by filaments at an altitude of several km [41], and plasma generation after laser beam propagation of 400 m [42] have been reported. These results show that the remote measurement using the filament plasma as a probe is possible. In addition, Kasparian *et al.* observed the increase of the cloud-cloud discharges using filaments, which shows the presence of plasma at the cloud position [43].

3.2. Advantages of FIBS

Ultrashort laser pulses have several advantages for application to LIBS. The breakdown energy threshold is reduced with decreasing laser pulse duration [3,4]. In addition, filaments can be generated for a long distance, as described in section 3.1. Since the laser intensity, on the order of up to 10^{14} W/cm^2, is higher than the threshold for producing plasmas in water droplets [44,45], filaments can break down a target placed far from the laser. Therefore, filaments are very useful for remote LIBS. Recently, LIBS using filaments, called filament-induced breakdown spectroscopy (FIBS), has been applied to metallic samples [46-49] and biological materials [50,51].

Xu *et al.* obtained an electron density of 8×10^{17} cm^{-3} and a plasma temperature of 6794 K using a lead target for FIBS [46]. The obtained electron density is similar to that obtained in LIBS using a nanosecond laser system [52], while the plasma temperature is about half [53]. Such a high electron density leads to a high signal intensity, and the low plasma temperature leads to low white-light noise, resulting in high SNR.

3.3. Measurement of Metal Targets

FIBS has been demonstrated for metal samples, such as copper [47,48], steel [47], aluminum [48,49], and lead [46]. In one experiment [47], laser pulses of energy 250 mJ, pulse width 80-800 fs, repetition rate 10 Hz were collimated to a diameter of 30 mm and irradiated on a target located 20-90 m from the laser system. Thirty filaments were observed in the cross section of the laser beam. The emission from the target was collected by a telescope with a diameter of 200 mm located near the laser system, and analyzed by a spectrometer with an ICCD camera. The spectra of copper were clearly measured at a distance of 90 m. In addition, the range corrected signal intensity was found to be constant up to the measurement length of 90 m. In the case of the focused beam of a normal nanosecond laser, the efficiency of ionization decreases with increasing distance, because the beam diameter increases along with focusing length. Rohwetter *et al.* obtained plasma line emission from an aluminum target at distances up to 180 m from the laser by FIBS [48]. FIBS has an advantage for the remote measurement, up to kilometer range [49], because the high intensity laser beam can propagate for a long distance.

3.4. Measurement of Bio-Materials

Remote detection and identification of biological agents has recently gained strong interest in security applications. FIBS measurements of bio-materials have been demonstrated [50,51], in which a laser beam of energy 0.5-12 mJ and pulse width 45 fs was focused on a target using a lens with focal length of 1-5 m [51]. The emission from the target was collected by a spherical mirror with a focal length of 1.5 m and a diameter of 300 mm, analyzed by a spectrometer with an ICCD camera. Barley, corn, and wheat were used for targets. In addition to atomic spectra of Si, C, Mg, Al, Na, Ca, Mn, Fe, Sr, K, the emission from molecules such as C_2 and CN was detected. By comparing the ratios of several elements such as Mg/Si, Al/Si, Al/Mn, and Si/Mn, the possibility of identification of bio-materials was demonstrated.

3.5. Lidar Measurement of Microparticles

Remote detection and identification of microparticles in air are desirable for monitoring toxic materials distributed in the atmosphere. It is also useful for monitoring saltwater particles, which cause deterioration of reinforced concrete structures such as buildings and bridges, and of performance of insulators in substations and power transmission lines. The effect is especially large in coastal regions with maritime winds. Moreover, remote identification of the chemical composition of atmospheric aerosols and clouds is attractive for the study of acid rain and for understanding cloud-aerosol interactions, which are important in developing global climate models [54].

Lidar is a powerful tool for remote measurement of microparticles in the atmosphere, in terms of parameters such as extinction coefficient, particle size distribution, and shape. However, measurement of the constituents of microparticles has not been achieved by ordinary lidar. FIBS has a great advantage for remote measurement of the constituents of microparticles in air, because a large number of microparticles can be ionized along a bundle of filaments in air over a long distance, which can realize lidar measurements. The first demonstration of the remote sensing of the constituents of microparticles in air using FIBS was reported by the author [55].

The experimental setup is shown in Fig. **(15)**. Artificial saltwater aerosols were generated from saltwater (300 g/L) using an ultrasonic humidifier, and were introduced into a tube with inner diameter 20 cm, length 6 m, and open ends. Laser pulses from a Ti:Al$_2$O$_3$ laser, with energy 130 mJ and duration 70 fs, were focused inside the tube using a spherical mirror. The image of the laser beam near the entrance of the tube is shown in Fig. **(16)**. Although the photograph was taken as an accumulation of several shots, the propagation of each filament is clearly observed.

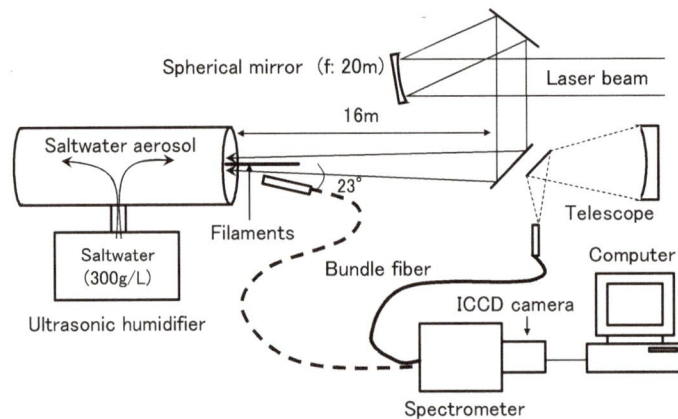

Fig. (15). Experimental setup for *in situ* and lidar measurements of Na emission from artificial saltwater aerosols using femtosecond laser pulses [55].

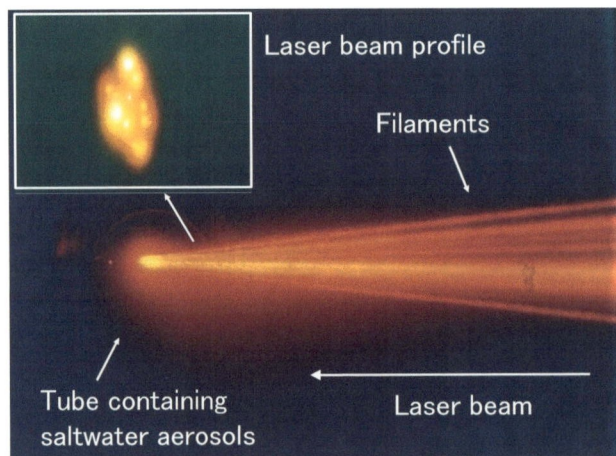

Fig. (16). Laser beam propagating around the entrance of the tube. Laser beam profile showing multiple filaments at the entrance of the tube is shown in the inset [55].

The laser pulses were focused using a spherical mirror of 20 m focal length, and the distance between the spherical mirror and the entrance of the tube was 18 m. The filamentation started 2-4 m before the entrance of the tube. Emission from the saltwater aerosols was collected by a bundled fiber located at the entrance of the tube (*in situ* measurement), or by a Newtonian telescope with 318 mm diameter, connected to the bundled fiber at 16 m from the entrance of the tube (lidar measurement). The collected light was fed into a spectrometer with an ICCD camera.

Fig. **(17)** shows Na emission spectra from saltwater aerosols in the *in situ* measurement, obtained at delay times of 20 ns, 40 ns, and 120 ns with respect to laser irradiation. The D_1 and D_2 lines of Na were clearly observed at 20 ns delay, at which the white light noise had sufficiently decayed. The Na emission intensity was lower at 40 ns delay, and almost no emission was observed at 120 ns delay. The white light noise and signal intensity decayed much faster compared to LIBS using nanosecond laser pulses.

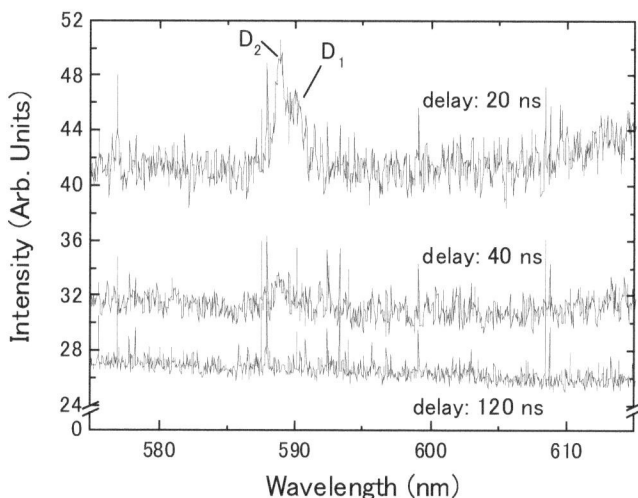

Fig. (17). Spectra of Na emission from saltwater aerosols irradiated by femtosecond laser pulses with delay times of 20, 40, and 120 ns from laser irradiation [55].

Fig. **(18)** shows Na emission spectra from saltwater aerosols in the lidar measurement. The reference emission spectrum of a spirit lamp containing salt is also shown. The gate width of the ICCD camera was set at 20 ns, and the gate delay was set at 15 ns. D_1 and D_2 lines of Na were clearly observed. These results demonstrate the possibility of remote sensing of the constituents of microparticles in air by FIBS.

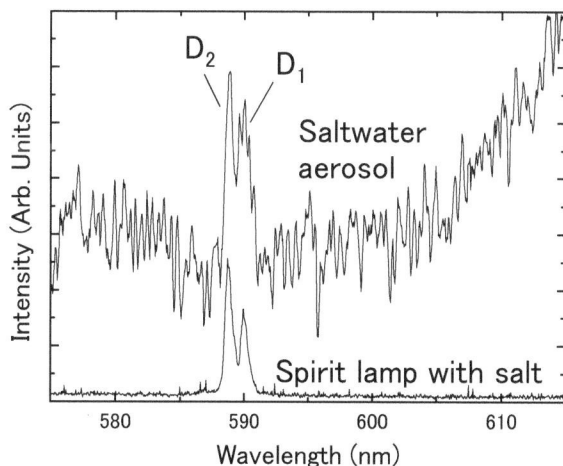

Fig. (18). Spectra of Na emission from saltwater aerosols irradiated by femtosecond laser pulses measured by lidar, along with emission of spirit lamp containing salt [55].

After the pioneering work presented above, the improvement of sensitivity and measurement distance has been reported [56]. Daigle *et al.* experimentally demonstrated ppm-level sensitivity and measurement distance of 70 m, showing a potential for kilometer-range application by extrapolation. They also demonstrated multiconstituent detection in contaminated aerosol clouds [57]. These results show the potential of lidar using filaments for remote detection of minor constituents.

4. NEW APPLICATION OF LIBS: ELECTRIC FIELD MEASUREMENT

4.1. Remote Electric Field Measurement

So far, LIBS has been applied mainly for the detection and identification of trace constituents, as described in the former sections. Recently, the author proposed and demonstrated a new application of LIBS: remote measurement of the electric field [10,11].

Remote, nondestructive, and time-resolved measurement of the atmospheric electric field may significantly advance the study of atmospheric electricity, such as lightning events, and the technologies for the design, diagnosis and protection of electric facilities against lightning strikes. However, conventional methods of detecting atmospheric electricity [58] are not comprehensive because of their effects on the field distribution, and remote measurement has been difficult.

A plasma is required to detect the electric field. A plasma in an electric field radiates in the ultraviolet (UV) region. The stronger the electric field, the higher the power of UV radiation. Seeding electrons are necessary for the plasma to emit UV light. There are various kinetic processes in air involving seeding electrons: ionization, excitation, recombination, and attachment [59]. The electric field increases the ionization and excitation rates, whereas the recombination rate decreases. Theoretically, the energy of UV emission and its duration can be used to obtain an exact value of the electric field in the space occupied by a bounded plasma. The key issues are how to make a suitable plasma source of UV emission that is sensitive to the electric field strength.

The filament plasma appears to be a good candidate for the source. Although dissociative recombination rapidly decreases the electron density, there is self-UV radiation of filament plasma even in the absence of an electric field [60]. In the presence of an electric field, the recombination process becomes slower, the electrons are heated by collisions, and the filament plasma may emit more powerful and longer UV radiation.

4.2. Experimental Results

A schematic diagram of the experimental setup is shown in Fig. **(19)**. Laser pulses from a Ti:Al$_2$O$_3$ laser, of wavelength 800 nm, pulse width 50 fs, and energy 84 mJ, were focused by a concave mirror of 10 m focal length. The optical axis of the laser beam was set at distances of 5 mm from the high voltage electrode (HVE). A negative high voltage was applied on a spherical HVE of 250 mm diameter placed at a the distance of 10.4 m from the focusing mirror. The negative voltage was varied between 0 and -400 kV. The field E was nonuniform and reached its maximum on the surface of the spherical HVE, as shown by the equation $E=U_0D/(2R^2)$, where U_0 is the applied voltage, D is the diameter of the spherical electrode, and R is the distance from the center of the electrode. From the equation, the field strength at the filament position under the HVE was calculated to vary from $E=0$ kV/cm to a maximum of $E = 29.6$ kV/cm, which was close to the corona threshold field in air.

At each voltage, the emission intensity in the spectral range near wavelength 337.1 nm was measured. This wavelength corresponds to the (0,0) band of the Second Positive system of N$_2$ (C$^3\Pi_u$-B$^3\Pi_g$). The emission from the filament plasma was collected by a telescope of diameter 152 mm located 20 m from the HVE and 1.3 m from the laser axis. The emission spectra were measured by a spectrometer with an ICCD camera. The variation of the emission spectra versus ICCD camera gate delay time from the laser shot was investigated.

The emission spectra of N$_2$ molecules, integrated over the gate width of 500 ns, were measured for filament plasma as shown in Fig. **(20)**. A major increase in the emission compared to self-radiation was observed at a voltage of -300 kV. The radiating part of the filament plasma was small because of the strong non-uniformity of the electric field in the vicinity of the laser filaments. The peak signal height of the N_2 emission near 337.1 nm in Fig. **(20)** is shown in Fig. **(21)**. Again, significant growth of the emission intensity was observed at absolute voltages of over 200 kV. In the absolute voltage range 200-400 kV, the signal growth was nonlinear. These results show that the filament plasma allows detection of very small changes in potential due to the exponential dependence.

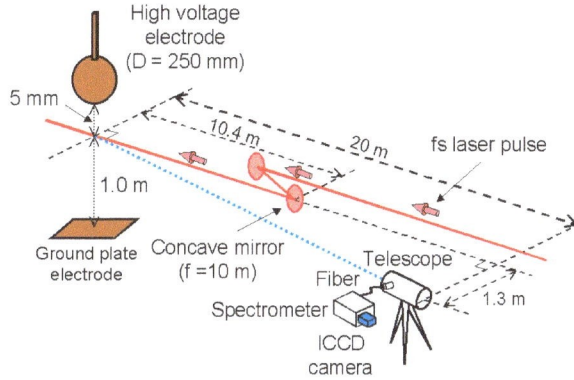

Fig. (19). Experimental setup for remote measurement of electric field.

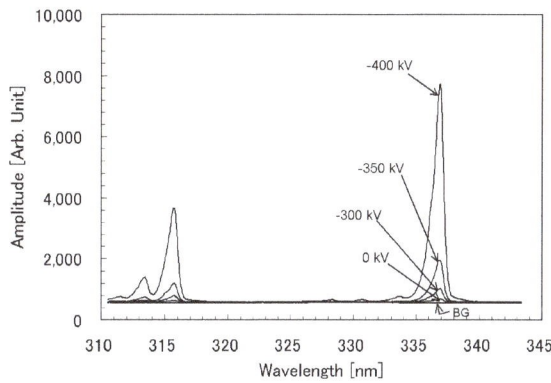

Fig. (20). Emission spectra of N_2 molecules at several applied voltages.

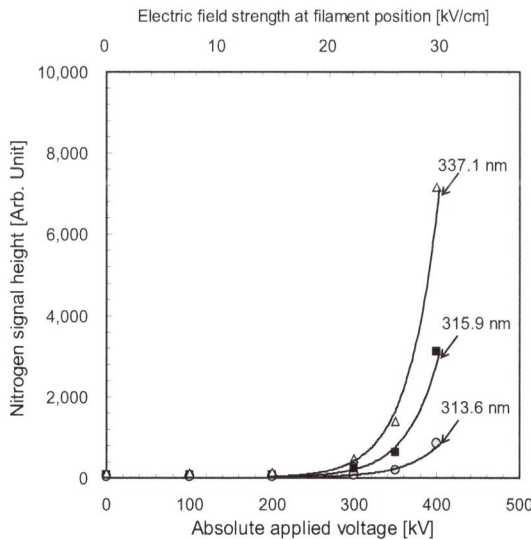

Fig. (21). Peak signal height of the N_2 emission over background spectra as a function of applied voltage.

This technique may open a very convenient and useful way for measuring the electric field and potential distributions in the atmosphere. The emission of excited nitrogen molecules in filament plasma produced in an external field is sensitive to electric field strength. The recombination time and the emission intensity of the filament plasma can be efficiently used for remote measurement. The filament plasma has many advantages for the remote measurement of the electric field. Having a high spatial resolution, the method enables atmospheric tomography as well as the rapid detection of the field dynamics. Low-cost and portable laser systems could easily make this method practical. Moreover, filaments have been demonstrated to propagate well in adverse atmospheric conditions such as turbulence [61], foggy [62] or rainy [63] atmospheres, partly due to the self-healing effect. However, precise control of the filament plasma position, which may depend on atmospheric conditions, has yet to be achieved. The necessary conformity of laboratory measurements and atmospheric measurement requires further development.

5. SUMMARY

Laser-induced breakdown spectroscopy (LIBS) and some of its applictions have been described.

LIBS has been shown to be an effective method to measure the chlorine concentration in concrete samples. A linear relationship between the spectral intensity of chlorine emission at wavelength 837.59 nm and chlorine concentration was verified. The signal-to noise ratio was higher than 2 for the sample with a chlorine concentration of 0.18 kg/m^3, which is below the threshold chlorine concentration of 0.6 kg/m^3 which causes corrosion of reinforcing bars in concrete structures. Therefore, LIBS is effective for quantitative measurement of chlorine concentration in concrete with high sensitivity.

LIBS using ultrashort laser pulses was also described. The use of ultrashort laser pulses have several advantages, such as reduction in the breakdown energy threshold and reduction in white light noise. In addition, filaments, which are generated as a result of the equilibrum between Kerr lens focusing and plasma defocusing, can be used for LIBS at long range. Filaments can be generated over a distance of more than several hundreds meters, and they have sufficiently high intensity to produce plasmas for various targets. LIBS using filaments, called filament-induced breakdown spectroscopy (FIBS), is useful for remote measurement. FIBS using multiple filaments is an attractive tool for remote measurement of microparticles in air, because a large number of microparticles can be ionized along a bundle of filaments in air for a long distance.

The application of FIBS to remote detection and identification of microparticles in air was demonstrated for saltwater particles. Na emission was observed at a distance of 16 m by lidar measurement using FIBS. The results show the possibility of remote measurement of the constituents of atmospheric microparticles, such as aerosols, clouds, and toxic materials, using filaments induced by ultrashort laser pulses.

As a new application of LIBS, remote measurement of electric field using the filament plasma was demonstrated. The dependence of radiation from a filament plasma on the external electric field was investigated using the 337.1 nm line, corresponding to the (0,0) band of the Second Positive system of N_2 ($C^3\Pi_u$-$B^3\Pi_g$). The results, which showed exponential dependence of the emission intensity on the field strengths, demonstrates the possibility of remote electric field measurement.

REFERENCES

[1]   A. Miziolek, V. Palleschi, I. Schechter, eds., *Laser-Induced Breakdown Spectroscopy*, Cambridge University Press, Cambridge, 2006.

[2]   D. Cremers, L. Radziemski, *Handbook of Laser-Induced Breakdown Spectroscopy*, John Wiley & Sons, West Sussex, 2006.

[3]   J. Noack, A. Vogel, "Laser-induced plasma formation in water at nanosecond to femtosecond time scales: calculation of thresholds, absorption coefficients, and energy density", *IEEE Journal of Quantum Electronics*, Vol. 35, pp. 1156-1167, 1999.

[4]   P. Kennedy, S. Boppart, D. Hammer, *et al.,* "A first-order model for computation of laser-induced breakdown threshold in ocular and aqueous media: part II - comparison to experiment", *IEEE Journal of Quantum Electronics*, Vol. 31, pp. 2250-2257, 1995.

[5]   A. Braun, G. Korn, X. Liu, *et al.,* "Self-channeling of high-peak-power femtosecond laser pulses in air", *Optics Letters*, Vol. 20, pp. 73-75, 1995.

[6] S. Chin, S. Hosseini, W. Liu, *et al.,* "The propagation of powerful femtosecond laser pulses in optical media : physics, applications, and new challenges", *Canadian Journal of Physics*, Vol. 83, pp. 863-905, 2005.

[7]   A. Couairon, A. Mysyrowicz, "Femtosecond filamentation in transparent media", *Physics Reports*, Vol. 441, pp. 47–189, 2007.

[8]   L. Bergé, S. Skupin, R. Nuter, *et al.,* "Ultrashort filaments of light in weakly ionized, optically transparent media", *Reports on Progress in Physics*, Vol. 70, pp. 1633-1713, 2007.

[9]   J. Kasparian, J.-P. Wolf, "Physics and application of atmospheric nonlinear optics and Filamentation", *Optics Express*, Vol. 16, pp. 466-493, 2008.

[10]   K. Sugiyama, T. Fujii, M. Miki, *et al.,* "Laser-filament-induced corona discharges and remote measurements of electric fields", *Optics Letters*, Vol. 34, pp. 2964-2966, 2009.

[11]   K. Sugiyama, T. Fujii, M. Miki, *et al.,* "Submicrosecond laser-filament-assisted corona bursts near a high-voltage electrode", *Physics of Plasmas*, Vol. 17, pp. 043108, 2010.

[12]   C. Aragon, J. Aguilera, "Characterization of laser induced plasmas by optical emission spectroscopy: a review of experiments and methods", *Spectrochimica Acta Part B*, Vol. 63, pp. 893-916, 2008.

[13]   A. Whitehouse, J. Young, I. Botheroyd, *et al.,* "Remote material analysis of nuclear power station power generation tubes by laser-induced breakdown spectroscopy", *Spectrochimica Acta Part B*, Vol. 56, pp. 821-830, 2001.

[14]   B. Smith, A. Quentmeier, M. Bolshov, *et al.,* "Measurement of uranium isotope ratios in solid samples using laser ablation and diode laser-excited atomic fluorescence spectrometry", *Spectrochimica Acta Part B*, Vol. 54, pp. 943-958, 1999.

[15]   A. Quentmeier, M. Bolshov, K. Niemax, "Measurement of uranium isotope ratios in solid samples using laser ablation and diode laser-atomic absorption spectrometry", *Spectrochimica Acta Part B*, Vol. 56, pp. 45-55, 2000.

[16]   H. Telle, D. Beddows, G. Morris, *et al.,* "Sensitive and selective spectrochemical analysis of metallic samples: the combination of laser-induced breakdown spectroscopy and laser-induced fluorescence spectroscopy", *Spectrochimica Acta Part B*, Vol. 56, pp. 947-960, 2000.

[17]   H. Kondo, N. Hamada, K. Wagatsuma, "Determination of phosphorus in steel by the combined technique of laser induced breakdown spectrometry with laser induced fluorescence spectrometry", *Spectrochimica Acta Part B*, Vol. 64, pp. 884-489, 2009.

[18]   F. Hilbk-Kortenbruck, R. Noll, P. Wintjens, *et al.,* "Analysis of heavy metals in soils using laser-induced breakdown spectrometry combined with laser-induced fluorescence", *Spectrochimica Acta Part B*, Vol. 56, pp. 933-945, 2001.

[19]   M. Nakane, A. Kuwako, K. Nishizawa, *et al.,* "Analysis of trace metal elements in water using laser-induced fluorescence for laser-breakdown plasma", *Proc. SPIE*, Vol. 3935, pp. 122-131, 2000.

[20]   K. Sugiyama, T. Fujii, T. Matsumura, *et al.,* "Detection of chlorine with concentration of 0.18 kg/m^3 in concrete by laser-induced breakdown spectroscopy", *Applied Optics*, Vol. 49, pp. C181-C190, 2010.

[21]   D. Cremers, L. Radziemski, "Detection of chlorine and fluorine in air by laser-induced breakdown spectrometry", *Analytical Chemistry*, Vol. 55, pp. 1252-1256, 1983.

[22]   L. St-Onge, E. Kwong, M. Sabsabi, *et al.,* "Quantitative analysis of pharmaceutical products by laser-induced breakdown spectroscopy", *Spectrochimica Acta Part B*, Vol. 57, pp. 1131-1140, 2002.

[23]   G. Wilsch, F. Weritz, D. Schaurich, *et al.,* "Determination of chloride content in concrete structures with laser-induced breakdown spectroscopy", *Construction and Building Materials*, Vol. 19, pp. 724-730, 2005.

[24]   R. Gronlund, M. Lundqvist, S. Svanberg, "Remote imaging laser-induced breakdown spectroscopy and remote cultural heritage ablative cleaning", *Optics Letters*, Vol. 30, pp. 2882-2884, 2005.

[25]   F.-C. De Lucia Jr. *et al.,* "Double pulse laser-induced breakdown spectroscopy of explosives: Initial study towards improved discrimination", *Spectrochimica Acta Part B*, Vol. 62, pp. 1399-1404, 2007.

[26]   J.-L. Gottfried *et al.,* "Double-pulse standoff laser-induced breakdown spectroscopy for versatile hazardous materials detection", *Spectrochimica Acta Part B*, Vol. 62, pp. 1405-1411, 2007.

[27]   G. Méchain *et al.,* "Long-range self-channeling of infrared laser pulses in air: a new propagation regime without ionization", *Applied Physics B*, Vol. 79, pp. 379-382, 2004.

[28]   P. Béjot, J. Kasparian, S. Henin, *et al.,* "Higher-Order Kerr Terms Allow Ionization-Free Filamentation in Gases ", *Physical Review Letters*, Vol. 104, pp. 103903 , 2010.

[29]   J. Kasparian, R. Sauerbrey, S. Chin, "The critical laser intensity of self-guided light filaments in air", *Applied Physics B*, Vol. 71, pp. 877-879, 2000.

[30]   A. Becker, N. Aközbek, K. Vijayalakshmi, *et al.,* "Intensity clamping and re-focusing of intense femtosecond laser pulses in nitrogen molecular gas", *Applied Physics B*, Vol. 73, pp. 287-290, 2001.

[31]   T. Fujii, M. Miki, N. Goto, *et al.,* "Leader effects on femtosecond-laser-filament-triggered discharges ", *Physics of Plasmas*, Vol. 15, pp. 013107, 2008.

[32]   A. Braun, G. Korn, X. Liu, *et al.,* "Self-channeling of high-peak-power femtosecond laser pulses in air", Optics Letters, Vol. 20, pp. 73-75, 1995.

[33]   E. Nibbering, P. Curley, G. Grillon, *et al.,* "Conical emission from self-guided femtosecond pulses in air", *Optics Letters*, Vol. 21, pp. 62-65, 1996.

[34]   B. Fontaine, F. Vidal, Z. Jiang, *et al.,* "Filamentation of ultrashort pulse laser beams resulting from their propagation over long distances in air", *Physics of Plasmas*, Vol. 6, pp. 1615-1621 (1999).

[35]   H. Yang, J. Zhang, Y. Li, *et al.,* "Characteristics of self-guided laser plasma channels generated by femtosecond laser pulses in air", *Physical Review E*, Vol. 66, pp. 016406, 2002.

[36]   F. Théberge, W. Liu, P. Simard, *et al.,* "Plasma density inside a femtosecond laser filament in air: Strong dependence on external focusing", *Physical Review E*, Vol. 74, pp. 036406, 2006.

[37]   J. Bernhardt, W. Liu, F. Théberge, *et al.,* "Spectroscopic analysis of femtosecond laser plasma filament in air ", *Optics Communications*, Vol. 281, pp. 1268–1274, 2008.

[38]   M. Mlejnek, E. Wright, J. Moloney, "Dynamic spatial replenishment of femtosecond pulses propagating in air", *Optics Letters*, Vol. 23, pp. 382-384, 1998.

[39]   W. Liu, F. Théberge, E. Arévalo, *et al.,* "Experiment and simulations on the energy reservoir effect in femtosecond light filaments", *Optics Letters*, Vol. 30, pp. 2602-2604, 2005.

[40]   S. Eisenmann, J. Penano, P. Sprangle, *et al.,* "Effect of an Energy Reservoir on the Atmospheric Propagation of Laser-Plasma Filaments", *Physical Review Letters*, Vol. 100, pp. 155003, 2008.

[41]   M. Rodriguez, R. Bourayou, G. Méjean, *et al.,* "Kilometer-range nonlinear propagation of femtosecond laser pulses", *Physical Review E*, Vol. 69, pp. 036607, 2004.

[42]   G. Méchain, C. D Amico, Y.-B. André, *et al.,* "Range of plasma filaments created in air by a multiterawatt femtosecond laser", *Optics Communications*, Vol. 247, pp. 171–180, 2005.

[43]   J. Kasparian, R. Ackermann, Y.-B. André, *et al.,* "Electric events synchronized with laser filaments in thunder clouds", *Optics Express*, Vol. 16, pp. 5757-5763 (2008).

[44]   F. Theopold, J.-P. Wolf, L. Wöste, "DIAL revisited: BELINDA and white-light femtosecond lidar", in *Lidar: Range-Resolved Optical Remote Sensing of the Atmosphere*, C. Weitkamp, ed., Springer, New York, pp. 399-443, 2005.

[45]   C. Favre, V. Boutou, S. C. Hill, *et al.,* "White-light nanosource with directional emission", *Physical Review Letters*, Vol. 89, pp. 35002, 2002.

[46]   H. Xu, J. Bernhardt, P. Mathieu, *et al.,* "Understanding the advantage of remote femtosecond laser-induced breakdown spectroscopy of metallic targets", *Journal of Applied Physics*, Vol. 101, pp. 033124, 2007.

[47]   K. Stelmaszczyk, P. Rohwetter, G. Méjean, *et al.,* "Long-distance remote laser-induced breakdown spectroscopy using filamentation in air" , *Applied Physics Letters*, Vol. 85, pp. 3977-3979, 2004.

[48]   Ph. Rohwetter, K. Stelmaszczyk, L. Wöste, *et al.,* "Filament-induced remote surface ablation for long range laser-induced breakdown spectroscopy operation", *Spectrochimica Acta Part B*, Vol. 60, pp. 1025-1033, 2005.

[49]   W. Liu, H. Xu, G. Méjean, *et al.,* "Efficient non-gated remote filament-induced breakdown spectroscopy of metallic sample", *Spectrochimica Acta Part B*, Vol. 62, pp. 76-81, 2007.

[50]   H. Xu, W. Liu, S. Chin, "Remote time-resolved filament-induced breakdown spectroscopy of biological materials", *Optics Letters*, Vol. 31, pp. 1540-1542, 2006.

[51]   H. Xu G. Méjean, W. Liu, *et al.,* "Remote detection of similar biological materials using femtosecond filament-induced breakdown spectroscopy", *Applied Physics B*, Vol. 87, pp. 151-156, 2007.

[52]   M. Sabsabi, P. Cielo, "Quantitative analysis of aluminum alloys by laser-induced breakdown spectroscopy and plasma characterization", *Applied Spectroscopy*, Vol. 49, pp. 499-507, 1995.

[53]   Y. Lee, S. Samuel, P. Sawan, *et al.,* "Interaction of a Laser Beam with Metals. Part II: Space-Resolved Studies of Laser-Ablated Plasma Emission", *Applied Spectroscopy*, Vol. 46, pp, 436-441, 1992.

[54]   S. Borrmann, J. Curtius, "Lasing on a cloudy afternoon", *Nature*, Vol. 418, pp. 826-827, 2002.

[55] T. Fujii, N. Goto, M. Miki, *et al.,* "Lidar measurement of constituents of microparticles in air by laser-induced breakdown spectroscopy using femtosecond terawatt laser pulses", *Optics Letters*, Vol. 31, pp. 3456-3458, 2006.

[56] J.-F. Daigle, G. Méjean, W. Liu *et al.,* "Long range trace detection in aqueous aerosol using remote filament-induced breakdown spectroscopy", *Applied Physics B*, Vol. 87, pp. 749-754, 2007.

[57] J.-F. Daigle, P. Mathieu, G. Roy, *et al.,* "Multi-constituents detection in contaminated aerosol clouds using remote-filament-induced breakdown spectroscopy", *Optics Communications*, Vol. 278, pp. 147–152, 2007.

[58] D. MacGorman, W. Rust, *The Electrical Nature of Storms*, Oxford University Press, London, 2006.

[59] Yu. Raizer, *Gas Discharge Physics*, Springer, Berlin, 1991.

[60] Q. Luo, W. Liu, S. L. Chin, "Lasing action in air induced by ultrafast laser filamentation ", *Applied Physics B*, Vol. 76, pp. 337–340, 2003.

[61] R. Salamé, N. Lascoux, E. Salmon, *et al.,* "Propagation of laser filaments through an extended turbulent region", *Applied Physics Letters*, Vol. 91, pp. 171106, 2007.

[62] G. Méjean, J. Kasparian, J. Yu, *et al.,* "Multifilamentation transmission through fog", *Physical Review E*, Vol. 72, pp. 026611, 2005.

[63] G. Méchain, G. Méjean, R. Ackermann, *et al.,* "Propagation of fs-TW laser filaments in adverse atmospheric conditions", *Applied Physics B*, Vol. 80, pp. 785–789, 2005.

CONCLUDING REMARKS

This book represents a compilation of recent advances in laser remote sensing at close ranges, in the order of m to tens of m, and applications to industry, environment, and public safety. Since the applicability of laser remote sensing to these disciplines is very diverse, this compilation represents only a small part of the possible applications. The table on the following page is a summary of the applications of the various lidar systems: Mie, absorption (and differential absorption), Raman, fluorescence, Doppler, as well as 3D laser radar, remote laser ultrasound, and remote laser-induced breakdown spectroscopy. The applications covered in this book are indicated in blue, with the corresponding chapter number.

Regrettably, no application of laser remote sensing to manufacturing was included in this book. Possible applications include metrology and/or flaw testing for quality control, and temperature measurement in extreme conditions (such as inside furnaces). The 3D laser radar, which was applied to traffic safety systems, may well be adapted to metrology of large structures such as ships, oil or gas tanks, and furnaces. Temperature measurement in combustion fields using coherent Anti-Stokes Raman spectroscopy (CARS) is a well-established technique, but its extension to larger measurement ranges is still to be realized. Laser ultrasonics, whose application to detection of defects in concrete was included in this book, is also applicable to metals and alloys, and should be effective for flaw detection (both surface and inner flaws) in situations in which remote sensing is required, such as high temperature environments.

Laser remote sensing of gases is applicable to the energy industry, as well as for public safety and maintenance of infrastructure (such as gas pipelines, hydrogen fueling stations, concerete structures). Another important application, which was not covered in this book, is the detection of toxic/hazardous gases and substances, of both natural and human origin. The former includes volcanic gases containing hydrogen sulfide (H_2S), which should be detected from a safe distance to prevent accidental deaths. The latter includes poisonous gases and radioactive substances, which must also be detected from a safe distance. Combination with Doppler lidar may provide accurate forecast of the movement of toxic/hazardous gases and/or substances.

It is unnecessary to mention the importance of laser remote sensing in atmospheric and environmental studies, as lidar systems are routinely used in these disciplines. Future need for development may arise for monitoring of greenhouse gases. Differential absorption lidar systems for measurement of atmospheric CO_2 have been developed, but their use is still limited to academic and scientific applications. Development of a more robust, economical, compact system, which can be used in industrial applications, may be called for in the near future.

The authors and editors hope that this book will arise interest in laser remote sensing in the industrial and scientific communities, and look forward to wider applications of laser remote sensing.

Tetsuo Fukuchi
Editor

INDUSTRIAL APPLICATIONS OF LIDAR SYSTEMS

(applications covered in this book are indicated in blue)

	Manufacturing	Energy	Infrastructure, Public Safety	Environment
Mie lidar			storm forecasting, storm warning	atmospheric monitoring (Ch.1)
				air quality monitoring (Ch.2)
Absorption lidar, Differential absorption lidar	gas monitoring	methane leak detection (Ch.3)		atmospheric monitoring
		flue gas monitoring (Ch.3)	toxic/volcanic gas detection	air pollution monitoring (Ch.3)
Raman lidar, CARS	temperature measurement	hydrogen leak detection (Ch.4)		atmospheric monitoring
Fluorescence lidar, remote LIF	surface flaw testing	oil leak detection in power plants	water quality, oil spill detection, pollutant monitoring (Ch.5)	
				vegetation monitoring (Ch.6)
Doppler lidar		wind profiling for energy farm (Ch.7)	aircraft wake detection at airports	atmospheric monitoring
			storm forecasting, storm warning	
3D laser radar (range finder)	metrology and surface profiling		traffic safety systems (Ch.8)	
	surface flaw testing		surveying, structural flaw testing	
Remote laser ultrasound	surface flaw testing		concrete structure testing (Ch.9)	
	inner/backwall flaw testing			
Remote laser-induced breakdown spectroscopy	element analysis		concrete structure testing (Ch.10)	aerosol detection (Ch.10)
				water quality, soil contamination monitoring

compiled by Tetsuo Fukuchi

Subject Index